# 2024 38th Symposium on Microelectronics Technology and Devices (SBMicro 2024)

**Joao Pessoa, Brazil**
**2-6 September 2024**

IEEE Catalog Number: CFP24SBO-POD
ISBN: 979-8-3315-4064-7

**Copyright © 2024 by the Institute of Electrical and Electronics Engineers, Inc.
All Rights Reserved**

*Copyright and Reprint Permissions*: Abstracting is permitted with credit to the source. Libraries are permitted to photocopy beyond the limit of U.S. copyright law for private use of patrons those articles in this volume that carry a code at the bottom of the first page, provided the per-copy fee indicated in the code is paid through Copyright Clearance Center, 222 Rosewood Drive, Danvers, MA 01923.

For other copying, reprint or republication permission, write to IEEE Copyrights Manager, IEEE Service Center, 445 Hoes Lane, Piscataway, NJ 08854. All rights reserved.

***\*\*\* This is a print representation of what appears in the IEEE Digital Library. Some format issues inherent in the e-media version may also appear in this print version.***

| | |
|---|---|
| IEEE Catalog Number: | CFP24SBO-POD |
| ISBN (Print-On-Demand): | 979-8-3315-4064-7 |
| ISBN (Online): | 979-8-3315-4063-0 |
| ISSN: | 2832-420X |

**Additional Copies of This Publication Are Available From:**

Curran Associates, Inc
57 Morehouse Lane
Red Hook, NY 12571 USA
Phone:     (845) 758-0400
Fax:     (845) 758-2633
E-mail:     curran@proceedings.com
Web:     www.proceedings.com

# 2024 38th Symposium on Microelectronics Technology and Devices (SBMicro 2024)

**Joao Pessoa, Brazil**
**2-6 September 2024**

**IEEE Catalog Number:** CFP24SBO-POD
**ISBN:** 979-8-3315-4064-7

# TABLE OF CONTENTS

Love Wave Acoustic Sensor as a Multisensing Device in Liquids .................................................................... 1
Ollivier Tamarin, Asawari Choudhari, Maxence Rube, Martine Sebeloue, Idris Sadli,
Dominique Rebiere, Corinne Dejous

Electrical Channel Length in the Subthreshold Operation of Junctionless Nanowire Transistors ...................... 5
Renan Trevisoli, Everton M. Silva, Rodrigo T. Doria

Two-Stage Transconductance Operational Amplifier Designed with VFET Experimental Data ...................... 9
Arllen D. R. Ribeiro, Vanessa C. P. Silva, Joao A. Martino, Anabela Veloso, Naoto Horiguchi,
Paula G. D. Agopian

Evolution of the Content and the Approach of Microelectronics Training to Regain Skills and
Competences .................................................................................................................................................... 13
Olivier Bonnaud

Analysis of a Low-Dropout Voltage Regulator Designed Using Omega-Gate Nanowire Transistors
Experimental Data ............................................................................................................................................ 17
Pedro H. Penna Da Silva, Joao A. Martino, Paula G. D. Agopian

Visible Light Tuning with Tridoped TeO2-ZnO Vitreous Samples for Photonics .................................... 21
Daniel Kendji Kumada, Beatrice Sayuri Kato, Raphael De Carvalho Gonçalves, Camila Dias
Da Silva Bordon, José Augusto Martins Garcia, Luciana Reyes Pires Kassab

Integrating STEAM and Maker Education in High School to Introduce the Microchip
Manufacturing Process ..................................................................................................................................... 25
Priscila Costa, Lucas Paiva Dias, Eduardo Ceretta Moreira

Reversible Memory Operation of Al/TeO$_2$-ZnO-Au/ TeO$_2$-ZnO/p-Si MIS Structures ...................... 29
José Augusto Martins Garcia, Leonardo Bontempo, Daniel Keij Kumada, Luciana Reyes
Pires Kassab, Sebastião Gomes Dos Santos Filho

A Simple Method of Fabrication of the Stainless Steel/Copper Oxide Nanoparticles Hybrid
Structure for Sensing Applications ................................................................................................................... 33
Andrei Alaferdov, Carolina Carvalho Previdi Nunes, Matheus Dias Sousa, Igor Fernandes
Namba, Fabio Domingues Caetano, Fernando Idalirio De Lima Leite

Effect of Interface Traps on the Different Conduction Mechanisms of MISHEMT from 200 K to
450 K ................................................................................................................................................................ 37
Welder F. Perina, Joao A. Martino, Paula G. D. Agopian

Influence of the Temperature on the Operational Transconductance Amplifier Designed with Triple
Gate SOI FinFETs ............................................................................................................................................ 41
Henrique Hilkner, Paula Ghedini Der Agopian, Joao Antonio Martino

Impact of Ionizing Radiation and Temperature on the Performance of pMOSFETs with Different
Layouts ............................................................................................................................................................. 45
Guilherme Inácio Grandesi, Paulo R. Garcia, Alexis V. Boas, Renato Giacomini, L. E. Seixas,
Marcilei A. Guazzelli

A Frugal Integrated Circuit Packaging for Non-Planar Surface-Conformant Electronics
Applications ...................................................................................................................................................... 49
Leonardo Shimizu Yojo, Favero Guilherme Santos, Louise Patron Etcheverry, Carlos R. P.
Dos Santos Junior, Fagnaldo Braga Pontes, Elvio C. Dutra E Silva

Development of a Double Pulse Test Plataform for Switching Loss Investigation in Emerging SiC MOSFET Technology .......... 53
Denison Rodrigo Ferreira Silva, Joel Felipe Guerreiro, Lucas B. Spejo, Marcos V. Puydinger Dos Santos

A Semi-Automatic Tool for the Extraction of RF-Plasmas Parameters by Langmuir Probe .......... 57
Rodrigo Cicareli, Giuseppe A. Cirino

Green Synthesis of Anatase Titanium Dioxide Nanoparticles from Joannesia Princeps Extract for Enhanced Photovoltaic Performance .......... 61
Felipe S. C. Portes, Adhimar F. Oliveira, Maria E. L. González

$SF_6/O_2$ Plasma for ICP/RIE SiC Etching .......... 64
R. R. César, M. Mederos, F. H. Cioldin, R. M. Beraldo, R. C. Teixeira, R. A. Minamisawa, J. A. Diniz

Experimentally Exploring the Performance of MOSFET Devices at Deep Cryogenic Temperatures .......... 68
Lucas Stucchi-Zucchi, Marcelo Pavanello, Francisco Rouxinol, José Alexandre Diniz, Francisco Brito

Optical Properties of Extended Inverted Pyramids Arrays for Enhanced Light Absorption in Silicon Solar Cells Simulated with Comsol Multiphysics .......... 72
Marcel Castilho Batista De Carvalho, Sebastiao Gomes Dos Santos Filho

Transfer of InGaP/GaAs Thin-Film to Unprecedented Flexible Polymeric Bases of PVC:PMMA:DOP Modified with EG for Solar Cell Applications .......... 76
Graciana S. Sousa, Luciana D. Pinto, Fabiele C. Tavares, Guillermo J. N. Soares, Rudy M. S. Kawabata, Rogério Valaski, Alexander. M. Silva, Maurício P. Pires, Roberto Jakomin, Patrícia L. Souza

Development of Epoxy Bonding Techniques for III-V on Silicon Tandem Solar Cells .......... 80
Willian M. M. Bazilio, Rudy M. S. Kawabata, R. T. Mourão, Guilherme M. Torelly, Patricia L. De Souza

Impact of Ionizing Radiation on the Behavior of Pseudo-Resistors with Temperature-Dependent Analysis .......... 84
Antonio Aurélio De Sousa Gomes, Cleiton Felix Pereira, Marcilei A. Guazzelli, Alexis C. Vilas Boas, Ricardo Germano Stolf, Renato Camargo Giacomini

Smaller Bond Pad for Device Reliability .......... 88
Ng Hong Seng, Lee Kuan Fang, Eddie Chaim Tau Tat, Florinna Sim, Jerald Sim Mong Joo, Deborah Debbie Anak Philip

Micromachined Passive Waveguide Fabrication with Fs Laser in Ag-Doped $GeO_2$–PbO Glasses for Photonics: Straight, Curved and Y Shaped Configurations .......... 92
Thiago Vecchi Fernandes, Camila D. S. Bordon, Niklaus U. Wetter, Wagner De Rossi, Luciana R. P. Kassab

Determining Neutron-Based Static Cross-Section of a SRAM-Based FPGA in a Simplified Setup .......... 96
Julia Willow Benvenutti, Fábio Benevenuti, Lívia Streit, Fernanda Kastensmidt

Experimental Comparison of Junctionless and Inversion-Mode Nanowire SOI MOSFETs Down to Cryogenics Temperatures .......... 100
Jefferson Almeida Matos, Flávio Enrico Bergamaschi, Michelly De Souza, Sylvain Barraud, Mikael Cassé, Olivier Faynot, Marcelo Antonio Pavanello

Single Nanofabrication Step of Low Series Resistance Nanowire-Based Devices for Giant Piezoresistance Characterization ............ 104
  *Kung Shao Chi, Lucas Barroso Spejo, Renato. A. Minamisawa, Marcos V. Puydinger Dos Santos*

Influence of Disorder on the Structural Analysis of a Quantum Bragg Mirror Detector ............ 108
  *Germano Maioli Penello, Pedro Henrique Pereira, Guilherme Monteiro Torelly, Rudy Massami Sakamoto Kawabata, Lucas Andrade Teixeira De Souza, Sérgio Luiz Morelhão, Alain André Quivy*

Influence of Extraction Methods on the Threshold Voltage Variability Results in SOI Nanosheets ............ 111
  *Vinícius Rodrigues Prates, Jaime Calçade Rodrigues, Marcelo Antonio Pavanello, Michelly De Souza*

Optimizing Broadband InGaAs/InP Photodetectors for the SWIR Range ............ 115
  *Marcelo G. Rua, Rudy M. S. Kawabata, Mauricio P. Pires, Carlos L. Ferreira, Guilherme M. Torelly, Patrícia L. Souza*

Neutron-Induced Effects on a Commercial GaN High Electron Mobility Transistor ............ 118
  *Alexis Cristiano Vilas Bôas, Saulo Gabriel Alberton, Paulo Roberto Garcia, Nilberto H. Medina, Vitor Ângelo P. Aguiar, Marco Antônio A. Melo, Roberto Baginski B. Santos, Renato C. Giacomini, Tássio V. Cavalcante, Luis Eduardo Seixas, Saulo Finco, Francisco Rogelio Palomo Pinto, Marcilei Guazzelli*

TCAD-Based Performance Evaluation of Dual-Gate UTBB SOI Junctionless ISFET for pH Detection ............ 122
  *Claudio Villela Moreira, Marcelo Antonio Pavanello*

Proposal of ᴮᴱSOI MOSFET Source Sensing Region for pH Monitoring Applications ............ 126
  *Pedro H. Duarte, Ricardo C. Rangel, Joao A. Martino*

Zero Temperature Coefficient Study Regarding the Half-Diamond Layout Style for MOSFETs ............ 130
  *M. A. P. Peixoto, M. P. Braga De Lima, E. H. S. Galembeck, M. M. Correia, L. M. Camillo, S. P. Gimenez*

Characterization of AlGaN/GaN HEMTs with Different Manufacturing Characteristics ............ 134
  *E. C. Panzo, J. Candido, N. Graziano Júnior, E. Simoen, M. G. C. Andrade*

Proposal for an Ion-Sensitive Floating-Gate MOSFET with Tunable Sensitivity and Memory Properties ............ 138
  *Henrique L. Carvalho, Ricardo C. Rangel, Joao A. Martino*

**Author Index**

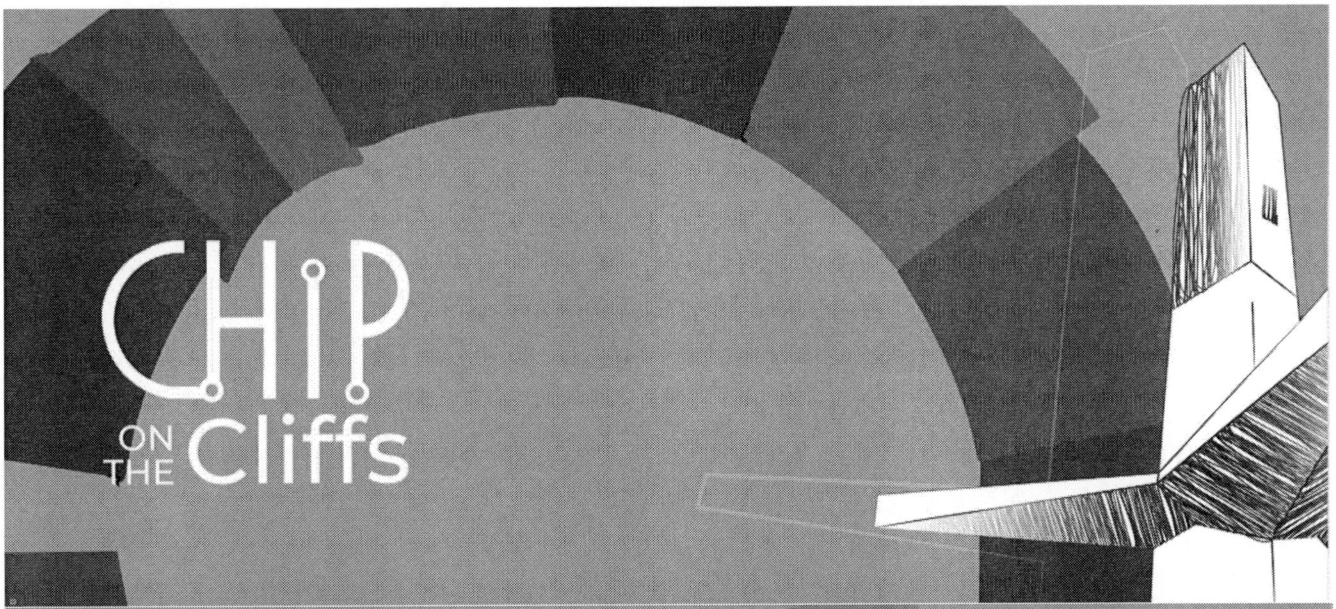

# SBMicro 2024

## 38TH SYMPOSIUM ON MICROELECTRONICS TECHNOLOGY AND DEVICES

### PROCEEDINGS

João Pessoa, Paraíba, Brazil

September 02 to 06, 2024

Organized by

UFCG · UFPB · INSTITUTO FEDERAL Paraíba

Sponsored by

## Organizing Committee

**General Chair:**
- Cleonilson Protasio de Souza, UFPB, Brazil

**Co-General Chair:**
- Marcelo Soares Lubaszewski, UFRGS, Brazil

**Program Chairs:**
- Paula Ghedini Der Agopian, UNESP, Brazil
- João Antonio Martino, USP, Brazil

**Tutorial Chair:**
- Rodrigo Trevisoli Doria, FEI, Brazil
- José Alexandre Diniz, UNICAMP, Brazil

**Publication Chair:**
- André Augusto Mariano, UFPR, Brazil

**Publicity Chair:**
- Linnyer Beatrys Ruiz Aylon, UEM, Brazil

**Local Organization Committee:**
- Ademar Virgolino da S. Netto UFPB, Brazil
- Alisson Brito, UFPB, Brazil
- Cleumar Moreira, IFPB, Brazil
- Ewerton Monteiro Salvador, UFPB, Brazil
- Juan Mauricio Villanueva, UFPB, Brazil
- Raimundo Carlos S. Freire, UFCG, Brazil
- Rômulo Calado Pantaleão Camara, UFPB, Brazil
- Suellen Finizola Dantas Maia, FAPESQ-PB, Brazil
- Verônica Maria Lima Silva, UFPB, Brazil
- Waslon Terllizzie A. Lopes, UFPB, Brazil

**Finance Chairs:**
- Juan Mauricio Villanueva, UFPB, Brazil
- Vimar Villela Ravagnani, SBMICRO, Brazil

# SBMicro 2024
## 38TH SYMPOSIUM ON MICROELECTRONICS TECHNOLOGY AND DEVICES

*List of Contents*

## Devices Physics and Characterization

Experimental Comparison of Junctionless and Inversion-Mode Nanowire SOI MOSFETs Down to Cryogenics Temperatures

*Almeida Matos, Jefferson; Bergamaschi, Flávio Enrico; Barraud, Sylvain; Cassé, Mikael; Faynot, Olivier; Pavanello, Marcelo A; de Souza, Michelly.*

Exploring the Performance of 180 nm MOSFET Devices at Deep Cryogenic Temperatures for Enhanced Cryo-Control Applications

*Stucchi-Zucchi, Lucas; Pavanello, Marcelo A; Rouxinol, Francisco; Diniz, José A; Brito-Filho, Francisco A.*

**Influence of Extraction Methods on the Threshold Voltage Variability Results in SOI Nanosheets**
*Prates, Vinicius\*; Rodrigues, Jaime; Pavanello, Marcelo A; de Souza, Michelly*

**Electrical Channel Length in the Subthreshold Operation of Junctionless Nanowire Transistors**
*Trevisoli, Renan\*; Silva, Everton; Doria, Rodrigo T.*

**Zero Temperature Coefficient Point Results for a Half-Diamond Hybrid Gate Layout Style MOSFET**
*Peixoto, Marco Aurélio P\*; Braga de Lima, Marcos Paulo; Galembeck, Egon; Correia, Marcelo; Gimenez, Salvador Pinillos; Camillo, Luciano.*

**Analysis of a low-dropout voltage regulator designed using omega-gate nanowire transistors experimental data.**
*Penna da Silva, Pedro Henrique\*; Martino, Joao Antonio; Agopian, Paula Ghedini Der*

**Effect of interface traps on the different conduction mechanisms of MISHEMT from 200 K to 450 K**
*Perina, Welder Fernandes; Martino, Joao Antonio; Agopian, Paula Ghedini Der*

**Development of a double pulse test plataform for switching loss investigation in emerging SiC MOSFET technology**
*Ferreira Silva, Denison Rodrigo; Guerreiro, Joel; Spejo, Lucas B; Puydinger dos Santos, Marcos*

**Neutron-induced effects on a commercial GaN High Electron Mobility Transistor**
*Vilas Bôas, Alexis Cristiano\*; Alberton, Saulo Gabriel; Garcia Jr., Paulo Roberto; Medina, Nilberto; Aguiar, Vitor Ângelo P.; Melo, Marco Antonio A.; Santos, Roberto Baginski B.; Giacomini, Renato C; Cavalcante, Tássio V.; Seixas Junior, Luis Eduardo; Finco, Saulo; Palomo Pinto, Francisco R; Guazzelli, Marcilei A.*

**Impact of Ionizing Radiation on the Behavior of Pseudo-Resistors with Temperature-Dependent Analysis**
*de Sousa Gomes, Antonio Aurélio\*; Felix Pereira, Cleiton; Guazzelli, Marcilei A; Vilas Boas, Alexis C.; Germano Stolf , Ricardo; Giacomini, Renato C.*

**Evolution of the content and the approach of Microelectronics training to regain skills and competences**
*Bonnaud, Olivier.*

Two-Stage Transconductance Operational Amplifier designed with VFET experimental data.
*Silva, Vanessa; Ribeiro, Arllen Dos Reis\*; Martino, Joao Antonio; Horiguchi, Naoto; Veloso, Anabela; Agopian, Paula Ghedini Der*

Characterization of AlGaN/GaN HEMTs with Different Gate Materials.
*Panzo, Eduardo C; Candido, Josué; Graciano, Nilton ; Simoen, Eddy ; Caño de Andrade, Maria Glória.*

Impact of Ionizing Radiation and Temperature on the Performance of pMOSFETs with Different Layouts.
*Grandesi, Guilherme I; Garcia, Paulo; Vilas Bôas, Alexis Cristiano; Giacomini, Renato C; Seixas, Luis; Guazzelli, Marcilei A.*

Influence of the Temperature on the Operational Transconductance Amplifier designed with triple gate SOI FinFETs.
*Hilkner, Henrique\*; Martino, Joao Antonio; Agopian, Paula Ghedini Der.*

## **Optoelectonics, Photonics and Photovoltaic**

Micromachined passive waveguide fabrication with fs laser in Ag-doped GeO2–PbO glasses for photonics: straight, curved and Y shaped configurations
*Fernandes, Thiago Vecchi\*; Bordon, Camila ; Wetter, Niklaus; de Rossi, Wagner; Kassab, Luciana R. Pires*

Visible light tuning with tridoped TeO2-ZnO vitreous samples for photonics
*Kumada, Daniel Kendji\*; Kato, Beatrice ; Gonçalves, Raphael ; Kassab, Luciana R. Pires; Garcia, José Augusto Martins; Bordon, Camila.*

Optimizing broadband InGaAs/InP photodetectors for the SWIR range
*Rua, Marcelo Gomes\*; Kawabata, Rudy; Pires, Mauricio; Ferreira, Carlos; Torelly, Guilherme; Souza, Patricia L.*

Green Synthesis of Anatase Titanium Dioxide Nanoparticles from Joannesia Princeps Extract for Enhanced Photovoltaic Performance
*Portes, Felipe Sievert; Oliveira, Adhimar F; González, Maria Elena*

Optical properties of extended inverted pyramids array for enhanced light absorption in silicon solar cells simulated with Comsol Multiphysics

*Batista de Carvalho, Marcel C\*; dos Santos Filho, Sebastião Gomes.*

Influence of disorder on the structural analysis of a quantum Bragg mirror detector

*Maioli Penello, Germano; Torelly, Guilherme Monteiro; Pereira, Pedro Henrique; Kawabata, Rudy M. S.; Morelhão, Sérgio Luiz; Quivy, Alain Andre; Andrade Teixeira de Souza , Lucas.*

Development of Epoxy Bonding Techniques for III-V on Silicon Tandem Solar Cells

*Bazilio, Willian M. M.; Kawabata, Rudy M. S.; Mourão, Renato T.; Torelly, Guilherme Monteiro; Souza, Patricia L .*

## **Fabrication Process  and Application**

SiO2 films obtained by PECVD for applications in photonic devices based on LiNbO3 thin films.

*Mederos Vidal, Melissa\*; Reigota César, Rodrigo; Hummel Cioldin, Frederico; Teixeira, Ricardo C; Silva Barbosa, Felippe Alexandre*

Single nanofabrication step of low series resistance nanowire-based devices for giant piezoresistance characterization

*Chi, Kung Shao; Spejo, Lucas Barroso; Minamisawa, Renato. A; Puydinger dos Santos, Marcos Vinicius*

A Simple Method of Fabrication of the Stainless Steel/ Copper Oxide Nanoparticles Hybrid Structure for Sensing Applications

*Alaferdov, Andrei\*; Carvalho Previdi Nunes, Carolina; Dias Sousa, Matheus ; Fernandes Namba, Igor; Domingues Caetano, Fabio; Idalirio de Lima Leite, Fernando*

Transfer of InGaP/GaAs thin-film to unprecedented flexible polymeric bases of PVC:PMMA:DOP modified with EG for solar cell applications

*de Sousa, Graciana dos Santos\*; Dornelas Pinto, Luciana; C. Tavares, Fabiele; J. N. Soares, Guillermo; M. S. Kawabata, Rudy; Valaski, Rogério; M. Silva, Alexander; P. Pires, Maurício; Jakomin, Roberto;  Souza, Patricia L*

A frugal integrated circuit packaging for non-planar surface-conformant electronics applications

Yojo, Leonardo S*; Santos, Fávero; Etcheverry, Louise; dos Santos Junior, Carlos; Pontes, Fagnaldo; Dutra, Elvio

A Semi-Automatic Tool for the Extraction of RF-Plasmas Parameters by Langmuir Probe

Cicareli, Rodrigo; Cirino, Giuseppe A.

SF6/O2 plasma for ICP/RIE SiC Etching.

Cesar, Rodrigo Reigota; Mederos, M.; Cioldin, F. H.; Beraldo, R. M.; Teixeira, R. C.; Minamisawa, R. A.; Diniz, José Alexandre.

Integrating STEAM and Maker Education in High School to Introduce the Microchip Manufacturing Process.

Costa, Priscila; Dias, Lucas Paiva; Moreira, Eduardo Ceretta.

Smaller Bond Pad for Device Reliability.

Ng, Hong Seng; Fang, Lee Kuan; Tat, Eddie Chaim Tau; Sim, Florinna; Joo, Jerald Sim Mong; Philip, Deborah Debbie Anak.

## Memory

Proposal for an Ion-Sensitive Floating-Gate MOSFET with Tunable Sensitivity and Memory Properties

Carvalho, Henrique Lanfredi; Rangel, Ricardo Cardoso; Martino, Joao Antonio

Reversible Memory Operation of Al/TeO2-ZnO-Au/ TeO2-ZnO/p-Si MIS Structures

Garcia, José Augusto Martins; Bontempo, Leonardo; Kumada, Daniel Kendji; Kassab, Luciana R. Pires; dos Santos Filho, Sebastião Gomes.

Determining Neutron-based static cross-section of a SRAM-based FPGA in a simplified setup

Benvenutti, Julia Willow; Benevenuti, Fábio; Streit, Lívia; Kastensmidt, Fernanda.

## Sensors and Biosensors

Proposal of BE SOI MOSFET source sensing region for pH monitoring applications.
*Duarte, Pedro H\*; Rangel, Ricardo C; Martino, Joao Antonio.*

TCAD-based Performance Evaluation of Dual-Gate UTBB SOI Junctionless ISFET for pH Detection
*Moreira, Claudio V; Pavanello, Marcelo A*

Love wave acoustic sensor as a multisensing device in liquids
*TAMARIN, Ollivier\*; CHOUDHARI, Asawari; RUBE, Maxence; LACHAUD, Jean Luc; SADLI, Idris; SEBELOUE, Martine; REBIERE, Dominique; DEJOUS, Corinne*

# Love wave acoustic sensor as a multisensing device in liquids

Ollivier TAMARIN[*†], Asawari CHOUDHARI[*†], Maxence RUBE[*], Martine SEBELOUE[*], Idris SADLI[*],
Dominique REBIERE[†], Corinne DEJOUS[†]

[*] *Univ. Guyane, Espace-Dev, UMR 228*, Cayenne, F-97300
[†] *Univ. Bordeaux, CNRS, Bordeaux INP, IMS, UMR 5218*, Talence, F-33400

*Abstract*—This paper presents a new measurement approach with Love wave sensor delay line. We demonstrate that by positioning a liquid on the entire surface of sensor, based on the piezoelectricity of the substrate, it is possible to distinguish mechanical parameters as well as electrical parameters of liquid sample. This method called "multiphysic approach" is possible *in situ* thanks to recent open loop readout electronic circuit and onboard micro controller.

*Index Terms*—Love wave sensor, liquid sample, electrical parameters, readout electronic circuit, real time detection

## I. INTRODUCTION

In Amazonian area, environmental contamination must be contained or even anticipated. For that, environmental studies are generally carried out using remote sensing and in laboratory biochemical analysis of samples taken on site. These above methods are not compatible with the need to have the most exhaustive water quality data possible on a large spatial and temporal scale and at low cost. An alternative approach, consists in using *in-situ* biochemical sensors which can be "easily" deployable and/or usable.

In the field of detection in a liquid medium, optical, electrochemical, and acoustic sensors are mainly studied and used.

Optical sensors are widely used for the estimation of physical parameters (temperature, pressure, liquid level, vibration, etc.) [1]. They have also developed greatly for estimation of biochemical species in liquids based on fluorescence properties and multi (or mono) chromatic absorption [1]. Even if several advantages argue in their favour (great sensitivity, wide range of measurement, multiplexing capacity, insensitivity to electromagnetic interference), one of the main disadvantages is the difficult (if not impossible) operability in highly turbid liquids due the "light" which interfere with suspended matters.

Electrochemical sensors allow multiple applications ranging from the characterisation of a liquid, especially for measurements of electrical properties (pH, dissolved oxygen [2]), to the detection of biochemical species requiring electrode fonctionalization. Nevertheless, with electrochemical sensors, only electrical energy can be used as sensing mechanism.

Surface acoustic wave (SAW) sensors are good candidates for compact, communicating *in-situ* detection systems [3]. As they are used as filters in the field of telecommunications since the 1980s [3], they can be mass produced at low cost with a very small size which allow a good adaptation to integration

into a communication system based on the concept of the Internet of Things [4].

Thus, transduction by acoustic waves could be complementary (or even better) than optical transduction in Amazonian waters, in particular, by avoiding the absorption and/or diffraction of light by suspended particles present in turbid waters. Moreover, acoustic wave devices present good performance and practicality as sensors in gaseous and liquid environments, such as good sensitivity, ease of functionalization thanks to their sensitive surface [5], and good accessibility for circuit integration microfluidics in aqueous environments [6].

Finally, compared other transduction, in the family of SAW sensors, Love wave (LW) sensors are essentially multiphysics as the generation of the acoustic wave is based on piezoelectricity. Thus, in transmission, there are two energy flows : acoustic (resulting from a mechanical displacement along the piezoelectric substrate), and electromagnetic (propagating between the input and output Inter Digitated Transducers) which can be used for detection. Furthermore, as presented in Fig. 1, the structure of the Inter Digitated Transducer (IDT) on the substrate, which could be compared to an impedimetric sensor, can also provide information on the detection mechanism.

The objective of this paper consists in present how LW sensors can be a possible multisensing device in liquids, based on a "multiphysic" approach. The first section presents the LW sensor principle and the classical (historic) use based on acoustic energy as "single" sensing mechanism for biochemical detection or liquid mechanical parameters estimation. Thus by taking advantage of the latest development of onboard microcontrollers, "low cost" Radio Frequency (RF) instrumentation tools as well as the piezoelectric property of the LW sensor, we will demonstrate how we can use both acoustic and electric energy flow for multiphysic sensing mechanism. Before the conclusion, some results highlighting the possibility to realize a multisensing tool in liquids will be presented.

## II. CLASSICAL USE OF LOVE WAVE SENSOR

### A. Love wave sensor acoustic response

LW transducers are based on the generation of a Shear Horizontal (SH) wave propagating in a thin layer deposited on a piezoelectric substrate. These transducers can operate in liquid media without excessive energy loss [5]. The detection mechanism is essentially based on the mass effect (quantity of

979-8-3315-4064-7/24 $31.00 © 2024 IEEE

mass deposited on the surface of the sensor) but also on the mechanical properties of the liquid medium, in particular its density and its viscosity [7]. For this, the strategy "classicaly" used consists in minimizing the influence of the dielectric parameters of the solution on the response of the sensor by localizing the liquid sample only on the acoustic path oh the LW sensor as presented on Fig. 1 (left). Thus, only acoustic energy is deliberately favored as the sensitive mechanism of interest, which positions LW sensors as a complementary (or additional) transducer to other types of sensors. Even if this approach offers numerous advantages in particular by favoring excellent sensitivity and promoting to a certain extent specificity to a mass effect [5], its deprives other information, in particular associated with electrical perturbation. This "non used" energy, considered as noise, can enquire about the physical characteristics of the liquid medium where the detection occurs.

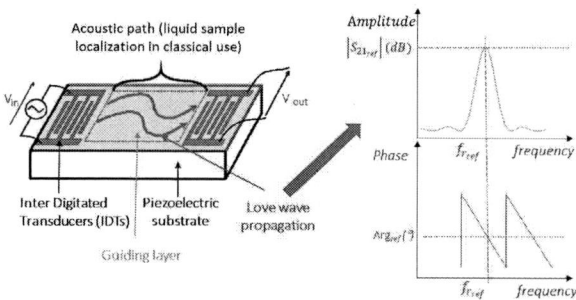

Fig. 1: (Left) LW sensor in delay line configuration for classical use. (Right) LW wave amplitude and phase.

The device we are currently using, and described in [8] is a double LW delay line based on a piezoelectric AT quartz substrate of $500\mu m$ thickness, with a guiding layer of $SiO_2$ of 4 to $6.4\mu m$ (see Fig. 1 (left)). The operating principle of the LW sensor is based on the generation of an acoustic wave by the reverse piezoelectric effect of the quartz. Indeed, as can be seen on Fig. 1 (left), by applying an electrical signal to the metallic contact of the interdigitated electrodes (IDTs), a mechanical deformation is generated and propagates along the acoustic path between the IDTs. Thus, the detection mechanism is based on the disturbance of the propagation parameters of the acoustic wave by a target or a liquid medium on the surface which will modify the amplitude and the velocity (phase) of the LW (see Fig. 1 (right)). In the the case of a biochemical detection of a molecule, a sensitive coating is necessary and allow specific interaction between the molecule and the surface of the LW sensor (see Fig. 2 left).

### B. Oscillator readout electronic circuit for "in situ" sensing

For in laboratory experiments, the dedicated instrumentation tool is the Vector Network Analyser (VNA). Due to the cost, the space and specific manipulation required for this system, one of the first approach used to bring the LW sensor on site consisted in insert it in an oscillation loop which is very

well known in the RF domain. This readout circuit delivers "spontaneously", without command, a periodic signal whose frequency can vary. In the case of their application to SAW delay line transducers, they are mounted in a test cell [7] and placed in the feedback loop of an amplifier, thus forming a closed loop system in which the signal oscillates. A small part of the signal is recorded using a coupler to measure the oscillation frequency which varies when the system (the surface of LW delay line) is disturbed. On the electronic level, to oscillate at a determined frequency "f0", the closed loop electronic chain must satisfy the Barkhausen conditions both on the gain and on the phase as presented in Fig. 2 (right).

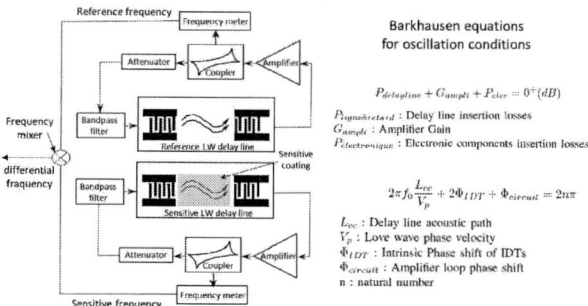

Fig. 2: (Left) Dual LW sensor in oscillator loop for specific biochemical detection. (Right) Barkhausen oscillation conditions.

Typically, during the detection mechanism, the LW phase velocity variation along the delay line will result in a phase shift. Thus, the frequency of the oscillator loop will change so as to maintain the phase condition of Barkhausen equation.

Experimentally, we obtained with our oscillation loops very good short-term stability of only a few $Hz.s^{-1}$ when the resonance frequency of the sensor is around $100MHz$. This property strongly contributes to the good resolution or detection limit of the sensor with the oscillator circuit. However, only the oscillation frequency which is the image of the phase of the sensor is accessible. This constraint does not allow to observe in real time the "quality" of the generation and transmission of the acoustic signal (*i.e.* LW sensor amplitude and phase fequency response as presented in Fig. 1 (right)), and even less to extract a variation in electrical impedance of the IDT.

### III. MULTIPHYSIC USE OF LOVE WAVE SENSOR

#### A. Low cost open loop instrumentation tools

For around one decade, low-cost and commercial open loop electronic interrogation circuits have emerged and are now widely used in the "RF" community. Based on the "open source" principle, both from a hardware and software point of view, these circuits are called "nano VNAs" and are based on the analysis of the transmission ($S_{21}$) and/or reflexion ($S_{11}$) signal of a device under test (in our case, the LW transducer). In this paper, we present two devices used by

979-8-3315-4064-7/24 $31.00 © 2024 IEEE

our team : i) an homemade one which consists in an electronic board controlled by an Arduino Due microcontroller (Arduino, Scarmagno, Italy) with a direct digital synthesizer (AD 9954), which can send a sinusoidal electronic signal to the transducer with a frequency range between 100 MHz and 120 MHz. The second circuit is an I/Q demodulator (AD 8302) which allows the input and output signals of the LW sensor to be compared; ii) the second one is a is a commercial nano VNA built around an ARM® microcontroller (see Fig. 3).

Although the performance of nano VNAs is not comparable to that of instrumentation VNAs (frequency range, number of points, measurement of the 4 S parameters, fineness of calibration), they offer good level performance for applications in the frequency ranges of our Love wave transducers.

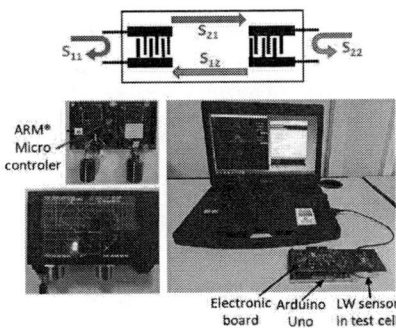

Fig. 3: (Top) LW sensor S parameters. (Bottom left) Commercial nanoVNA. (Bottom right) Home made nanoVNA.

The FEMTO-ST laboratory in Besançon, at the origin of the Open-Loop (OL) electronic circuit from which we took inspiration, demonstrated theoretically that the OL and closed loop (CL) approaches associated with a delay line present equal detection limits for a wide range of frequencies and delay times [9]. On a practical level with our devices, an experimental comparison of the behavior of our delay lines inserted in our OL and CL circuits confirms this observation. However, the open loop approach is very interesting because it frees us from the problems of "dropout" of oscillation conditions due, among other things, to experimental protocol, drop in transmission level (amplitude), phase jumps. But still associated with actual on-board computers, as we will show in the following subsection, the OL approach allows us to extract much more information than the CL approach.

*B. Multiphysic approach with "enriched protocol"*

In order to introduce the multiphysic approach mentioned in the introduction taking into account the piezoelectricity of the substrate, it is necessary to (re)define the different energy flows existing during the classic operation of a LW transducer.

In Fig. 4, electrical and electromagnetic energies are represented in blue and acoustic energies in red. More precisely, as detailed in [10], by studying the $S_{21}$ parameter, the electromagnetic coupling between the input and output of the

sensor which has an almost instantaneous transition time ($\Delta t \approx 0s$) is sensitive to the electrical parameters of the medium of detection. The $S_{21}$ parameter allow also to study the acoustic wave whose transit time depends on the geometry of the delay line and the materials constituting it ($\Delta t \approx 2\mu s$ for our sensor as can be seen in Fig. 5). This acoustic energy is mainly sensitive to the mechanical parameters of the medium. Finally, from the sensor input impedance calculable from the $S_{11}$ parameter we can extract the resistance and capacitance which are sensitive to the electrical parameters of the LW sensor. These two parameters are sensitive to electrical parameters, mainly at lower frequency compared to resonance [10].

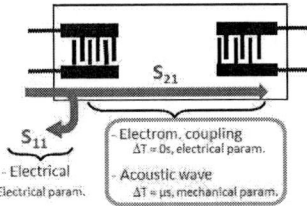

Fig. 4: LW energy flow chart for multiphysical approach.

On table I, by using the multiphysic approach and the LW sensor frequency response as presented on Fig.1, we synthesize the signals in both frequency and time domains that we can exploit in order to realize a biochemical detection, but also following the physical (electrical and mechanical) parameters of the medium of detection.

To manage to use this multiphysical approach and obtain an enriched response of the LW sensor in liquids the conventional protocol was updated by localyzing the liquid sample on the acoustic path as well as on the input and output IDTs. For that a "new" microfluidic PDMS chip called "long chamber PDMS chip" was realized as described in [11], contrarily to the "short chamber PDMS chip" used in classical approach.

TABLE I: Signals of interest from a LW sensor in multiphysical approach

| Domain | $S_{11}$ (reflection) | $S_{21}$ (transmission) |
|---|---|---|
| time | non applicable | LW EM amplitude (dB) |
| | | LW acoustic amplitude (dB) |
| | | LW acoustic echo (dB) |
| | | acoustic transit time of LW (s) |
| | | acoustic transit time of LW echo (s) |
| frequency | Low frequency Impedance | Resonance frequency (Hz) |
| | * Resistance (h) | Phase (°) |
| | * Capacitance (F) | Insertion losses (dB) |
| | * Inductance (H) | |

## IV. PRELIMINARY RESULTS BY USING LOVE WAVE MULTISENSING APPROACH

*A. Love wave sensor response in electrical conductive liquids*

The first experiments we carried out with multiphysic approach consist in testing the behaviour of a LW sensor without sensitive coating in a liquid with several electrical conductivity. Whether the effect of mechanical properties of

979-8-3315-4064-7/24 $31.00 © 2024 IEEE

a liquid (viscosity and density) on the LW sensor response is well known since a few decades by using $S_{21}$ parameter, as far we know, LW sensors are not widely used to estimate electrical parameters of liquids without sensitive coating. Thus, we prepared 6 turbid solutions ranging from 0 FTU (deionized water) to 1000 FTU which correspond to liquid samples having electrical conductivity from 0 to 3684 μS/cm [11].

As we indicated previously, in order to be able to extract maximum information from the LW sensor response, the sample must be positioned over the entire surface of the transducer (IDTs and acoustic path). However, in this case, as presented in Fig. 5 (left), the $S_{21}$ response of the sensor is significantly degraded due to the strong electromagnetic coupling between IDTs.

Fortunately, nowadays, according to powerful onboard microcontroller, and the OL readout electronic circuit, "a simple" method which consist in a time gating filtering allow to obtain the "pure" acoustic response of the sensor. Fig. 5 (right) presents the inverse Fourier transform of the $S_{21}$ response to pass into the time domain, where the electromagnetic (t ≈ 0 s) and acoustic (t ≈ 2μs in the black box) energies are clearly identified. Thus, after a Fourier transform of the time domain acoustic energy, Fig. 6 (right) represents the $S_{21}$ "pure" acoustic wave response of the LW sensor.

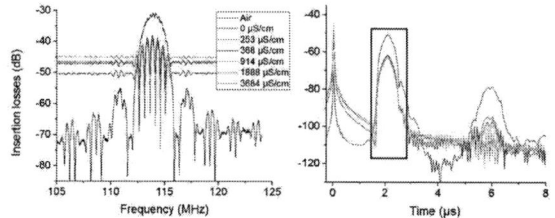

Fig. 5: Raw LW sensor response in Formazine solutions. Frequency (left) and time (right) domain responses

### B. Time gated multiphysical results

On Fig. 6, $S_{21}$ acoustic and $S_{11}$ magnitude responses are presented around the resonance frequency. The acoustic response of the sensor exhibit that the influence of the presence of a liquid in relation to air is significant when DeIonized water (DI) water is introduced (around -10 dB). Only by considering the liquids, the influence of the electrical conductivity is about -2 dB shift from DI water to the solution at 3684 μS/cm. Thus the relative magnitude shift is about -4% from the lowest to highest electrical conductivity.

Nevertheless, the interesting and new observation is the behaviour of the LW sensor by studying the $S_{11}$ response. As can be seen on Fig. 6 left, $S_{11}$ response show a decrease of the magnitude response from -1,2 dB to -1,67 dB from DI water to the highest conductive liquid sample which corresponds to a relative shift of 39%. As the viscosity and density of the Formazine solutions do not vary a lot contrarily to the electrical conductivity, both $S_{21}$ and $S_{11}$ parameters can distinguish mechanical from electrical parameters.

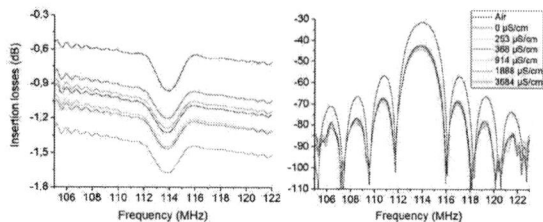

Fig. 6: Time gated LW sensor response in Formazine solutions. $S_{11}$ (left) and $S_{21}$ "acoustic" (right) parameter

## V. CONCLUSION

In this paper, we present a multiphysical approach to use LW sensors . If preprocessing of the LW raw signal is necessary when positioning the liquids on the entire surface of the device (IDTs + acoustic path), onboard microcontroller and OL readout electronic circuit allow do distinguish the electrical parameters of a complex liquids from mechanical parameters. Further work are now undertaken to use the LW sensor as a multiparameter probe unit which can realize a biochemical detection when estimating the electrical parameters of the liquid medium.

### ACKNOWLEDGMENT

This work was partly supported by "CARTEL" French Guiana FEDER Project under Grant SYNERGIE - GY0015845. The authors also want to warmly thank Serge DESTOR for his technical implication.

### REFERENCES

[1] D. A. Parande *et al*, "optical sensors and their applications," Journal of Scientific Research and Reviews, vol. 1, no. 5, pp. 60–68, 2012.

[2] Y. Wei *et al*, "Review of Dissolved Oxygen Detection Technology: From Laboratory Analysis to Online Intelligent Detection," Sensors, vol. 19, no. 18, p. 3995, 9 2019.

[3] C. C. W. Ruppel *et al*, "Acoustic Wave Filter Technology - A Review," IEEE Transactions on Ultrasonics, Ferroelectrics, and Frequency Control, vol. 64, no. 9, pp. 1390–1400, 2017.

[4] Z. Tang *et al*, "SAW Delay Line Based IoT Smart Sensing in Water Distribution System," 2018 IEEE 20th International Conference on High Performance Computing and Communications, (HPCC/SmartCity/DSS), pp. 1474–1478, 2018.

[5] M. Rocha-Gaso *et al*, "Love Wave Biosensors: A Review," State of the Art in Biosensors - General Aspects, pp. 277–310, 2013.

[6] A. M. Nightingale *et al* "Trends in microfluidic systems for in situ chemical analysis of natural waters," Sensors and Actuators, B: Chemical, vol. 221, pp. 1398–1405, 2015.

[7] V. Raimbault *et al*, "Acoustic Love wave platform with PDMS microfluidic chip," Sensors and Actuators A: Physical, vol. 142, no. 1, pp. 160–165, 2008.

[8] N. Moll *et al*, "A Love wave immunosensor for whole E. coli bacteria detection using an innovative two-step immobilisation approach," Biosensors and Bioelectronics, vol. 22, no. 9-10, pp. 2145–2150, 2007.

[9] P. Durdaut *et al*, "Equivalence of Open-Loop and Closed-Loop Operation of SAW Resonators and Delay Lines," Sensors, vol. 19, no. 1, 2019.

[10] M. Rube *et al*, "A Dual Love wave and Impedance-based Sensor: Response Enrichment," in 2020 IEEE SENSORS. IEEE,10 2020, pp. 1–4.

[11] A. Choudhari *et al*, "Love wave acoustic sensor response in high turbidity liquid environment," in 2022 IEEE Sensors. IEEE, 10 2022, pp. 1–4.

# Electrical Channel Length in the Subthreshold Operation of Junctionless Nanowire Transistors

Renan Trevisoli[1,2], *Senior Member, IEEE*, Everton M. Silva[3], and Rodrigo T. Doria[3], *Senior Member, IEEE*

[1]Pontifícia Universidade Católica de São Paulo, PUC-SP–São Paulo, Brazil
[2]Insper Instituto de Ensino e Pesquisa–São Paulo, Brazil
[3]Centro Universitário FEI, Electrical Engineering Department–São Bernardo do Campo, Brazil
e-mail: rtdoria@pucsp.br

*Abstract*— **In this work, the electrical length of Junctionless Nanowire Transistors biased in the subthreshold condition is analyzed. An analytical model for the lateral depletion regions is proposed and validated using numerical simulations. A method for extracting the effective channel length based on the occurrence of short-channel effects is also presented, validated with the simulations, and applied to experimental devices.**

*Keywords– Junctionless Transistors; Effective Channel Length, Subthreshold Operation.*

## I. INTRODUCTION

Junctionless Nanowire Transistors (JNTs) have been developed aiming at a simpler fabrication process with respect to conventional inversion mode devices [1-2]. JNTs are composed of a silicon nanowire, which is surrounded by the gate stack. The nanowire presents a heavy constant doping concentration from source to drain, such that no p-n junctions are needed.

When the device operates in the off-state condition, a depletion region is induced in the silicon layer due to the difference between the work functions of the gate and the silicon. In this condition, there is a potential barrier that inhibits the carriers' flow from source to drain [3]. Due to the absence of junctions, the depletion region induced by the gate can extend towards the source/drain regions [3], with an effective channel length ($L_{eff}$) longer than the mask length ($L$) in the subthreshold condition as demonstrated in [4,5,6]. This behavior is opposite from the one observed in inversion-mode transistors, where the effective channel length is smaller than $L$ owing to the dopants' diffusion from source/drain regions into the channel [7]. Therefore, gate control over the depletion charge is improved in JNTs, reducing the occurrence of short-channel effects [8].

Different works have considered the longer $L_{eff}$ of junctionless devices as an advantage for ultimate technological nodes [9,10]. Therefore, some works have recently been published aiming to evaluate $L_{eff}$ in JNTs. In [6], an extraction method for the effective channel length of JNTs through the capacitance curves is proposed. However, the capacitance measurement requires long and/or wide devices, limiting its application. Thus, this work aims to explore the subthreshold swing as a way to extract the effective channel length of the devices. In Section II, three-dimensional simulation results are presented in order to analyze quantitatively the electrical channel length, similarly as done in [4]. An analytical model for the calculation of $L_{eff}$ is proposed in Section III, where it is validated for devices with different characteristics and biases. In Section IV, a method for extracting the effective channel length based on the subthreshold slope ($S$) is proposed, validated against simulated results, and applied to experimental data. Finally, Section V presents the conclusions of this work.

## II. SIMULATION RESULTS

The numerical simulations of triple-gate Junctionless devices on Silicon-on-Insulator technology were performed using Synopsys tools [11]. The models used in the simulations account for the drift-diffusion transport mechanism, bandgap narrowing, doping-dependent generation/recombination, and mobility dependence on the lateral electric field. In the simulated devices, the nanowire width ($W$) and height ($H$) have been varied between 10 and 20 nm, the effective gate oxide thickness ($t_{ox}$) between 1 and 3 nm, the source/drain lengths ($L_{SD}$) between 1 and 40 nm, the doping concentration ($N_D$) between $5 \times 10^{18}$ cm$^{-3}$ and $10^{19}$ cm$^{-3}$. The simulated devices present a channel length of 30 nm.

The simulated drain current ($I_D$) is shown in Fig. 1 as a function of the gate voltage ($V_{GS}$) for transistors with different doping concentrations and source/drain lengths biased with a drain-to-source voltage ($V_{DS}$) of 1 V. An increase in the subthreshold slope and in the off-state current ($I_{OFF}$ − extracted at $V_{GS} = V_{TH} − 0.3$ V, been $V_{TH}$ the threshold voltage) can be observed when $L_{SD}$ is reduced for both doping concentrations. When $L_{SD}$ is very

---

[1]This work was supported by National Council for Scientific and Technological Development grants #427975/2016-6, #406193/2022-3, and #311892/2023-0, financed in part by the Coordenação de Aperfeiçoamento de Pessoal de Nível Superior - Brasil (CAPES) - Finance Code 001, and in part by PUC-SP - PIPEq - AuxP under Grant 29509.

979-8-3315-4064-7/24 $31.00 © 2024 IEEE

short, there is no physical space for the gate-induced lateral depletion region. For devices with longer $L_{SD}$, the gate-induced depletion region can extend towards the source/drain regions, increasing the electrical channel length and reducing the short-channel effects occurrence.

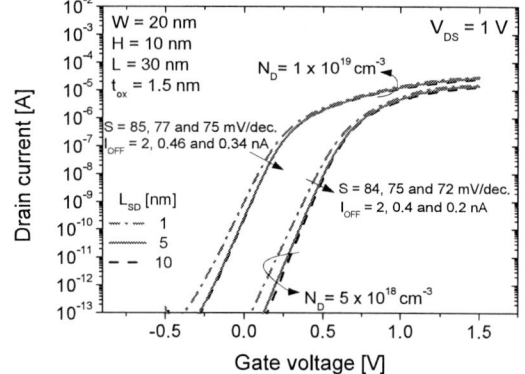

Fig.1. Simulated drain current in logarithmic scale as a function of the gate voltage for junctionless devices of different doping concentrations and source/drain lengths.

In order to analyze the effective electrical length of the JNTs, the electron density ($n$) has been extracted from the simulations, similarly as done in [4,6], for JNTs, and in [12], for inversion mode devices. In Fig. 2, the electron density is presented along the channel length for devices with different doping concentrations, indicating the source, channel, and drain regions. The cutline was performed along the channel length direction near the center of the nanowire cross-section (at the position $W/2$ and $H/2$) for $V_{GS} = 0$ and different drain biases ($V_D$). The cutline position was defined based on the JNT operation. As the center of the nanowire is the last region that becomes depleted when the gate voltage is reduced below the threshold voltage ($V_{TH}$), this region presents a shorter $L_{eff}$, being responsible for the short-channel effects occurrence.

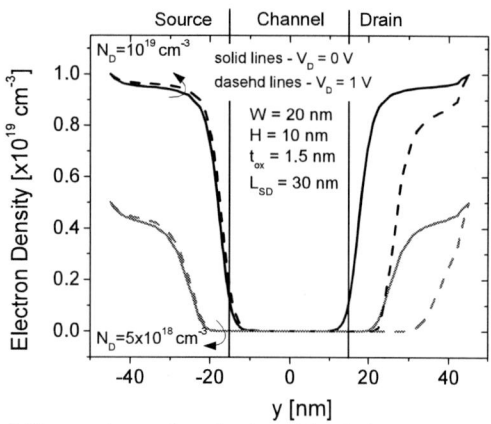

Fig.2. Electron density along the channel length direction for devices of different doping concentrations and drain biases.

From Fig. 2, the electron density is extremely reduced in the channel region, indicating that it is depleted. It can also be noted that the plateau region related to the depletion penetrates in the source/drain regions, indicating an effective channel length longer than $L$ [4]. The $L_{eff}$ increase is higher for the devices with the lower doping concentration. One can observe that the increase in the drain voltage raises the depletion region towards the drain side, increasing $L_{eff}$. To provide a quantitative analysis of the effective channel length, the derivative of the electron density presented in Fig. 2 has been calculated, as shown in Fig. 3. The minimum and the maximum points of the curve represent the maximum absolute variations of the concentration. These points can be understood as the extensions of the depletion region towards the source/drain.

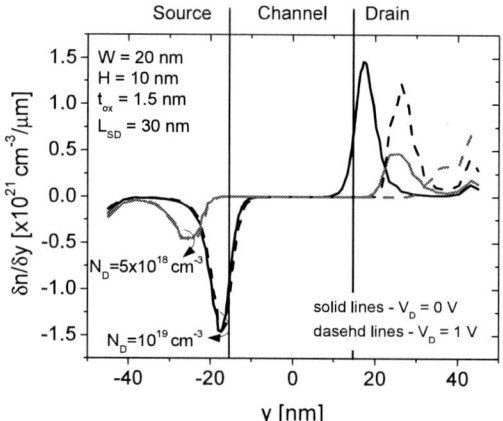

Fig.3. Derivative of the electron density towards the channel length direction for JNTs of different doping concentrations and drain biases.

III.     ANALYTICAL MODEL

As demonstrated in Section II and [4], in the subthreshold condition, the lateral depletion varies with the device characteristics and the applied biases. In order to understand the role of each physical parameter and bias in $L_{eff}$, a simplified approach has been considered.

The depletion at the source/drain regions are induced laterally by the influence of the gate voltage on the channel potential. Therefore, a simple form of calculating the depletion lengths inside source (drain) region ($L_{S(D)}$) is to consider the solution of the one-dimensional Poisson equation, given by

$$L_{S(D)} = \sqrt{\frac{2\varepsilon_{Si}}{qN_D}V_{S(D)}}, \qquad (1)$$

where $V_{S(D)}$ is the potential at the beginning (end) of the channel, which is responsible for the generation of the lateral depletion, $q$ and $\varepsilon_{Si}$ have their usual meanings.

When the device is biased in the on-state condition, there is no gate-induced lateral depletion, such that $V_{S(D)}$ should be zero. When the gate voltage is reduced below the threshold voltage in the case of an n-type device, $V_{S(D)}$ should increase with a reduction in $V_{GS}$ due to lateral depletion. For a long junctionless device, the increase in the potential with the gate voltage reduction is linear.

However, for short-channel devices, the drain bias can affect the potential in the channel, so a correction is needed. In [13], the correction is performed by adding the variation of the minimum potential in the channel due to the influences of the source and drain regions on the gate voltage. In the present work, a similar correction is used. The potentials for the source/drain regions can be obtained by

$$V_S = -(V_{GS} - \Delta V_{GS} - V_{TH} - \varphi_{Vt}), \qquad (2)$$
$$V_D = -(V_{GS} - \Delta V_{GS} - V_{TH} - \varphi_{Vt} - V_{DS}), \qquad (3)$$

where $\Delta V_{GS}$ is the variation of the minimum potential in the channel, and $\varphi_{Vt}$ is the surface potential in the threshold condition given by (4), according to [14]. It is worth mentioning that the difference $V_{GS} - V_{TH}$ is used since the lateral depletion occurs only when the device is operating in the subthreshold condition. Even when the JNT is biased in the threshold, there is a surface potential $\varphi_{Vt}$, which does not generate the lateral depletion such that it is also subtracted in (2) and (3).

$$\varphi_{Vt} = -\frac{qN_D}{\varepsilon_{Si}}\left(\frac{WH}{2H+W}\right)^2 \qquad (4)$$

To obtain $\Delta V_{GS}$, the three-dimensional Poisson equation should be solved in the channel region. A simplified approach for this solution is to consider the superposition principle, where only the solution of a 3D Laplace equation is needed. Following [13,15], the variation of the minimum potential can be obtained by

$$\Delta V_{GS} = \frac{V \sinh\left(\frac{x_{min}}{\lambda}\right) + U \sinh\left(\frac{L-x_{min}}{\lambda}\right)}{\sinh\left(\frac{L}{\lambda}\right)}, \qquad (5)$$

where $x_{min}$ is the position of the minimum potential in the channel, $V = V_{GS} - V_{TH} + \varphi_{Vt}$, $U = V_{GS} - V_{TH} + \varphi_{Vt} - V_{DS}$ [13] and $\lambda$ is the characteristics length. The position of the minimum potential can be obtained by [15].

$$x_{min} = \frac{\lambda}{2}\ln\left[\frac{U\exp\left(\frac{L}{\lambda}\right)-V}{V-U\exp\left(\frac{-L}{\lambda}\right)}\right]. \qquad (6)$$

Using (1) to (6), the length of the lateral depletion regions can be calculated. However, (2) and (3) are only valid in the subthreshold regime since $V_{S(D)}$ should be zero in the on-state. To provide a continuous function for the calculation of the lateral depletion length, a smooth function similar to [16] can be used.

The electrical length of a junctionless device can be obtained by

$$L_{eff} = L + \sqrt{\frac{2\varepsilon_{Si}}{qN_D}V_S} + \sqrt{\frac{2\varepsilon_{Si}}{qN_D}V_D}. \qquad (7)$$

In Fig. 4, the effective channel length calculated using (4) to (7) is validated by the comparison with simulated results, where it is presented as a function of the gate overdrive voltage. Devices with different widths, heights, gate oxide thicknesses, and doping concentrations have been considered. It is clear from all the curves of Fig. 4 that the model adequately predicts the extension length of lateral depletion of JNTs in the subthreshold operation, with a maximum error of 2.5 nm (< 8%).

## IV. ELECTRICAL CHANNEL LENGTH EXTRACTION

As already mentioned, the increase in the electrical channel length when the junctionless device is operating

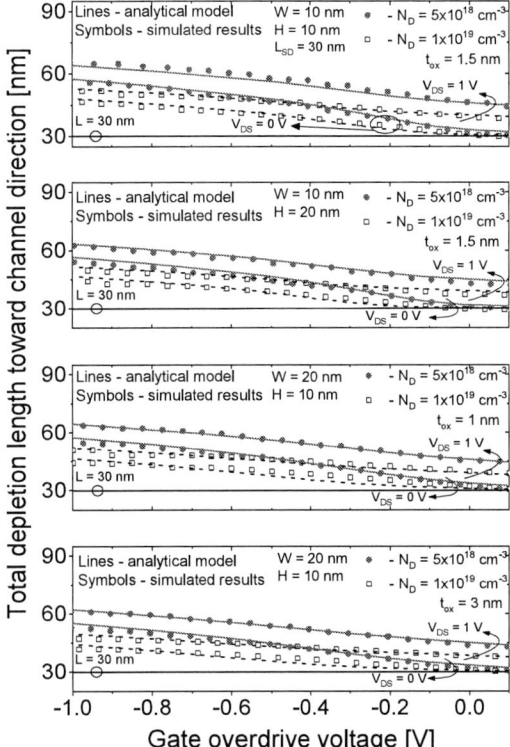

Fig. 4. Total depletion length towards the channel as a function of the overdrive gate voltage comparing the analytical model with numerical simulations for devices with different characteristics.

in the subthreshold condition affects the subthreshold swing. Therefore, $S$ variation can be used in order to obtain the channel length variation with the applied biases and device physical characteristics.

According to [13], the subthreshold swing obtained for a short-channel junctionless device with a low drain bias can be calculated by

$$S = \left(1 + 2\frac{\sinh(L/(2\lambda))}{\sinh(L/\lambda)}\right)\frac{kT}{q}\ln(10)n_{body}, \qquad (8)$$

where $k$ is the Boltzmann constant and $n_{body}$ is the body factor. Eq. (8) can be rewritten as

$$S = \left(1 + 2\frac{\sinh(L/(2\lambda))}{\sinh(L/\lambda)}\right)S_{long}, \qquad (9)$$

where $S_{long}$ is the subthreshold slope for a long JNT, i.e. $L \gg \lambda$. If $L = 10\lambda$, the hyperbolic term in (9) is approximately 0.01. By isolating the $L$ in (9), the electrical channel length in the subthreshold condition can be obtained by

$$L_{eff} = 2\lambda \cosh^{-1}\left(\frac{S_{long}}{S-S_{long}}\right). \qquad (10)$$

In order to analyze the validity of (10), the effective channel length was extracted from the subthreshold slope extracted at $V_{GS} = V_{TH} - 0.3\,\text{V}$ for the devices presented in Fig. 4 and compared to the values extracted through the derivative of the electron concentration in Table I. The extracted values are relatively close to the ones obtained in the simulations. The error is lower for the wider devices, which are the ones most susceptible to the short-channel effects. As the extraction method uses the SCEs

979-8-3315-4064-7/24 $31.00 © 2024 IEEE

occurrence to predict the effective length in the subthreshold regime, it is expected that the precision of the method is improved for shorter devices.

*Table I. Comparison between the effective channel length extracted from the simulated electron density and the one obtained from (10) for $V_{DS} = 50\ mV$.*

| W [nm] | H [nm] | $t_{ox}$ [nm] | $N_D$ [cm⁻³] | Effective channel length [nm] | | Error [%] |
|---|---|---|---|---|---|---|
| | | | | Simulated | Eq. (10) | |
| 10 | 10 | 1.5 | $5 \times 10^{18}$ | 40 | 35 | 12.5 |
| 10 | 10 | 1.5 | $1 \times 10^{19}$ | 36 | 33 | 8.3 |
| 10 | 20 | 1.5 | $5 \times 10^{18}$ | 40 | 35 | 12.5 |
| 10 | 20 | 1.5 | $1 \times 10^{19}$ | 35 | 32 | 8.6 |
| 20 | 10 | 1.0 | $5 \times 10^{18}$ | 41 | 40 | 2.4 |
| 20 | 10 | 1.0 | $1 \times 10^{19}$ | 36 | 36 | 0.0 |
| 20 | 10 | 3.0 | $5 \times 10^{18}$ | 39 | 38 | 2.6 |
| 20 | 10 | 3.0 | $1 \times 10^{19}$ | 34 | 32 | 5.9 |

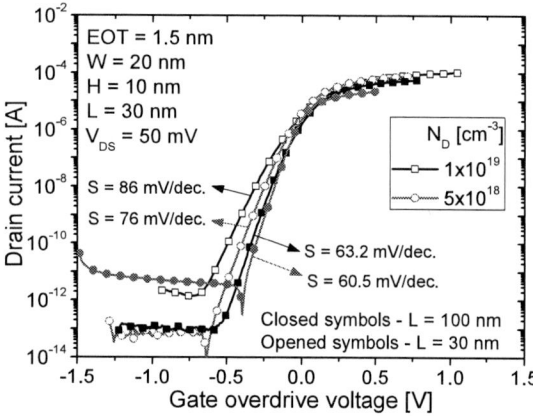

*Fig.5. Drain current as a function of the gate overdrive voltage for experimental devices of different doping concentrations, indicating S.*

Experimental devices composed of 50 parallel nanowires fabricated in CEA-LETI following [9] were measured. The JNTs were fabricated on a Silicon-on-Insulator wafer with a 145 nm-thick buried oxide and present width of 20 nm, height of 10 nm, and channel length of 30 nm. Devices with different doping concentrations were measured: $N_D = 5 \times 10^{18}$ cm⁻³ and $10^{19}$ cm⁻³. The gate stack is composed of HfSiON/TiN/Polysilicon with an effective oxide thickness (EOT) of 1.5 nm.

The drain current measured from the experimental short-channel devices with different doping concentrations is presented in Fig. 5 as a function of the gate overdrive voltage. The extracted subthreshold slope is also indicated in the figure. It is clear that the heavily doped devices exhibited a higher subthreshold slope, which indicates that these devices present a higher off-state current. From the $S$ values and using (10), the effective channel length of the experimental devices can be calculated. The obtained electrical lengths are $L_{eff} = 40$ nm for $N_D = 5 \times 10^{18}$ cm⁻³ and $L_{eff} = 33$ nm for $N_D = 1 \times 10^{19}$ cm⁻³, indicating that, similarly to the simulated data, the devices with lower doping concentrations present a longer electrical length.

## V. CONCLUSIONS

This work has presented an analysis of the electrical length of the Junctionless Nanowire Transistors biased in the subthreshold condition. It has been demonstrated that the electrical length can be significantly larger than the gate length due to the gate-induced lateral depletion regions towards the source/drain, reducing the occurrence of short-channel effects. A method for the extraction of the effective length based on the short-channel effects occurrence in the $I_D$ versus $V_{GS}$ curves is also proposed and validated. It has been shown that the method precision is higher for devices most susceptible to short-channel effects. The electrical length method is applied to experimental devices, also indicating that lower-doped devices present a longer electrical length.

### REFERENCES

[1] J.P. Colinge, C.W. Lee, A. Afzalian et al., "SOI gated resistor: CMOS without junctions," In: IEEE Int. SOI Conf., 2009, pp.1-2.

[2] J.P. Colinge, C.W. Lee, A. Afzalian, et al., "Nanowire transistors without junctions," Nature Nanotech, vol. 5, pp. 225-229, 2010.

[3] J.P. Colinge, A. Kranti, R. Yan et al., "Junctionless Nanowire Transistor (JNT): Properties and design guidelines", Solid-State Electronics, vol. 65-66, pp. 33-37, 2011.

[4] R. Trevisoli, R.T. Doria, M. de Souza, M.A. Pavanello, "Effective channel length in Junctionless Nanowire Transistors", In: 30th SBMicro, 2015.

[5] R. Trevisoli, R.T. Doria, M. de Souza, M.A. Pavanello, "Lateral Spacers Influence on the Effective Channel Length of Junctionless Nanowire Transistors," 2017 IEEE S3S Conf., 2017.

[6] E.M. Silva, R. Trevisoli, R.T. Doria, "Junctionless Nanowire Transistors Effective Channel Length Extraction through Capacitance Characteristics", Solid-State Electron., vol. 208, pp. 108734, 2023.

[7] V. Trivedi, J.G. Fossum, M.M. Chowdhury, "Nanoscale FinFETs with gate-source/drain underlap", IEEE Trans. Electron Devices, vol. 52, pp. 56-62, 2005.

[8] S. Sahay, M.J. Kumar, Junctionless Field-Effect Transistors: Design, Modeling, and Simulation, John Wiley & Sons, 419p., New York, 2019.

[9] D. Bosch, J. P. Colinge, G. Ghibaudo et al., "All-Operation-Regime Characterization and Modeling of Drain Current Variability in Junctionless and Inversion-Mode FDSOI Transistors", 2020 IEEE Symp. VLSI Techn., pp. 1-2, 2020.

[10] C.W. Lee, A. Afzalian, R. Yan et al., "Performance Estimation of Junctionless Multigate Transistors", Solid-State Electronics, vol. 54, pp. 97-103, 2010.

[11] Sentaurus Device User Guide, Synopsys, USA, 2023.

[12] R. Narayanan, A. Ortiz-Conde, J. J. Liou et al., "Two-dimensional numerical analysis for extracting the effective channel length of short-channel MOSFET", Solid-State Electronics, vol. 38, pp. 1155-1159, 1995.

[13] R.D. Trevisoli, R.T. Doria, M. de Souza, and M.A. Pavanello, "Drain Current and Short Channel Effects Modeling in Junctionless Nanowire Transistors," Journal of Integrated Circuits Systems, vol. 8, 2013, pp. 116-124.

[14] R.D. Trevisoli, R.T. Doria, M. de Souza, and M.A. Pavanello, "Threshold voltage in junctionless nanowire transistors," Semicond. Science and Technology, vol. 26, pp. 105 009, 2011.

[15] N. Lakhdar, and F. Djeffal, "A two-dimensional analytical model of subthreshold behavior to study the scaling capability of deep submicron double-gate GaN-MESFETs," J. Comput. Elec., vol. 10, pp. 382-387, 2011.

[16] B. Iniguez, L.F. Ferreira, B. Gentinne et al., "A physically based C∞-continuous fully depleted SOI MOSFET model for analog applications," IEEE Trans. Elec. Dev., vol. 43, pp. 568-575, 1996.

979-8-3315-4064-7/24 $31.00 © 2024 IEEE

# Two-Stage Transconductance Operational Amplifier designed with VFET experimental data

Arllen D. R. Ribeiro[1], Vanessa C.P. Silva[2], Joao A. Martino[1], Anabela Veloso[3],
Naoto Horiguchi[3] and Paula G. D. Agopian[1,2]
[1]LSI/PSI/USP, University of Sao Paulo, Sao Paulo, Brazil
[2]UNESP, Sao Paulo States University, Sao Joao da Boa Vista, Brazil
[3]Imec, Leuven, Belgium.
email: arllen@usp.br

*Abstract—* **In this paper, the design of a Two-Stage Transconductance Operational Amplifier (OTA) was studied, using vertical nanowire experimental data. The vertical nanowire device was investigated in both its forward and reverse mode. After analyzing the main electrical parameters of the devices such as Threshold Voltage (V$_{TH}$), Drain Current (I$_{DS}$), Transconductance (gm), Output Conductance (gd), Early Voltage (V$_{EA}$), and Intrinsic Voltage Gain (Av) of the device, it was concluded that, that, for analog integrated circuit applications, the best configuration is in the forward mode thanks to presenting a higher intrinsic voltage gain. Once the best mode for the vertical nanowire was chosen, three OTA circuits were designed, varying the quantity of vertical nanowires among them, to observe how the area affects circuit performance by analyzing its key operating parameters such as voltage gain (Av), gain-bandwidth product (GBW), phase margin (PM), and power dissipation (PD). As a result, it was observed that for a larger area, there is a higher GBW reaching 2GHz; however, the integrated circuit has a higher power dissipation, reaching 4000µW due to the increase in current caused by the number of added parallel devices.**

*Keywords— transistors with gate-all-around; vertical nanowires; transconductance operational circuit*

## I. INTRODUCTION

Aiming to find new devices to continue technological scaling, coupled with their good performance and eliminating or reducing short-channel effects, gate-all-around devices have become quite recurrent for the next step in technological evolution [1]. Gate-all-around devices have become capable of operating at nanoscale technology nodes, still offering excellent performance without suffering from short-channel effects, due to their strong electrostatic coupling [2].

Among gate-all-around devices, vertical nanowire devices (VFETs) have been studied. These devices have shown themselves to be quite distinctive and promising [3]. Recent studies show that these devices exhibit better electrostatic coupling, offer the possibility of working with even smaller technological nodes like 3 to 5nm, demonstrate better control of short-channel effects, and have higher switching speeds [4]. Another interesting characteristic of VFET devices is that they can be designed with two different configurations, one with the source as the bottom electrode (forward mode) and the other with the source as the top electrode (reverse mode), and depending on the configuration, VFET devices will exhibit slightly different electrical behaviors [5].

This work focuses on analyzing the performance of a Two-Stage Transconductance Operational Amplifier circuit, using VFET experimental data in the forward mode configuration, by varying the number of nanowires in parallel, showing the influence of area on the main parameters of the OTA circuit, such as voltage gain (A$_V$), gain bandwidth (GBW) and power dissipation.

## II. DEVICE CHARACTERISTICS

The device used in this work is a vertical nanowire device (VFET) fabricated in Imec/Belgium. This device is of p-type VFET and was fabricated on a silicon-on-insulator (SOI) wafer. The device presents the following main physical characteristics: buried oxide thickness of 145nm, gate length of 50nm, channel diameter of 20nm, and a gate dielectric with an equivalent oxide thickness of 0.9nm. Further information about the device and its fabrication can be found in reference [4]. The VFET mode chosen for operation was the forward mode, which will be justified in the Results and Analysis section. The Fig. 1 below shows a cross-section of the vertical nanowire device.

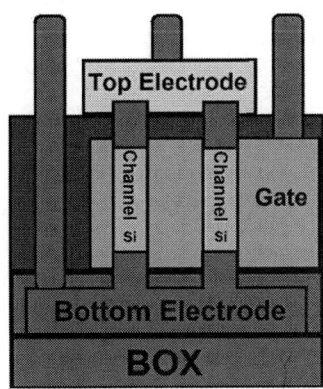

Fig.1- Schematic of the vertical nanowire device (VFET).

## III. RESULTS AND ANALYSIS

The first step in the development of this study was the extraction of the device's transfer curves, both for the forward mode and for the reverse mode. The pMOS transfer curves of the VFETs were obtained using the B1500A from Keysight®. The gate bias (V$_{GS}$) ranged from 0.5V to -1.0V in increments of -10mV, while the drain voltage (V$_{DS}$) varied from -50mV to -700mV, covering the triode and saturation regions, respectively, at room temperature. Output characteristics were measured with a drain bias (V$_{DS}$) ranging from 0V to -1.0V in -10mV increments, but only for a gate overdrive voltage (V$_{GT}$=V$_{GS}$ – V$_T$, where V$_T$ is the threshold voltage) of 200mV.

979-8-3315-4064-7/24 $31.00 © 2024 IEEE

After all measurements, a scan was performed on the device, extracting its main electrical parameters such as Threshold Voltage (VTH), it would be Saturation Drain Current (IDS), Transconductance (gm), Output Conductance (gd), Early Voltage (VEA), and intrinsic voltage gain (Av) of the device.

Through parameter extraction, some differences were observed between the forward and the reverse configuration. Table 1 shows, in quantitative values, the results of the extraction of the device's main electrical parameters.

Table 1 – Electrical parameters extracted from the VFET.

| Electrical Parameters | VFET Mode | |
|---|---|---|
| | Forward | Reverse |
| Threshold Voltage (V) | 0.15 | 0.15 |
| Drain Current (μA) | 7.89 | 8.75 |
| Transconductance (mS) | 1.61 | 2.85 |
| Output Conductance (mS) | 0.0281 | 0.202 |
| Early Voltage (V) | 6.13 | 2.09 |
| Intrinsic Voltage Gain (Av) | 34.87 | 22.93 |

Through Table 1, it is observed that for both configurations forward and reverse, there is a $V_{TH}$ of 0.15V. The fact that the threshold voltage is lower is linked to the small diameter of the device, increasing the electrostatic coupling, and consequently resulting in a lower threshold voltage [5].

Analyzing the drain current, in the forward mode, there is a lower drain current compared to the reverse mode. This decrease in drain current occurs because in the forward mode there is a higher access resistance compared to the reverse mode [6]. This difference in drain current will influence other important parameters of the device.

The next analyzed parameter is the transconductance. It is already known from the literature that devices with good electrostatic coupling, such as in the case of the VFET, exhibit higher transconductance [7]. For the VFET studied in this work, the device in reverse mode exhibited a higher transconductance than in forward mode due to the lower access resistance [6].

Since the focus of this work is to design an analog building block, another fundamental parameter analyzed was the output conductance. The study of output conductance is crucial when working with analog integrated circuits designed with specific devices because the output conductance directly affects the device's electrical behavior within the circuit it is placed in. A higher output conductance can lead to lower amplifier efficiency as it may introduce additional load on the output circuit, impacting linearity and frequency response [8]. It was observed that in the forward mode, the device exhibited a smaller output conductance than the device in the reverse mode. This indicates that in the reverse mode, there is a higher variation between the drain and source currents compared to the forward mode.

Finally, the last electrical parameter analyzed was the intrinsic voltage gain of the device. This electrical parameter

is inversely related to the output conductance of the device. and the Early voltage. Since the device in the forward mode exhibited smaller output conductance and a higher Early voltage compared to the device in the reverse mode, the intrinsic voltage gain is higher when the device is biased in the forward mode.

After analyzing the electrical parameters of the devices, it was concluded that since the intrinsic voltage gain is better when the devices are biased in the forward mode, this is the ideal configuration to be worked on in VFET devices to design an amplifier block. The OTA simulations were performed using Cadence Virtuoso Analog Design Environment based on Verilog-A model with experimental lookup table [9].

For the circuit simulation, it was considered that n-type and p-type devices are symmetrical. Figure 2 shows the IDS as a function of the VGS output curve and the IDS as a function of the VDS output characteristics, respectively, taking int account the measured and simulated data, for different bias conditions.

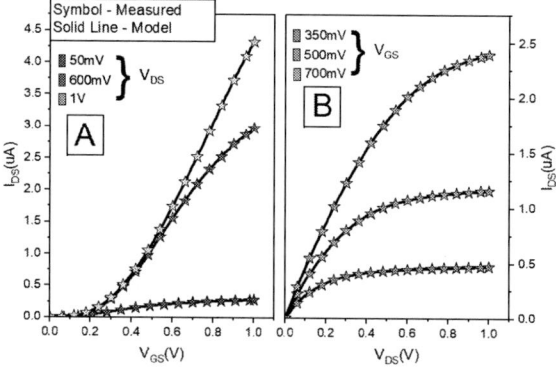

Fig. 2- Drain current as a function of gate voltage (A) and drain voltage(B).

The Figures 2A and 2B not only show the device's drain current curves but also validate the modeled device in the simulator with experimental measurements, exhibiting a perfect fit between the measured and modeled curves. The error between model and measurements is smaller than 1%. This indicates that when OTA will be designed in the simulator, the circuit is evaluated considering the real behavior of devices. The Figure 3 shows the schematic of a Two-Stage OTA.

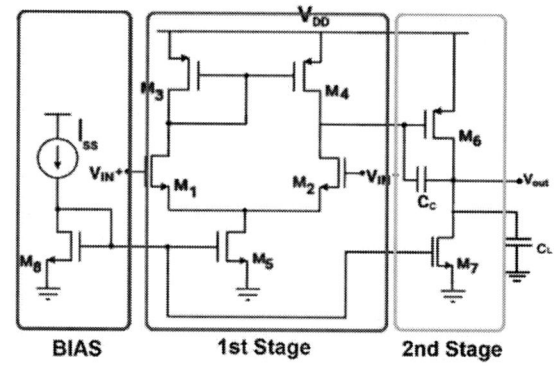

Fig. 3- OTA circuit Schematic.

979-8-3315-4064-7/24 $31.00 © 2024 IEEE

To design the circuit, it was considered that the nMOS device behavior is the same as the pMOS device behavior. The circuit presents the following distribution: the first stage comprises devices M1 and M2, compose a differential pair circuit with an active load represented by devices M3 and M4; the second stage contains a common-source amplifier represented by device M6; the current mirror is composed of devices M8, M5, and M7. The circuit also includes a compensation capacitor (Cc) and a capacitive load represented by $C_L$.

As one of the main advantages of using VFET is area reduction, three different designs were implemented for circuit modeling, varying the number of parallel nanowires (nNW) in each device. This experimental analysis allowed us to observe the benefits and drawbacks that increasing the area, that is, a greater number of parallel nanowires, can bring to an OTA circuit. The first design features only one nanowire for M1 and M2 (differential pair) devices, representing the minimum area value for project development. In the second design, an intermediate increase in the number of parallel nanowires was proposed to observe the effect of area enlargement versus circuit performance. Finally, the third design represents the maximum number of transistors in parallel obtained by the device, resulting in a significantly larger area.

The Table 2 shows the arrangement of devices within the circuit and the quantity of nanowires in parallel designed in each case.

Table 2 – Electrical parameters extracted from the VFET.

| OTA | Electrical parameters | | nNW | |
|---|---|---|---|---|
| | | | M1 - M2 | 1 |
| 1st Project | Iss (µA) | 0.37 | M3 -M 4 | 1 |
| | $V_{DD}$ (V) | 1.8 | M5 | 2 |
| | $V_{CM}$ (mV) | 940 | M6 | 9 |
| | gm/$I_{DS}$ ($V^{-1}$) | 8 | M7 | 10 |
| | $C_L$ (fF) | 200 | M8 | 1 |
| OTA | Electrical parameters | | nNW | |
| | | | M1 - M2 | 4 |
| 2nd Project | Iss (µA) | 1.45 | M3 -M 4 | 4 |
| | $V_{DD}$ (V) | 1.8 | M5 | 8 |
| | $V_{CM}$ (mV) | 940 | M6 | 36 |
| | gm/$I_{DS}$ ($V^{-1}$) | 8 | M7 | 40 |
| | $C_L$ (fF) | 200 | M8 | 4 |
| OTA | Electrical parameters | | nNW | |
| | | | M1 - M2 | 400 |
| 3rd Project | Iss (µA) | 580 | M3 -M 4 | 400 |
| | $V_{DD}$ (V) | 1.8 | M5 | 800 |
| | $V_{CM}$ (mV) | 940 | M6 | 3600 |
| | gm/$I_{DS}$ ($V^{-1}$) | 8 | M7 | 4000 |
| | $C_L$ (fF) | 200 | M8 | 400 |

It is worth noting that the number of parallel nanowires is distributed per device, considering the project specifications. For this work, the adopted specification was to operate the devices in the strong inversion condition, with a transistor efficiency (gm/$I_{DS}$) chosen as 8$V^{-1}$. Another consideration made for the distribution of the number of parallel nanowires is to increase the effective width of the device and achieve the current of each stage. For the circuit simulation, additional specifications were proposed, such as a supply voltage ($V_{DD}$) of 1.8V and a load to be sustained of 200$pF$.

The Table 3 presents all the main figure of merits of the three OTA circuits designed in this work.

Table 3 – Main results of the OTA circuits.

| Parameters Analyzed | 1st Project | 2nd Project | 3rd Project |
|---|---|---|---|
| Phase Margin (°) | 64 | 63 | 68 |
| Cc (fF) | 40 | 40 | 20 |
| GBW (MHz) | 12 | 46 | 2000 |
| 1° Stage Gain (dB) | 20 | 20 | 20 |
| 2° Stage Gain (dB) | 36 | 36 | 36 |
| Total Gain (dB) | 56 | 56 | 56 |
| Power (µW) | 10 | 39 | 4000 |

It is known in the literature that for a circuit to achieve stability, its Phase Margin must be at least 60° [10], a value that depends on the Compensation Capacitor coupled to the circuit. As shown in Table 3, for the three projects, this minimum value of phase margin was achieved, but for the third project, a much larger compensation capacitor was required to achieve this stability. Having a larger capacitor is not very viable, as it will also physically occupy a larger area in the circuit. Another important point is that the GBW is inversely proportional to $C_C$ and although for the third project a larger $C_C$ was required, this project reaches much higher GBW than that of the other two projects, thanks to the transconductance increase with effective width.

Analyzing the voltage gain of the circuit, all three projects exhibited equal voltage gains, indicating that the number of parallel nanowires was not a significant factor in this circuit parameter, as increasing the number of parallel nanowires proportionally increases other electrical parameters of the device as well. During the study of the circuit's voltage gain, it was observed that it exhibited a very high output conductance reducing the voltage gain. This observation was more pronounced in the first stage where the high output conductance of active load takes place. It is possible to achieve a higher gain, but the device would deviate from the desired biasing or the $V_{DS}$ would become asymmetric.

The Figure 4 shows two graphics displaying the Voltage Gain and Phase Margin for the three projects, as a function of frequency.

It is possible to observe the relatively higher GBW of the third project compared to the other two projects, while still presenting the same voltage gain and practically the same Phase Margin.

979-8-3315-4064-7/24 $31.00 © 2024 IEEE

In terms of power dissipation, the third project showed a higher value, due to the amount of current that should be draw from the power supply with increasing the number of parallel nanowires, as previously commented.

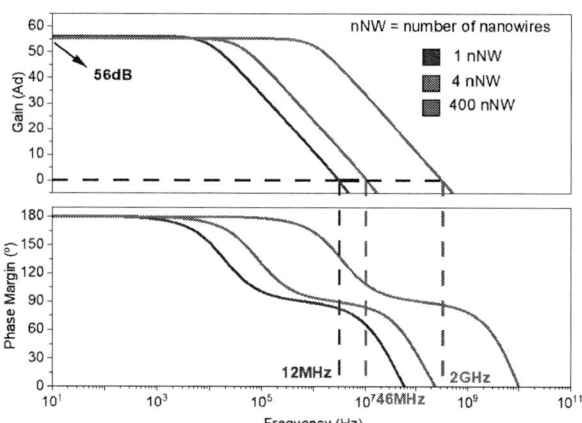

Fig. 4- Voltage Gain and Phase Margin as a function of Frequency.

## IV. CONCLUSIONS

In this work, a study of a Two-Stage Transconductance Operational Amplifier circuit designed with vertical nanowire devices (VFET) in the forward mode was conducted. Initially, a parametric analysis was performed with the device in its two configurations: forward mode and reverse mode, in order to analyze which configuration was best suited for working with analog circuits. It was concluded that the forward mode was preferable due to its higher intrinsic voltage gain in response to frequency.

Subsequently, three different circuits were designed with VFETs in the forward mode, with the only difference between them being the design area - the quantity of nanowires in parallel. It was observed that for all three projects, the circuit exhibited a voltage gain of 56dB. However, for the circuit designed with the highest number of nanowires in parallel, it had a relatively higher GBW reaching up to 2 GHz, compared to the others. Nonetheless, a larger compensation capacitor was required to achieve stability, and it exhibited higher power dissipation, reaching 4000µW.

This study has shown that vertical nanowire devices can yield promising results when applied in analog circuits, which is relevant for the next steps of technological evolution.

## ACKNOWLEDGMENT

The authors acknowledge CNPq, São Paulo Research Foundation - FAPESP (under grant #2020/04867-2) and Coordenação de Aperfeiçoamento de Pessoal de Nível Superior - Brasil (CAPES) - Finance Code 001 for the financial support.

## REFERENCES

[1] J. P. Colinge, Silicon-On-Insulator Technology: Materials to VLSI, Boston (MS): Kluwer Academic Publishers, 2004.
[2] R. Yadav, K. Goyal and A. Kaushik, "Impact of Work Function and Silicon Thickness on 3-D Analytical Model of Gate All Around Field Effect Transistor," 2022 4th International Conference on Smart Systems and Inventive Technology (ICSSIT), Tirunelveli, India, 2022, pp. 706-709, doi: 10.1109/ICSSIT53264.2022.9716517.
[3] D. Ghosh, P. Bhulania and S. Kumar, "An approach for analytical modeling and simulation of gate all around MOSFET for 50 nm technology," 2015 2nd International Conference on Signal Processing and Integrated Networks (SPIN), Noida, India, 2015, pp. 950-953, doi: 10.1109/SPIN.2015.7095381.
[4] Simoen E, Veloso A, Matagne P. Impact of the Channel Doping on the Low-Frequency Noise of Gate-All-Around Silicon Vertical Nanowire pMOSFETs. Solid-State Electronics 2023, pp. 108576 doi: 10.1016/j.sse.2022.108576.
[5] A. Veloso et al., "Vertical Nanowire and Nanosheet FETs: Device Features, Novel Schemes for Improved Process Control and Enhanced Mobility, Potential for Faster & More Energy Efficient Circuits," 2019 IEEE International Electron Devices Meeting (IEDM), San Francisco, CA, USA, 2019, pp. 11.1.1-11.1.4, doi: 10.1109/IEDM19573.2019.8993602.
[6] E. Simoen, A. Veloso and P. Matagne, "On the Asymmetry of the DC and Low-Frequency Noise Characteristics of Vertical Nanowire pMOSFETs with Bulk Source Contact," 2021 Joint International EUROSOI Workshop and International Conference on Ultimate Integration on Silicon (EuroSOI-ULIS), Caen, France, 2021, pp. 1-4, doi: 10.1109/EuroSOI-ULIS53016.2021.9560186.
[7] A. D. R. Ribeiro, G. V. Araujo, J. A. Martino and P. G. D. Agopian, "Trade-off between channel length and mechanical stress in the Operational Transconductance Amplifier designed with SOI FinFET," 2023 37th Symposium on Microelectronics Technology and Devices.
[8] Sedra, Adel S., and Smith, Kenneth C. "Microelectronic Circuits." Oxford University Press, 2014. (Chapter 4: "MOS Field-Effect Transistors (MOSFETs)").
[9] Cadence ® Verilog ® -A Language Reference, [S.l:s.n.], 2006.
[10] B. Razavi, Design of Analog CMOS Integrated Circuits, McGraw-Hill Education, 2016.

# Evolution of the content and the approach of Microelectronics training to regain skills and competences

Olivier Bonnaud
*Executive direction*
*GIP-CNFM and University of Rennes*
Grenoble France and Rennes France
ORCID: 0000-0002-7922-4917
olivier.bonnaud@univ-rennes.fr

*Abstract*— The international development of digital technology is creating numerous problems for the use of associated equipment based on electronics and microelectronics. Increasing the performance of equipment, minimizing the use of natural resources and reducing energy consumption call for a very high-level industry and many new skills. The only way to meet these challenges is to train microelectronics specialists at all levels, in order to counter the phenomenon of shortage jobs and meet the priority of improving microelectronics. In line with promising research results, all specialties are concerned, and all skill levels too. An inventory of technical priorities has enabled us to establish a strategy for training technicians, engineers and PhDs, who need to acquire the skills and know-how to be integrated into both companies and training structures. The attractiveness of the field is also part of the drive to attract more future graduate people with female representativity. The paper concludes with details of the strategy pursued by the French microelectronics training network to meet these objectives. (*Abstract*)

*Keywords—microelectronics training, jobs in shortage, global challenges, new skills, advanced practical activities*

## I. INTRODUCTION

On a global scale, we are in the midst of a digital revolution, with changes in societal behavior and the implementation of numerous new approaches. The development of the Internet and the use of digitized data have invaded our daily lives, whether for administrative procedures, financial management, citizen security, control of industrial production, media activities, controlled entertainment, advertising, commercial activities or education. Figure 1 shows this growth, which varies exponentially and should lead to storage capacities in excess of one zettabyte ($10^{24}$) in the next few years [1].

The second consequence is the need for reliable, increasingly high-performance objects and equipment, in line with the needs expressed by industry and the market for societal applications. The application domains involving connected objects [2] are more and more devoted to security, finance and environment monitoring for which the reliability is the priority [3]. To this end, the design of the integrated circuits must be optimized, tested and verified at all the steps from the concept to the layout and the technological processes must be reproducible without any defects and with a very high yield of fabrication. The third consequence concerns energy consumption, which is increasingly high [4-5].

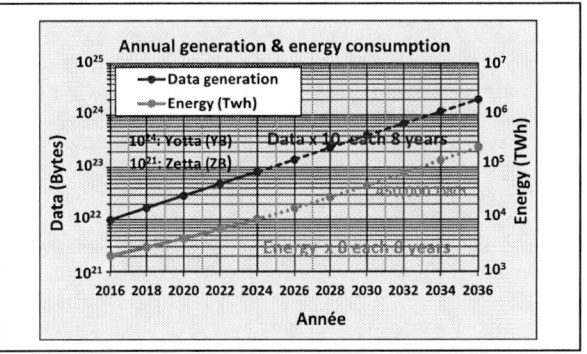

Fig. 1. Exponential growth of data and energy consumption. In less than one decade, the energy needed for digital equipement should pass over the global electrical production.

Serious forecasts show that consumption is growing exponentially, rising 8 times at every 8 years every 3 years (figure 1), and will lead the world to an energy impasse within a decade if drastic action is not taken. In this case, all electricity production would become insufficient to support digital activities. Since digital development is unlikely to slow down over the next few years, the most realistic solution is a significant improvement in the electronics associated with equipment dedicated to data generation, transmission and processing, whether for users, servers or data centers. A strategy is therefore needed to meet this demand, which can only be satisfied if the necessary human resources are available with the necessary skills.

Firstly, this document presents the workforce requirements for improvements in electronics and microelectronics, followed by human resources needs, highlighting jobs in shortage. Next, it looks at the actions taken in France by the national microelectronics and nanotechnology training network (CNFM) [6-7] to meet technical and technological needs in line with the results of available research, and to meet human resources needs by making the field more attractive. It then looks at the actions taken by the national microelectronics and nanotechnology training network in France to meet both technical and technological needs, in line with recent research results, and human resources needs, by focusing on the attractiveness of the field. The implementation of a 5-year innovation strategy project should boost this development. The name of this project is INFORISM for Strategic and Innovative Training Engineering in Microelectronics [8], included in the France 2030 plan [9].

979-8-3315-4064-7/24 $31.00 © 2024 IEEE

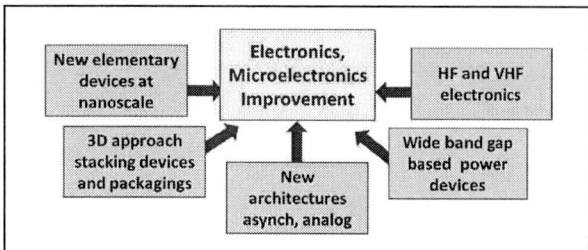

Fig. 2. Different approaches for improving the performance of microelectronics cicuit ans systems.

## II. MICROELECTRONICS IMPROVEMENT REQUIRED

To reduce energy consumption, several approaches have been proposed. On the one hand, it is necessary to reduce the power consumption of elementary microelectronic components, while improving data throughput and reducing dimensions. On the other hand, circuit design and architecture need to be reviewed in order to optimize the size and number of components per circuit, control active zones and minimize interconnection distances between the various electronic functions of the circuits. This applies to analog and digital circuits, as well as to power and frequency [10]. Figure 2 summarizes these points. At the level of nanoscale components, the primary objective is to reduce leakage currents, mainly due to tunneling in ultra-thin layers. Indeed, for circuits such as memories with several billion transistors, the sum of leakage currents becomes much greater than the processing current more especially for the technologies developed in new nodes lower than 7nm [11]. These technologies are expected to increase the flow of data, and the integration as well. To decrease the leakage current, the proposed solutions involve using materials with higher permittivity and wider bandgaps, but also introducing insulating zones, as in the case of SOI (Silicon on Insulator) technologies such as FDSOI. The development of wide-bandgap components such as SiC or GaN minimizes off-state currents and on-state resistance. This is particularly important for power supplies, which are very useful for all transmission and processing stages in data centers. Promising new results have been obtained with semiconducting oxides such as GIZO (Gallium Indium Zinc Oxide) [12], but also with organic semiconductors. The use of chiplets not only minimizes Joule effect losses, but also improves response times. Circuit packaging using chiplets also improves frequency and energy performance. For example, stacked MOSFET circuits may improve the memories integration and decrease the losses. The introduction of new functions, such as neuromorphic circuit modules, allows minimizing the size of integrated circuits, which is a step in the right direction in terms of power consumption and computing speed. The analog design is becoming much more useful for many electronics functions. The proportion of analog parts even in high integrated digital circuits is permanently increasing. In the future 7nm and smaller nodes, the importance of analog modules should represent more than 50% of the total circuit as predicted by W. Haensch [13]. This approach allows minimizing the leakage currents generated by the synchronous electronic functions that induces a permanent electrical energy consumption. In circuit architecture, one way of reducing power consumption is to introduce new standby modes that enable low-power operation by cutting power to the inactive chip, even when other chips in the same package are active. [14].

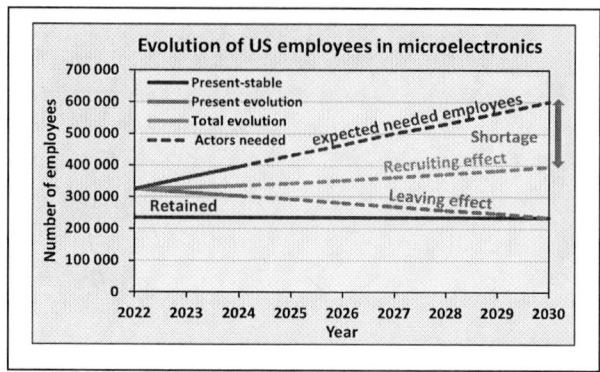

Fig. 3. Shortage in microelectronics employers in US expected by Intel. A deficit of two hundred thousands recruitement should not be satisfied.

This technique allows dropping power consumption significantly in standby mode. Looking ahead, several improvements should be achieved in the next few years if quantum electronics involving qubits is sufficiently integrated to cover the consumer market. These approaches mentioned above are among many others that can be consulted in the review article by J.R. Léquepeys *et al.* [15]. Clearly, the tasks are enormous, and new skills and competencies are needed to face these challenges.

## III. SKILLED HUMAN RESOURCES SHORTAGE

In fact, if the need for skills is growing, it's important to remember that in the field of information and communication technologies, the skills shortage is nothing new. Worldwide, and particularly in the USA, this deficit has been predicted for almost a decade [16].

Figure 3, which shows the evolution of jobs in this scientific field in the USA, highlights the gradual increase in the employee deficit up to 2030, considering departures (leaving people) and recruitment. These departures include the retirement of older employees, as well as people leaving for other, potentially more lucrative, sectors of the economy. Recruitment will not be able to compensate for these departures and to satisfy the projected needs. This shortfall could represent 200,000 employees [16] in this country. This trend is unfortunately similar for the Western countries, at least. The reasons for the departure of some skilled employees may be linked to the level of salaries in this field, which is highly competitive on a global scale. The competition with the huge number of companies in the Far East forces Western companies induces a limitation of the remuneration of even their most highly skilled and indispensable employees.

Figure 4 deduced from the analysis made by Khonexio company in 2022 on French salaries [17], shows this parameter. Analog electronics, device process manufacturing and power electronic devices and circuits pay less and are therefore more in shortage. The shortage of specialists in "embedded electronics" is due to the considerable increase in connected object applications, which call for more specialists, even with higher remuneration [2].

Figure 5 shows the impact of this parameter, among others, on the recruitment shortage derived from the French government website [18]. In 2022, an average of 40% of vacancies at all diploma levels failed to find a candidate.

979-8-3315-4064-7/24 $31.00 © 2024 IEEE

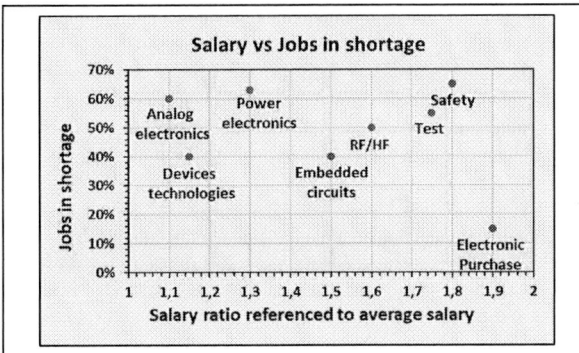

Fig. 4. Salary ratio by area of specialization. The current priority skills correspond to the lowest salaries. This is one explanation for the shortage in these specific fields. [From Khonexio analysis, 2022].

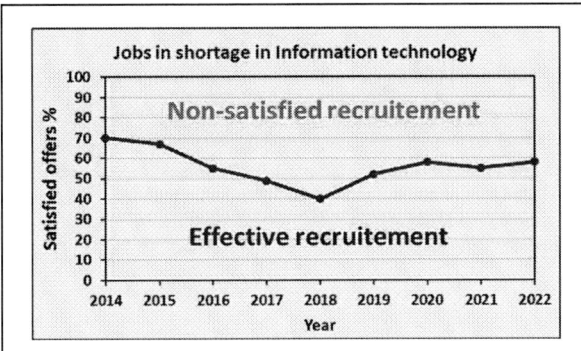

Fig. 5. Difficulty of recruitement in the information technology whatever the level, from technicians to engineers and doctors.

As a result, academic institutions need to increase the number of students in this field via attractiveness actions, whatever their level of graduation, and adapt the content of their teaching, particularly know-how, to meet the technical and energy challenges.

## IV. ACTIVITY OF THE FRENCH NETWORK TO PRODUCE SKILLS AND KNOW-HOW

The French microelectronics training network, GIP-CNFM [6], was set up over 40 years ago [7]. Its main mission is to develop and maintain technical and technological platforms on which students at all levels, from technician to engineer and doctor, can acquire the know-how essential to the technical professions in this engineering field. This network is made up of 12 academic establishments, which operate technical platforms that are locally inter-university, and the "ACSIEL Alliance electronique", an industry union with over 120 industrial members involved in the design and manufacture of microelectronic components, circuits and boards. This organization enables us to maintain a close link between the needs of industry and research, and the training strategy and development of technology platforms. In view of the skills shortages mentioned above and revealed in a "Comité Stratégique de Filière", set up in 2019 and steered by the industry union and representatives of the training network, the government has launched an action plan and opened calls for projects. It is in this context that the network's activities are being developed within the INFORISM project [8]. Every year, the platforms are used by around 20,000 students on platforms distributed throughout the 12 inter-university clusters.

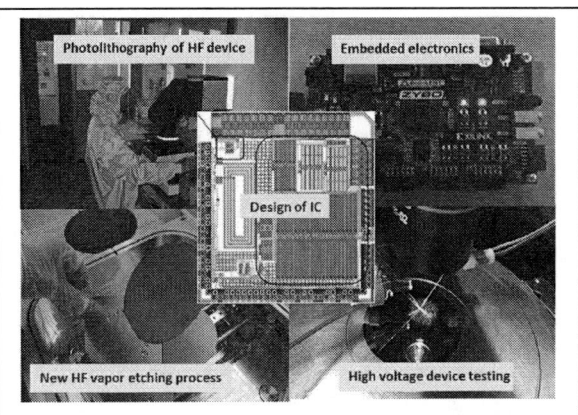

Fig. 6. Examples of realization by the students on several platforms of the main priority domains.

Figure 6 shows the achievements of students on different workstations, bearing in mind that to cover the whole discipline at network level, more than 100 platforms are offered to university training courses, but also to companies as part of vocational or lifelong training. 7 platforms are clean rooms in which students can build elementary components and circuits. Given that students may use several technical platforms in the course of their studies, these 20,000 students on platforms may in fact be coming from a physical community of around 16,000 people. The number of users graduating in bachelor's, master's, engineer's and doctorate degrees is around 5,000 annually, with degree courses lasting several years. The topics are directly connected to the priorities, knowing that more than a half of the students have practice in order to acquire the background that can be provided in all the interuniversity centers. At post-graduate level, the centers specialize in nanotechnologies at Paris-Saclay, Toulouse and Grenoble, CAD for advanced integrated circuits at Montpellier, Marseille and Paris, prototyping at Grenoble, HF electronics at the Lille and Limoges centers, electromagnetic compatibility (EMC) at Strasbourg, power electronics at Lyon and organic electronics at Bordeaux and Rennes. Figure 7 shows the distribution of activities for the majority of students in initial training. The five priorities appear in this survey. Maximum job offers from companies are in the specialties of integrated circuit design, testing and characterization, and embedded electronics. Computer-aided design activities are well managed by the network, thanks to the national software services of the GIP-CNFM network under the responsibility of the Montpellier center. All CAD software licenses are negotiated by colleagues at this center

Another joint activity is managed by the Grenoble center, enabling circuit prototyping for multi-chip projects. A part of the mission of the network consists to convince the higher education institutions (engineers, masters, PhD), to maintain a significant part of electronics in the menu of their students that can take benefit of these platforms to be better adapted to industry and research activities. This number of graduates is insufficient to meet needs, but they now have skills in priority areas [10]. To increase the pool of trainees, the network is carrying out awareness-raising and attractiveness campaigns. It should be noted that efforts have been made for several years to attract young women, as presented at SBMicro 2019 and published in 2020 [19]. In fact, in companies and universities alike, women account for less than 20% of technical employees at all levels from technician to doctorate.

979-8-3315-4064-7/24 $31.00 © 2024 IEEE

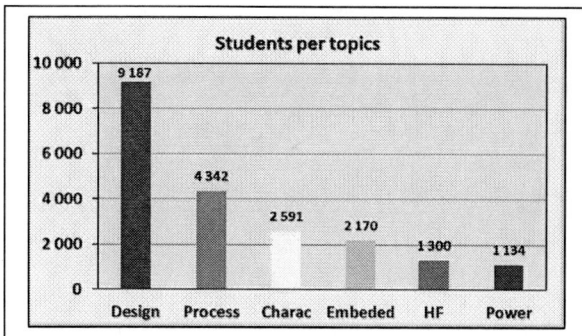

Fig. 7. Number of students per main training topic on the platofrms of the French network. The data are issued of the last academic year, corresponding to the beginning of the 5-years project. The are mainly involved in the priority topics.

This missing community presents a problem, as it leads to a 40% reduction in the skills pool compared to other industrial fields. Another way to increase the number of students consists to attract foreign students. Unfortunately, the worldwide political situation during the recent years has increased the difficulty to receive students coming from abroad, mainly at the level of master and PhD. However, the possibility to receive in the academic institution the foreign remain a priority for doctoral schools, at least in France. To meet these needs, we would also need to increase the number of teachers in the field. As already mentioned, the number of graduates, engineers and PhDs is below requirements, as is the case in higher education. The trend over the last ten years has been not to renew retirements. This has led to a reduction in the community of electronics and microelectronics educators to around 30%, with the recruitment decision moving from a national to a local organization by the academic institution [20]. Because they are increasingly financially independent, the strategy of universities and engineering schools is based on a clientelist approach to students, for whom the cost of studies is minimized. As a result, future engineers are less and less offered studies of a highly technological nature, leading to a sharp decline in the number of professors at these establishments. This trend naturally runs counter to the national strategy, and several joint working groups of professors and industrialists have been set up to convince government bodies to adapt the organization of higher education specifically to microelectronics.

## V. CONCLUSION

The meteoric evolution of societal needs has led to a galloping demand for digital tools for all kinds of applications, resulting in an exponential growth in the number of these objects, coupled with an exponential growth in energy consumption. It is urgent to counter this deleterious trend by significantly improving all equipment based on electronic components and circuits. This strategy is now spreading over the world with new national programs such as Chips Acts that are established all over the major countries involved in the microelectronics industry [21]. For this aim, new and more abundant skills are needed. Actions are being taken in this direction at French level, namely to provide new trainees with the skills for innovation and to increase the pool of trainees. Behind this, there is a problem linked to the media that are not interested to promote sciences and technologies, and to the higher education institutions that are preferring to attract students in other fields that are more lucrative with lower and

cheaper practice. This approach is not adapted to a long-term policy. However, it is clear that the strategy developed by the French network should be applied worldwide. This means a higher cooperation between international academic institutions to gather the efforts of all the global skills in order to avoid a tragic freeze in energy consumption.

## ACKNOWLEDGMENT

These activities are partially supported by grants from the French Ministry of Higher Education and Research and from the France 2030 program, AMI-CMA INFORISM: "Interest Call for Competences and Jobs in Shortage", Project ANR23-CMAS-0024. The author wants to thank their colleagues of the CNFM national network for their contribution in many works presented in this document. Special thanks to Lorraine Chagoya-Garzon for the technical support in the preparation of this presentation.

## REFERENCES

[1] "Tsunami of data could consume one fifth of global electricity by 2025," https://www.theguardian.com/environment/2017, Feb. 2018.

[2] O. Bonnaud, "The technological challenges of microelectronics for the next generations of connected sensorsIJPEST, 14, 2020, pp. 1-8.

[3] O. Bonnaud, L. Fesquet, "Innovation for Education on Internet of Things", PETI, vol. 9, 2018, pp. 1-8.

[4] E. Gelenbe and Y. Caseau, "The Impact of Information Technology on Energy Consumption and Carbon Emissions", ACM publication, pp. 1-15, June 2015, last access July 2023.

[5] M. Koot, F. Wijnhoven, Usage impact on data center electricity needs: A system dynamic forecasting model, Applied Energy, 2021, vol. 291.

[6] CNFM, "Coordination Nationale pour la formation en Microélectronique et nanotechnologies," htpps://www.cnfm.fr, 2024.

[7] O. Bonnaud et al., GIP-CNFM: a French education network moving from microelectronics to nanotechnologies, Proc. IEEE GEEC Conference; Amman, Jordan, 2011, pp. 122-127.

[8] AMI-CMA INFORISM: Interest Call for Competences and Jobs in Shortage, France 2030 program, ANR-23-CMAS-0024, https://cime.grenoble-inp.fr/fr/evenements-a-venir, April. 2024.

[9] O. Bonnaud, National recovery plan for the electronics industry in France; Proc. 30th EAEEIE, Prague (Tchek Republic), 2021, pp. 1-5.

[10] O. Bonnaud, The Five Priority Topics of Microelectronics Training to Meet Future Societal Challenges, IEEExplore, Proc. SBMicro'2023, Rio de Janeiro (Brazil), 2023.

[11] Advanced Process Technology 7nm Node and Below Market Size, Report: Embracing Growth Opportunities in 2024-2031, VMR, April 2024, https://www.linkedin.com June 2024.

[12] S.K. Moore, The First High-Yield, Sub-Penny Plastic Processor", IEEE Spectrum Journal, August 2022.

[13] W. Haensch, "Silicon CMOS devices beyond scaling", IBM J. Res. Dev. 339, 2006.

[14] Intel forum, "Intel promises 10TB+ SSDs thanks to 3D Vertical NAND flash memory", March 26, 2015.

[15] J-R. Lèquepeys, et al., "Overcoming the Data Deluge Challenges with Greener Electronics", in Proc. ESSDERC'2021, 2021, pp. 7-14.

[16] N. Flaherty, "Skills shortages hit US semiconductor sovereignty", eeNews Europe, Dec. 08, 2021, https://www.eenewseurope.com/news/

[17] Konexio analysis, https://vipress.net/ads/ETUDE-SALAIRES-2022.pdf , last access April 2024.

[18] French governement website (in French) https://dares.travail-emploi.gouv.fr/, last access April 2024.

[19] O. Bonnaud, "Weak presence of women in microelectronics: analysis and suggestion of the French training network to be more attractive., Journal of Integrated Circuits ans Systems, vol.15 (2), 2020, pp. 1-5.

[20] O. Bonnaud, Growing Demand of Educators in Electrical and Electronic Engineering, IEEE Education. CFP07MGD-CDR, ISBN 978-1-4244-1916-6, Nov 2007.

[21] The IEEE International Roadmap for Devices and Systems (IRDS) Emerges as a Global Leader for Chips Acts Visions and Programs, IEEE Computer Society, PRNewswire, Dec 12, 2023.

# Analysis of a low-dropout voltage regulator designed using omega-gate nanowire transistors experimental data.

Pedro H. Penna da Silva[1*], Joao A. Martino[2], Senior Member IEEE and
Paula G. D. Agopian[1], Senior Member IEEE

[1] UNESP, Sao Paulo State University, Sao Joao da Boa Vista, Brazil

[2] LSI/PSI/USP, University of Sao Paulo, Sao Paulo, Brazil

e-mail[*]: pedro.penna@unesp.br

*Abstract*— **In this work, the low-dropout voltage regulator (LDO) was designed using omega-gate nanowire transistors. These transistors were modeled using the Verilog-A description based on the experimental data. After the model validation LDO's were developed using gm/I$_D$ strategy considering transconductance over drain current ratio (gm/I$_D$) of 7V$^{-1}$ and 8V$^{-1}$ with load capacitances of 10pf and 100pf. The LDO was designed to provide a voltage of 1.5V and 100μA at the output. The LDO biased transistors with gm/I$_D$ equal to 8V$^{-1}$ presented very good results where the Loop Voltage Gain was 52.2dB, the gain bandwidth product (GBW) was 5.9MHz for load of 10pf, the load regulation was 27V/A and power supply rejection (PSR) was 41dB showing that Omega-gate nanowire transistors can be a good option for LDOs circuits.**

*Keywords—Low-Dropout Voltage Regulator, LDO, Nanowire device, Omega-gate nanowire, LDO Omega-gate nanowire.*

## I. INTRODUCTION

Over the years and with the advancement of technology, new technological nodes have emerged, thus enabling scaling and the emergence of new and smaller devices, following what was proposed by Gordon Moore in 1965 through Moore's Law. However, with this reduction in the size of the devices, they began to suffer unwanted degradations such as current leakage and the short channel effect (SCE) [1]. Aiming to reduce or mitigating these effects, new architectures have emerged such as multiple-gate devices which, when implemented through nanowire structures, achieve better electrostatic isolation between gate and channel resulting in greater immunity to SCE [2][3]. A promising multiple-gate device architecture is the Ω-gate nanowire that will be used in this work.

The Ω-gate nanowire fabricated in Silicon on Insulator (SOI) wafer appears as a possible evolution of the planar devices. In addition, its fabrication process is more simple than the gate-all-around (GAA) one [3], which has a similar technology. Its great coupling between gate and channel ends up guaranteeing a lower SCE, greater immunity to radiation, as well to being suitable for high gain and high frequency analog applications, all these benefits are reported in [3] – [7].

The low-dropout voltage regulator (LDO) is one of the recurring blocks in the design of integrated circuits, as this block reduces the variation introduced by the supply voltage, providing more constant voltage and current delivered to the load [8].

In this work, the Ω-gate nanowire device was modeled based on its experimental data and the low-dropout voltage regulator (LDO) was developed for two different channel inversion regime conditions, aiming to evaluate the application of Ω-gate nanowire transistor in analog built blocks.

## II. DEVICE CHARACTERISTICS

The studied transistor was fabricated at CEA-LETI,

France, on a SOI wafer with buried oxide thickness (t$_{oxb}$) of 145nm. The Ω-gate nanowire transistor has the following characteristics: the gate material is composed of Titanium Nitride (TiN) covered by Poly-Silicon, the gate oxide is composed of the combination of materials, Hafnium, silicon, oxygen and nitrogen (HfSiON) with an effective oxide thickness (EOT) of 1.3nm, the channel length (L) of 100nm, the fin height (H$_{NW}$) of 10nm and fin width (W$_{NW}$) of 10nm. Using this information, it is possible to calculate the effective width of the channel by subtracting the portion of the channel not covered by the gate (5nm) from the circumference of the circle, which provides an effective width (Weff) equal to 26.42 nm. Figure 1.A shows the schematic structure of the Ω-gate transistor in 3D, and Figure 1.B shows cross section of Ω-gate.

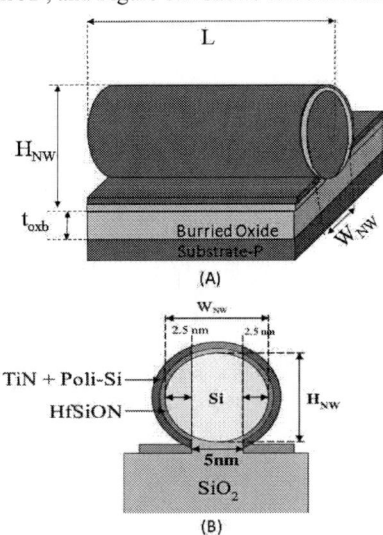

Figure 1.A 3D structure and Figure1.B a cross section of Ω-gate nanowire transistor [4].

## III. RESULTS

In order to design LDO with the Ω-gate device, the first step to be taken was to select an Ω-gate device that did not suffer with short channel effects. Using [4] as a reference, it was chosen to use a device with L ≥ 50nm, as devices larger than 50nm achieve greater immunity to SCE. Given this caveat in reference [4], an Ω-gate device with a channel length of 100nm will be used in this work.

In order to obtain an accurate Look Up Table, the transfer curves were measured with a gate step of 10 mV and drain step of 50 mV, and to model the device a C-V curve was also considered. The device modeling was performed using a Verilog-A language and Cadence Virtuoso software.

In order to verify whether the modeling is correct, two

transistor bias circuits were carried out and the transfer curves ($I_D$x$V_G$) and output curves ($I_D$x$V_D$) were measured for P-type and N-type. Figure 2 shows the drain current curves as a function of the gate voltage ($V_G$) for drain voltage ($V_D$) of 700 mV and the output characteristic ($I_D$ x $V_D$) for $V_G$ of 0.6V. When observing the Figure 2, it can be seen that the simulated result perfectly matches with model results for both the P and N-types. The obtained threshold voltages were symmetric for P and N-types with a |Vth| = 0.42V.

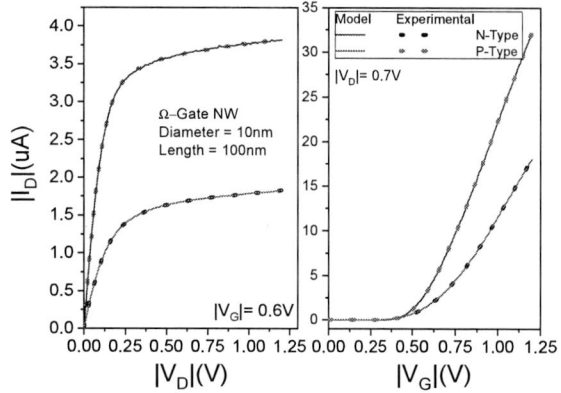

Figure 2. Drain current as a function of drain voltage (left) and gate voltage (right) of modelled and experimental P and N type Ω-gate nanowires transistors

The LDO circuit consists of an error amplifier (EA), which is powered by an ideal current source ($I_{SS}$) and a current mirror ($M_5$ and $M_6$).

The LDO is composed of two stages, where the first stage consists of a differential pair working as an error amplifier, which is biased by an ideal current source. The second stage is composed of a power transistor and the negative feedback loop formed by a resistive voltage divider. Figure 3 shows the schematic of the implemented LDO circuit, consisting of an input voltage source (VDD), a differential pair with active load formed by transistors $M_1$ to $M_4$, powered by current mirrors $M_5$, $M_6$, which reflect the current from the $I_{SS}$ voltage source. The transistor $M_P$ is utilized to supply the necessary current for the load, and resistors R1 and R2 constitute the feedback loop that provides the feedback voltage (Vfed), which will be compared with the reference voltage (VREF) to bias the gate of transistor $M_P$ and provide regulated output voltage and current. Finally, the load is simulated by a load capacitor ($C_L$) in parallel with a load resistor ($R_L$), supplying an output voltage ($V_{OUT}$). The compensation capacitor ($C_C$) is crucial for controlling the frequency response of the LDO circuit, ensuring regulator stability and minimizing supply voltage and current variations.

Firstly, the LDO circuit was developed using the gm/$I_D$ technique reported in [9], where the transistors of the differential amplifier are in the same bias. In this work, two different bias were developed for the transistors, which were gm/$I_D$ equal to 7V$^{-1}$ and 8V$^{-1}$. Due to the drain current difference between P-type and N-type transistors, it was necessary to use a relationship in the number of nanowires in parallel to have a properly effective width to maintain the same bias. Table 1 shows the list of necessary number of nanowires in parallel for the transistors to have the same bias. The LDO circuit was designed considering the output voltage of 1.5V and output current of 100uA as specifications. Another two different capacitive loads were considered 100pF

and 10pF for the same output specification. The number of multiple nanowires in of power transistor ($M_P$) was calculated aiming to provide 100uA for the load.

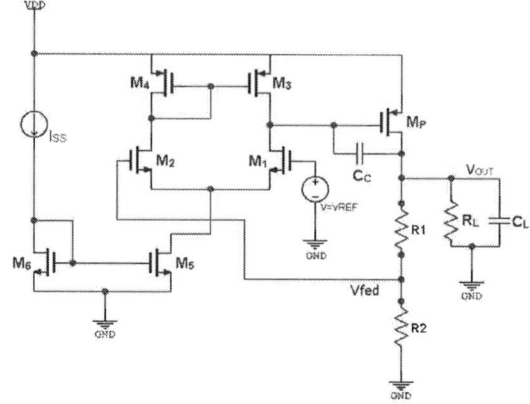

*Fig. 3 – Low-Dropout voltage circuit implemented.*

TABLE 1. NUMBER OF NANOWIRES IN PARALLEL FOR TWO DIFFERENT BIAS CONDITIONS (gM/$I_D$ = 7V$^{-1}$ AND 8V$^{-1}$).

| Device | Number of nanowires in parallel | |
|---|---|---|
| | gm/$I_D$ = 7 V$^{-1}$ | gm/$I_D$ = 8 V$^{-1}$ |
| $M_{1,2}$ | 2 | 2 |
| $M_{3,4}$ | 5 | 5 |
| $M_5$ | 4 | 4 |
| $M_6$ | 4 | 4 |
| $M_P$ | 51 | 65 |
| Channel Length | 100 nm | |

Table 2 shows the main obtained results for designed LDOs. The parameter gm$_{(MP)}$ represents the transconductance of the $M_P$ power transistor, while rds represents the output resistance related to the same transistor. The $A_{EA}$ and $R_D$ parameters can be explained together, as $A_{EA}$ refers to the error amplifier voltage gain, which directly depends on the error amplifier output resistance ($R_D$) and the transconductance of the transistor $M_1$ (gm$_{(M1)}$), where the formulation is given by equation (1). $C_{gg}$ represents the total gate capacitance, while $C_C$ represents the compensation capacitance used to guarantee that the Loop voltage gain reduces with a rate of 20 dB/dec along all amplifier frequency range and ensure a phase margin (PM) of 60°, aiming circuit stability [10].

Lastly, the feedback factor (β) is given by the ratio between the feedback loop resistors R1 and R2 or ratio between Vfed and $V_{OUT}$.

$$A_{EA} = gm_{(M1)} * R_D \qquad (1)$$

As mentioned earlier, the LDO was designed for two different biases (7V$^{-1}$ and 8V$^{-1}$) with two different load capacitances (10pF and 100pF). Tables 2 and 3 show the main obtained results.

It can be observed in Table 2 that for a higher capacitive load, it was necessary to increase the compensation capacitor value to ensure circuit stability for both bias conditions. Although the loop voltage gain remains the same for both capacitance considerations (Figures 4 and 5), the gain bandwidth product (GBW) was degraded when a higher compensation capacitor is needed, since this parameter is inversely proportional to this capacitance value.

979-8-3315-4064-7/24 $31.00 © 2024 IEEE

When the comparison is performed considering the different $gm/I_D$ condition, almost all LDO important parameters was changed.

TABLE 2. LDO PARAMETERS FOR LOAD EQUAL TO 100PF AND 10PF.

| LDO Parameters | $gm/I_D$ (V$^{-1}$) | | | |
|---|---|---|---|---|
| | 7 | | 8 | |
| | $C_L$ 100pf | $C_L$ 10pf | $C_L$ 100pf | $C_L$ 10pf |
| **gm$_{(MP)}$** (μS) | 696.74 | | 774.40 | |
| rds(KΩ) | 13.87 | | 14.56 | |
| A$_{EA}$ (dB) | 33.74 | | 36.89 | |
| R$_D$ (KΩ) | 630 | | 982 | |
| C$_{gg}$ (fF) | 1.71 | | 2.23 | |
| C$_c$ (pF) | 18 | 1.8 | 14 | 1.4 |
| β | 0.85 | | 0.85 | |

Table 2 also shows that the higher R$_D$ was obtained for higher $gm/I_D$ (8 V$^{-1}$), which contributes to obtaining the higher error amplifier voltage gain.

TABLE 3. LDO PERFORMANCE FOR LOAD EQUAL 100PF AND 10PF.

| LDO performance | $gm/I_D$ (V$^{-1}$) | | | |
|---|---|---|---|---|
| | 7 | | 8 | |
| | $C_L$ 100pf | $C_L$ 10pf | $C_L$ 100pf | $C_L$ 10pf |
| I$_{SS}$(μA) | 22.6 | | 17.44 | |
| I$_L$(μA) | 98 (~100) | | 98 (~100) | |
| C$_L$ (pF) | 100 | 10 | 100 | 10 |
| V$_{OUT}$ (V) | 1.5 | | 1.5 | |
| I$_Q$ (μA) | 45.20 | | 35.04 | |
| V$_{DO}$ (mV) | 300 | | | |
| Efficiency(%) | 57.05 | | 61.42 | |
| Load Regulation (V/A) | 35.08 | | 27 | |
| Line Regulation (mV/V) | 0.95 | | 1.08 | |
| Loop Voltage Gain (dB) | 47.83 | | 52.2 | |
| Power Consumption (μw) | 81.87 | | 63.06 | |
| PSR (dB) | -32.67 | | -41.13 | |
| GBW (MHz) | 0.501 | 5.31 | 0.620 | 5.85 |
| Phase Margin (°) | 64 | 64 | 64 | 64 |
| β | 0.85 | 0.85 | 0.85 | 0.85 |

Table 3 shows that the LDO designed for the biasing of 8V$^{-1}$ achieved higher efficiency than the one designed for biasing of 7V$^{-1}$, due to its lower quiescent current. Regarding load regulation, the best result was obtained for a biasing of 8V$^{-1}$, as load regulation depends on A$_{EA}$, and this parameter is higher for the biasing of 8V$^{-1}$ according to Table 2. For line regulation, results were very close for both biasing voltages, but there was a slight advantage for the lower biasing voltage (7V$^{-1}$). The loop voltage gain improved for the 8V$^{-1}$ biasing, as it directly depends on the A$_{EA}$ gain, which is higher for the 8V$^{-1}$ biasing.

The power consumption directly depends on the current flowing through the transistors M$_5$ and M$_6$, as well as the

current passing through R2 towards ground. The power consumption for both LDO designs is low due to the low current required by error amplifier.

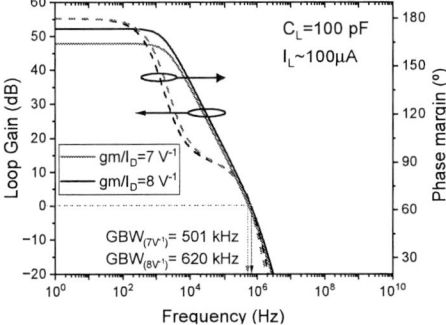

Figure 4. Loop Voltage Gain and phase LDO with load equal to 100pf.

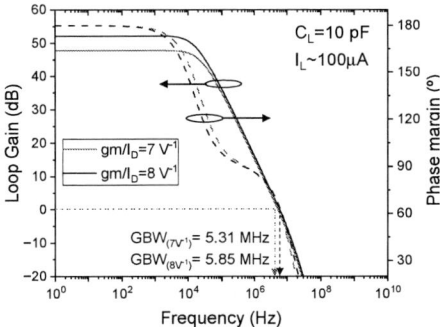

Figure 5. Loop Voltage Gain and phase LDO with load equal to 10pf.

The last analysis to be conducted is on the PSR, which can be observed in Figure 6, where it becomes a crucial parameter for the LDO, as the PSR suppresses the undesired ripple created by the input source. As can be seen in Table 3, the best PSR occurs for the 8V$^{-1}$ biasing.

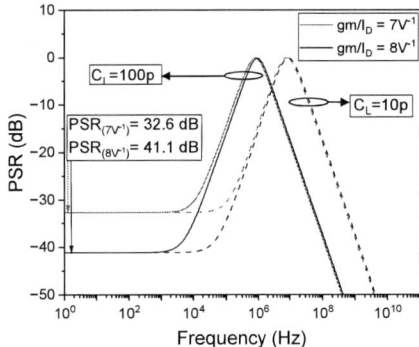

Figure 6. PSR's LDO with load equal to 100pF and 10pf.

The best results were achieved with the 8V$^{-1}$ biasing, showing a wider frequency bandwidth for a smaller load capacitance. Therefore, the load capacitance should be taken into account when the high frequency response is an important specification of the LDO project.

In order to make a comparison with a similar technology, the results from reference [11] were used, which pertains to an LDO developed with a vertical gate all around MOSFET nanowire (V-GAANW) transistors biased at $gm/I_D$ of 8V$^{-1}$ and with a load of 10pF.

979-8-3315-4064-7/24 $31.00 © 2024 IEEE

Table 4 compares the LDO parameters developed with V-GAANW MOSFETs and a 10pF load with two LDOs using $\Omega$-Gate nanowire transistors (this work), for capacitive load of 10pF. It is worth to highlight that the LDO designed with V-GAANW take into account an output current of 1μA for the same load, while the LDO designed in this work provide 100 μA to the load.

TABLE 4. COMPARISON OF A LDO DEVELOPED WITH A VERTICAL GATE ALL AROUND MOSFET NANOWIRE [11] AND ONE WITH $\Omega$-GATE NANOWIRE, BOTH WITH $gM/I_D$ EQUAL $8V^{-1}$.

| Technology | V-GAA-NW MOSFET | $\Omega$-Gate nanowire ($C_L$ = 10pF) |
|---|---|---|
| $I_L$ (μA) | 1 | 100 |
| $C_L$ (pF) | 10 | 10 |
| $V_{DO}$ (mV) | 300 | 300 |
| $V_{OUT}$ (V) | 1.5 | 1.5 |
| $I_Q$ (μA) | 1.46 | 35.04 |
| Efficiency(%) | 32.4 | 61.4 |
| Load Regulation (V/A) | 7370 | 27 |
| Line Regulation (mV/V) | 3.4 | 1.1 |
| Loop Voltage Gain (dB) | 41.4 | 52.2 |
| PSR (dB) | -48 | -41 |
| GBW (kHz) | 52 | 5900 |
| $C_C$ (pF) | 5.0 | 1.4 |
| $gm_{(MP)}$ (μS) | 7.6 | 774.4 |
| $A_{EA}$ (dB) | 27 | 37 |
| $R_{DS(MP)}$ (KΩ) | 2072 | 14 |
| $R_D$ (kΩ) | 9000 | 982 |
| $C_{gg}$ (fF) | 37.5 | 2.2 |
| β | 0.80 | 0.85 |

Although, usually the low quiescent current results in a high efficiency of LDO project and knowing that $I_Q$ of V-GAANW LDO (1.46 μA) is to much lower than the obtained $I_Q$ of $\Omega$-Gate nanowire one (35.04 μA), the efficiency of the LDO designed with $\Omega$-Gate nanowire is better (higher) due to the V-GAANW LDO output current to be very low (1 μA).

The load regulation was also significantly better in the LDOs developed with $\Omega$-Gate transistors (27V/A) compared to the one developed with V-GAANW (7370V/A), due to the fact that the transconductance and the error amplifier gain are higher than one obtained for V-GAANW LDO. The same analysis can be applied to the explanation of the better loop voltage gain value, which directly depends on $gm_{(MP)}$ and $A_{EA}$. Another parameter that $gm_{(MP)}$ has a strong impact is GBW, which is to much higher considering the $\Omega$-Gate nanowire LDO.

Although both LDO projects present a very good line regulation, the value obtained of LDO designed with $\Omega$-Gate transistors (1.1mV/V) is slightly better than one observed for V-GAANW LDO (3.4mV/V).

For the PSR, the best result was achieved by the V-GAANW LDO, with 48dB compared to 41dB for the one developed with $\Omega$-Gate. However, the $\Omega$-Gate LDO had a higher frequency.

In addition to all these factors, it should be emphasized that for the $\Omega$-Gate nanowire LDO with a 10pF load, a smaller compensation capacitor was used (1.4pF) compared to the one used for the nanowire MOSFET LDO (5pF), which takes up less area on the chip.

## CONCLUSIONS

This work initially presented LDOs developed for two different transistor biasing ($7V^{-1}$ and $8V^{-1}$) with loads of 10pF and 100pF. In this first presentation, it became evident that both LDOs exhibited excellent results, but the one developed with transistors biased at $8V^{-1}$ and $C_L$ equal to 10pF showed better loop voltage gain (52.2dB), GBW(5.9MHz), PSR (-41db), $I_Q$(35.04μA), and load regulation (27V/A) than the one developed with transistors biased at $7V^{-1}$. Subsequently, using data from an LDO developed with V-GAANW MOSFET transistors biased at $8V^{-1}$ and a 10pF load from reference [11], a comparison was made with the LDO developed with $\Omega$-Gate nanowire transistors (this work) with the same biasing ($8V^{-1}$) and load (10pF). This comparison showed that the LDO developed with $\Omega$-Gate nanowires transistors excelled in most parameters such as output current (100μA), efficiency (61.4%), line regulation (1.1mV/V), load regulation (27V/A), loop voltage gain (52.2dB), GBW (5.9MHz), and only fell short in PSR. All these factors favorable to $\Omega$-Gate nanowire make it an excellent option for the development of LDO regulators

## ACKNOWLEDGMENT

The authors acknowledge UNIVESP, CNPq, São Paulo Research Foundation - FAPESP (under grant #2020/04867-2) and Coordenação de Aperfeiçoamento de Pessoal de Nível Superior - Brasil (CAPES) - Finance Code 001 for the financial support. The authors would also thanks to CEA-LETI (France) for providing the $\Omega$-Gate device and Rodrigo do Nascimento Toledo for technical discussions.

## REFERENCES

[1] Colinge J-P 2008 FinFETs and Other Multi-Gate Transistors (Boston, MA: Springer US).

[2] J B. C. Paz et al., "Non-linearity analysis of triple gate SOI nanowires MOSFETS," 2016 31st Symposium on Microelectronics Technology and Devices (SBMicro), Belo Horizonte, Brazil, 2016, pp. 1-4, doi: 10.1109/SBMicro.2016.7731355.

[3] S. Barraud et al., "Performance of Omega-Shaped-Gate Silicon Nanowire MOSFET With Diameter Down to 8 nm," in IEEE Electron Device Letters, vol. 33, no. 11, pp. 1526-1528, Nov. 2012, doi: 10.1109/LED.2012.2212691.

[4] Perina, Welder & Martino, J.A. & Agopian, Paula. (2020). Analysis of Omega-Gate Nanowire SOI MOSFET Under Analog Point of View. Journal of Integrated Circuits and Systems. 15. 1-6. 10.29292/jics.v15i1.113.

[5] Silva V C P, Wirth G I, Martino J A and Agopian P G D 2019 A Negative-Bias-Temperature-Instability Study on Omega-Gate Silicon Nanowire SOI pMOSFETs 2019 34th Symposium on Microelectronics Technology and Devices (SBMicro) 2019 34th Symposium on Microelectronics Technology and Devices (SBMicro) pp 1–4

[6] Silva V C P, Martino J A and Agopian P G D 2018 Parasitic Conduction on $\Omega$-Gate Nanowires SOI nMOSFETs ECS Meet. Abstr. MA2018-01 1468–1468

[7] Vanessa Silva et al 2019 ECS J. Solid State Sci. Technol. 8 Q54

[8] J. Torres et al., "Low Drop-Out Voltage Regulators: Capacitor-less Architecture Comparison," in IEEE Circuits and Systems Magazine, vol. 14, no. 2, pp. 6-26, Secondquarter 2014, doi: 10.1109/MCAS.2014.2314263.

[9] F. Silveira, D. Flandre, and P. G. A. Jespers, "A gm/ID based methodology for the design of CMOS analog circuits and its application to the synthesis of a silicon-on-insulator micropower OTA," IEEE J Solid-State Circuits, vol. 31, no. 9, pp. 1314–1319, Sep. 1996,

[10] Razavi, B. Design of Analog CMOS Integrated Circuits. 2016, McGraw-Hill Higher Education. ISBN: 9780077496128.

[11] R. N. Tolêdo, J. A. Martino and P. G. D. Agopian, "Nanowire TFET with diferente Source Compositions Applied to Low-Dropout Voltage Regulator," in 2022 IEEE 36th Symposium on Microelectronics Technology, SBMICRO 2022, 16 September 2022.

# Visible light tuning with tridoped TeO2-ZnO vitreous samples for photonics

Daniel Kendji Kumada
Escola Politécnica da Universidade de São Paulo
Universidade de São Paulo
São Paulo, Brazil
https://orcid.org/0009-0007-4116-2785

Beatrice Sayuri Kato
Departamento de Ensino Geral
Faculdade de Tecnologia
de São Paulo
São Paulo, Brazil
beatricesayuri@gmail.com

Raphael de Carvalho Gonçalves
Departamento de Sistemas Eletrônicos
Faculdade de Tecnologia
de São Paulo
São Paulo, Brazil
raphaelcgkk@gmail.com

Camila Dias da Silva Bordon
Escola Politécnica da Universidade de São Paulo
Universidade de São Paulo
São Paulo, Brazil
https://orcid.org/0000-0002-3119-902X

José Augusto Martins Garcia
Escola Politécnica da Universidade de São Paulo
Universidade de São Paulo
São Paulo, Brazil
https://orcid.org/0009-0008-1022-6288

Luciana Reyes Pires Kassab
Departamento de Ensino Geral
Faculdade de Tecnologia
de São Paulo
São Paulo, Brazil
https://orcid.org/0000-0002-6795-5712

*Abstract*— This study focuses on the characterization of TeO$_2$-ZnO samples prepared with different concentrations of rare-earth ions (Tm$^{3+}$, Er$^{3+}$, and Yb$^{3+}$ ions) for potential application in photonics. Samples were produced with the melt-quenching technique and optical characterization, including luminescence and absorption measurements, was conducted, highlighting the influence of the rare-earth ion concentration on the emission intensities. The chromaticity diagram further illustrated the emitted colors, demonstrating light tuning capabilities. The sample with a higher concentration of Er$^{3+}$ and Yb$^{3+}$ ions exhibited consistent green emission for LED and display applications green light is observed for all excitation powers (with the highest purity for the lowest excitation power (14.4 W/cm$^2$), whereas the one with a lower concentration of both rare - earth ions showed a broader range of emission colors, from blue to green, indicating superior light tuning and more versatile applications across the visible spectrum. Results of luminescence intensity as a function of different laser powers indicated that two photons participate in the emission of green light (545 and 650 nm) and three photons are associated with the blue one (477 nm). The present results demonstrate a route to manage visible light emission and produce different photonic devices based on the efficiency frequency upconversion process of triply doped TeO$_2$-ZnO glasses with different concentrations of rare-earth ions.

*Keywords— Rare-earth ions, light tuning, luminescence spectroscopy, zinc-tellurite glasses, melt-quenching technique.*

## I. INTRODUCTION

The aim of this study is to produce TeO$_2$-ZnO glasses doped with Tm$^{3+}$, Er$^{3+}$, and Yb$^{3+}$ ions for light tuning in the visible spectra. The use of TeO$_2$-ZnO host is justified due to its excellent properties for photonic applications, such as high refractive index, good mechanical strength, large chemical durability, low melting temperature, and low phonon energy when compared to silicate and borate glasses, and excellent transmission from visible to mid-infrared region (400- 5000 nm) [1].

Among the diverse range of potential photonic applications of glasses based on TeO$_2$, notable examples include shielding for nuclear radiation [2], coatings for solar cells [3,4], laser functionality [5,6], memory devices [7], random lasers [8], temperature sensors [9], optical amplifiers using different configurations [10,11] and enhanced luminescence properties of rare- earth ions in the presence of silver nanoparticles [12].

The proposed rare-earth ions offer numerous technological prospects that drive the current research, owing to their light emissions across the blue, green, and red regions of the electromagnetic spectrum, crucial for RGB devices. This study presents opportunities for light tuning within the visible range. Although tellurite glasses demonstrated to be adequate hosts for white light generation [13], the present study intends to show a procedure in which different rare-earth ion concentrations can manage visible light emission. Then efficiency frequency upconversion process is reported based on triply doped TeO$_2$-ZnO glasses for photonic applications such as visible light tuning devices, LEDs, and displays.

## II. EXPERIMENTAL PROCEDURES

### A. Sample fabrication

The samples were fabricated using the melt-quenching technique. The starting composition (in wt %) was 85TeO$_2$-15ZnO. Two doped samples were prepared: 0.5Tm$_2$O$_3$/0.1Er$_2$O$_3$/2Yb$_2$O$_3$ (Sample 1), and 0.5Tm$_2$O$_3$/0.5Er$_2$O$_3$/3Yb$_2$O$_3$ (Sample 2). Fig. 1 shows the procedural flowchart employed for the sample preparation.

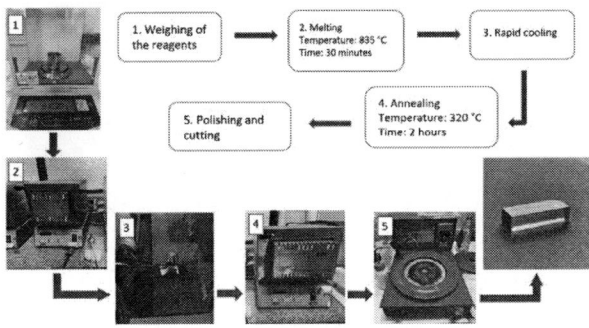

Fig. 1.   Flowchart of the vitreous sample fabrication.

The sample production initiates with weighing the reagents on a calibrated precise balance. The reagents are then mechanically mixed and transferred into a crucible made of pure platinum, which is subsequently placed inside a furnace. The crucible is maintained at 835°C for 30 minutes. After completion, the molten mixture is poured into a preheated brass mold and transferred into another furnace for heat treatment at 320°C for 2 hours to reduce internal stresses. After this step, the furnace is turned off, and the sample is gradually cooled over approximately 12 hours until reaching ambient temperature. Then the sample undergoes polishing and cleaving procedures.

### B. Characterization

For luminescence spectroscopy, it was used the experimental setup shown in Fig. 2, which included a diode laser (980 nm), lenses, mirrors, a monochromator, and a power meter (laser power ranged from 14.4 to 1076.5 W/cm$^2$). The optical absorption spectra were measured with the setup shown in Fig. 3, using the Ocean Optics (QE65PRO) spectrometer for visible spectra and the NIRQuest (NIRQ512) for the infrared one.

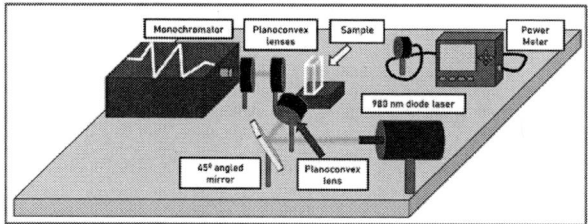

Fig. 2. Experimental setup used for luminescence measurements.

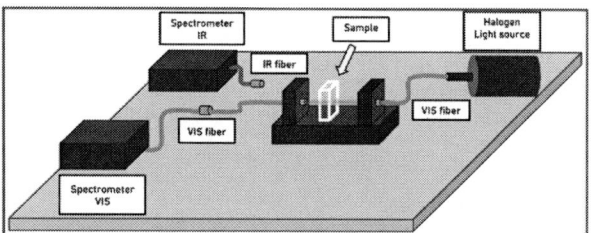

Fig. 3. Experimental setup used for absorption measurements.

### III. Results and Discussion

Fig. 4 shows the absorption bands in the visible and near-infrared regions and the corresponding electronic transitions associated with the rare-earth ions that corroborate the incorporation of trivalent rare-earth ions in both produced samples.

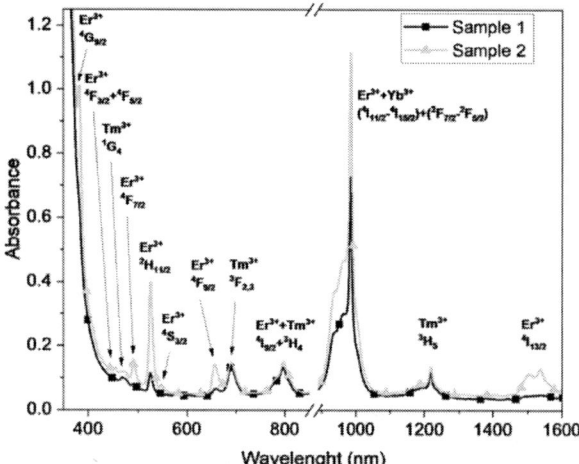

Fig. 4. Visible and near-infrared absorption spectra of the samples 1 and 2.

The higher concentration of Er$^{3+}$ and Yb$^{3+}$ ions present in Sample 2 with respect to Sample 1 explains the escalated intensity of the absorption bands attributed to Er$^{3+}$ and Yb$^{3+}$ ions.

The luminescence results of Samples 1 and 2 for different excitation powers at 980 nm, which is in resonance with the Yb$^{3+}$ transition ($^2F_{7/2} \rightarrow {}^2F_{5/2}$), are shown in Figure 5 and were obtained by the experimental setup of Fig. 2. The bands centered at 477 nm and 650 nm correspond to Tm$^{3+}$ transitions ($^1G_4 \rightarrow {}^3H_6$ and $^1G_4 \rightarrow {}^3F_4$, respectively) and those at 525 nm, 545 nm, and 650 nm to Er$^{3+}$ transitions ($^2H_{11/2} \rightarrow {}^4I_{15/2}$, $^4S_{3/2} \rightarrow {}^4I_{15/2}$, $^4F_{9/2} \rightarrow {}^4I_{15/2}$, respectively).

Notably, Sample 1, shows the prevalence of the emission in the blue region for higher powers and in the green region for lower powers. In contrast, for Sample 2 the emission predominantly occurs in the green region for all excitation powers.

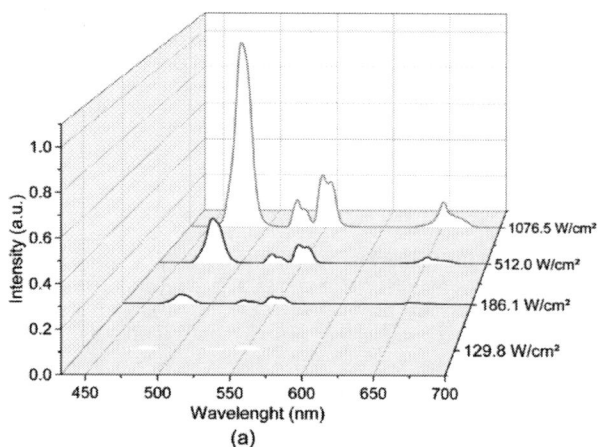

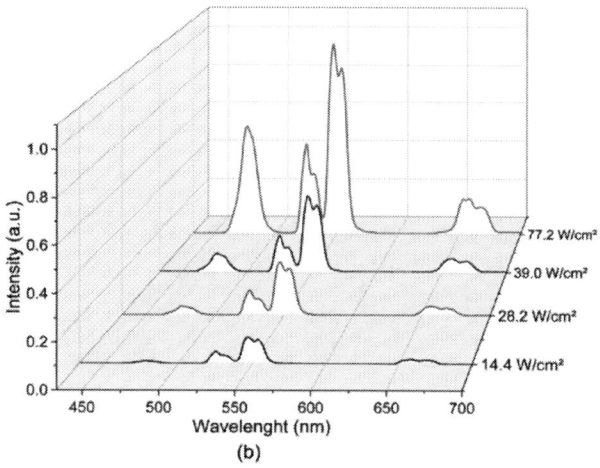

Fig. 5. Luminescence results of Sample 1 for excitation at 980 nm for different pump powers: (a) 129.8 – 1076.5 W/cm$^2$, and (b) 14.4 – 77.2 W/cm$^2$.

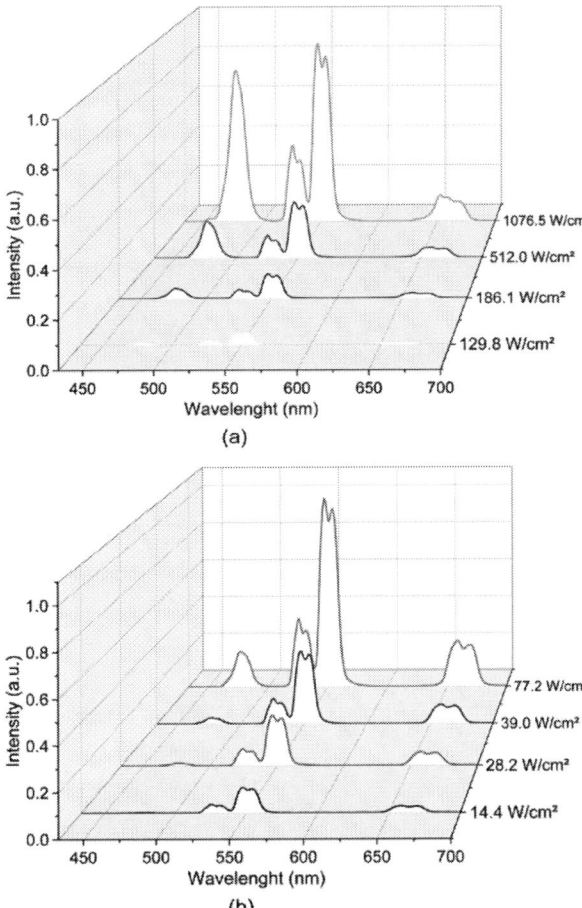

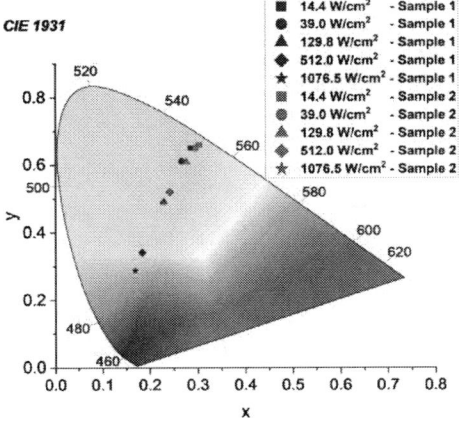

Fig. 6. Luminescence results of Sample 2 for excitation at 980 nm for different powers: (a) 129.8 – 1076.5 W/cm², and (b) 14.4 – 77.2 W/cm².

From the emission results, it is possible to obtain the coordinates (x, y) of the CIE chromaticity diagram, which indicate the emitted color for each excitation power. Thus, it is possible to observe light tuning. Fig. 7 shows the chromaticity diagram for both samples, where it is noticeable that Sample 1 exhibits greater light tuning in the visible region compared to Sample 2, ranging from the blue to the green region. On the other hand, Sample 2 exhibits emission only in the green region, as expected.

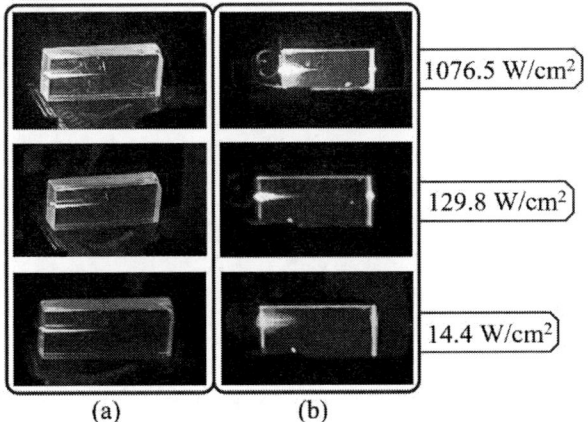

Fig. 7. CIE chromaticity diagram for samples 1 and 2 under 980 nm pump power sweep.

The images in Fig. 8 clearly show that Sample 1 emits blue light at higher powers and green light at lower powers, consistent with the findings presented in the CIE chromaticity diagram (Fig. 7). On the other hand, Sample 2 emits green light for all excitation powers; nonetheless, there is a noticeable shift in the green hue at lower powers, resulting in the emission of a purer green light. This different behavior is mainly attributed to the higher concentration of $Er^{3+}$ ions used to prepare Sample 2.

Fig. 8. Samples (a) 1 and (b) 2 emitting light under 980 nm excitation for the following powers: 1076.5 W/cm², 129.8 W/cm², and 14.4 W/cm².

The emission of sample 2 was also evaluated by calculating the purity of the emitted colors (in %), using Equation 1 [14], where (x,y) are the color coordinates of the sample, for each excitation power, $x_i = 0.333$, $y_i = 0.333$ are the coordinates of pure white light emission, and $(x_d, y_d)$ are the coordinates associated with the greenest emission, which occurs at 14.4 W/cm² ($x_d = 0.303$ and $y_d = 0.661$), as shown in Fig. 8. The results obtained are presented in table I and demonstrate that as the excitation power increases, the green emission becomes increasingly less pure, consistent with the results shown in Fig. 8.

$$Color\ Purity = \sqrt{\frac{(x - x_i)^2 + (y - y_i)^2}{(x_d - x_i)^2 + (y_d - y_i)^2}} \times 100\% \quad (1)$$

TABLE I. CHROMATICITY DIAGRAM COORDINATES FOR DIFFERENT POWERS AND THE RESPECTIVE PURITY OF THE GREEN LIGHT EMITTED (IN %) FOR SAMPLE 2.

| Power (W/cm²) | Coordinates | | Color purity (%) |
|---|---|---|---|
| | x | y | |
| 28.2 | 0.294 | 0.660 | 99.98 |
| 39.0 | 0.295 | 0.651 | 97.24 |
| 77.2 | 0.282 | 0.623 | 89.40 |
| 129.8 | 0.276 | 0.608 | 85.27 |
| 186.1 | 0.260 | 0.565 | 73.84 |
| 512.0 | 0.242 | 0.520 | 63.14 |
| 1076.5 | 0.228 | 0.489 | 57.09 |

Fig. 9 shows the dependence of the upconversion intensity ($I_{UC}$) versus laser power ($I_{exc}$) for sample 1, obtained by the results presented in Fig. 5. In the absence of saturation the upconversion emissions should satisfy the relation $I_{UC} \propto I_{ex}^n$

979-8-3315-4064-7/24 $31.00 © 2024 IEEE

where $n$ represents the number of photons that participate in the process. The slopes obtained from the log-log plot of the $I_{UC}$ versus $I_{exc}$ indicate that two photons participate in the production of the luminescence bands centered at 545 and 650 nm and three photons are associated with the generation of the 477 nm emission. The deviation from the expected slopes ($n$ = 2 and 3) is attributed to saturation of the $Yb^{3+}$ absorption transition [15]. The same behavior was observed for the sample 2.

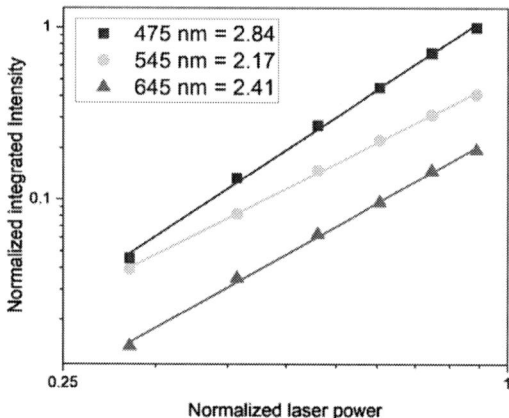

Fig. 9. Dependence of the upconversion intensity with the laser power.

## IV. Conclusion

The present investigation demonstrates that the efficiency frequency upconversion process of triply doped $TeO_2$-ZnO glasses can be used for photonic applications. In this context, it is shown that different concentrations of rare-earth ions can manage visible light emission. From the results obtained, it's apparent that Sample 2 possesses applications for displays and LEDs operating in the green region; green light is observed for all excitation powers (14.4 to 1076.5 $W/cm^2$), with the highest purity for the lowest excitation power (14.4 $W/cm^2$). This is attributed to the higher concentration of $Er^{3+}$ and $Yb^{3+}$ with respect to Sample 1. On the other hand, Sample 1 with a lower $Er^{3+}$ and $Yb^{3+}$ concentration demonstrates superior coverage across the visible spectrum (ranging from blue to green) and more versatile emission properties compared to Sample 2, making it more promising for photonic devices with applications that require broad light tuning. The slopes obtained from the log-log plot of the $I_{UC}$ versus $I_{exc}$ indicate that two photons participate in the production of the luminescence bands centered at 545 and 650 nm and three photons are associated with the generation of the 477 nm emission. The present results demonstrate a route to promote different photonic devices based on triply doped $TeO_2$-ZnO glasses produced with different concentrations of rare-earth ions.

## Acknowledgment

This study was financed in part by the Coordenação de Aperfeiçoamento de Pessoal de Nível Superior – Brasil (CAPES) – Finance Code 001. We acknowledge Conselho Nacional de Desenvolvimento Científico e Tecnológico— Grant 465.763/2014 (Instituto Nacional de Ciência e Tecnologia de Fotônica) and Grant 305745/2023-9.

## References

[1] G. Lozano, et al. "Cold white light emission in tellurite-zinc glasses doped with $Er^{3+}$–$Yb^{3+}$–$Tm^{3+}$ under 980 nm," J. Lumin., vol. 228, pp. 117538, December 2020.

[2] H. O. Tekin, et al. "Newly developed tellurium oxide glasses for nuclear shielding applications: an extended investigation," J. Non Cryst. Solids, vol. 528, pp. 119763, January 2020.

[3] J. A. M. GARCIA, et al. "Efficiency boost in Si-based solar cells using tellurite glass cover layer doped with $Eu^{3+}$ and silver nanoparticles," Opt. Mater., vol. 88, pp. 155-160, February 2019.

[4] B. C. Lima, L. A. Gómez-Malagón, A. S. L. Gomes, J. A. M. Garcia, & L. R. P. Kassab. "Plasmon-Assisted efficiency enhancement of Eu$^{3+}$-doped tellurite glass-covered solar cells," J. Electron. Mater, vol. 46, pp. 6750-6755, August 2017.

[5] M. J. V. BELL, et al. "Laser emission of a Nd-doped mixed tellurite and zinc oxide glass," JOSA B, vol. 31, Issue 7, pp. 1590-1594, 2014.

[6] L. M. MOREIRA, et al. "The effects of $Nd_2O_3$ concentration in the laser emission of $TeO_2$-ZnO glasses," Opt. Mater., vol. 58, pp. 84-88, August 2016.

[7] L. BONTEMPO, S. G. DOS SANTOS FILHO, L. R. P. KASSAB, "Conduction and reversible memory phenomena in Au-nanoparticles-incorporated $TeO_2$–ZnO films," Thin Solid Films, vol. 611, pp. 21-26, July 2016.

[8] J. G. Câmara, D. M. da Silva, L. R. P. Kassab, C. B. de Araújo, & A. S. Gomes, "Random laser emission from neodymium doped zinc tellurite glass-powder presenting luminescence concentration quenching," J. Lumin., vol. 233, pp.117936, May 2021.

[9] G. S. Bezerra, et al. "Influence of plasmonic and thermo-optical effects of silver nanoparticles on near-infrared optical thermometry in $Nd^{3+}$-doped $TeO_2$–ZnO glasses," J. Lumin., vol. 265, pp. 120222, January 2024.

[10] V. D. DEL CACHO, et al. "Fabrication of $Yb^{3+}$/$Er^{3+}$ codoped $Bi_2O_3$–$WO_3$–$TeO_2$ pedestal type waveguide for optical amplifiers," Opt. Mater., vol. 38, pp. 198-203, December 2014.

[11] E. S. MAGALHÃES, et al. "The influence of the different parameters used for the production of double line waveguides in $Nd^{3+}$ doped $TeO_2$-ZnO glasses by fs laser writing," Integrated Optics: Devices, Materials, and Technologies XXVI, SPIE, pp. 255-261, March 2022.

[12] T. A. A. DE ASSUMPCÃO, et al. "Frequency upconversion properties of $Tm^{3+}$ doped $TeO_2$-ZnO glasses containing silver nanoparticles," J. Alloy Compd., vol. 536, pp. S504-S506, September 2012.

[13] V. A. G. RIVERA, et al. "White light generation via up-conversion and blue tone in $Er^{3+}$/$Tm^{3+}$/$Yb^{3+}$-doped zinc-tellurite glasses," Opt. Mater., vol. 67, pp. 25-31, May 2017.

[14] V. D. S. de Souza, F. J. Caixeta, K. de Oliveira Lima, & R. R. Gonçalves, "Modulating white light emission temperature in $Ho^{3+}$/$Yb^{3+}$/$Tm^{3+}$ triply doped nanostructured $GeO_2$-$Nb_2O_5$ materials for WLEDs applications," J. Lumin., vol, 248, pp. 118978, August 2022.

[15] M. E. Camilo, E. D. O. Silva, L. R. Kassab, J. A. Garcia, & C. B. De Araújo. "White light generation controlled by changing the concentration of silver nanoparticles hosted by $Ho^{3+}$/$Tm^{3+}$/$Yb^{3+}$ doped $GeO_2$–PbO glasses". J. Alloys Compd., vol, *644*, pp. 155-158 September 2015.

# Integrating STEAM and Maker Education in High School to Introduce the Microchip Manufacturing Process

Priscila Costa
Campus Bage
Universidade Federal do Pampa
Bagé-RS-Brazil
priscilacosta.aluno@unipampa.edu.br

Lucas Paiva Dias
Campus Bage
Universidade Federal do Pampa
Bagé-RS-Brazil
lucaspd.aluno@unipampa.edu.br

Eduardo Ceretta Moreira
Campus Bage
Universidade Federal do Pampa
Bagé-RS-Brazil
eduardomoreira@unipampa.edu.br

*Abstract*—The rapid evolution of microchip technology has significantly transformed various global industries, necessitating innovative educational approaches to align with technological advancements. This article explores the integration of Science, Technology, Engineering, Arts, and Mathematics (STEAM) with Maker Education at Brazilian high schools, aiming to demystify the microchip manufacturing process. Through practical, hands-on activities and the utilization of 3D printed models, this study demonstrates a method for engaging students with the intricate concepts of microchip production and binary logic. The didactic tools developed for this purpose enable students to visualize and interact with the stages of microchip construction and function, fostering a deeper understanding and sparking interest in the field of microelectronics. This educational approach enhances student technical skills while simultaneously promoting problem-solving, critical thinking, and creativity, all essential competencies in the contemporary technological landscape. The methodology presented herein shows potential as a scalable model for secondary education, aiming to inspire future innovations in the integration of technology and interdisciplinary learning.

*Keywords— STEAM, Maker Education, microchip manufacturing, 3D printing, engineering education.*

## I. INTRODUCTION

Since the advent of the transistor in 1947, followed by significant advancements in integrated circuits, microchips have played a pivotal role in the digital revolution, substantially impacting the global economy. This technological evolution has not only transformed the technology industry but has also permeated various economic sectors. In Brazil, the semiconductor sector demonstrates robust performance but continues to face challenges such as dependence on imported chips and electronic components. This underscores the need for ongoing investments in research, development, and infrastructure to strengthen the national industry.

Internationally, initiatives like the CHIPS and Science Act of 2022 in the United States, and in Brazil, the National Institute of Science - INCT NAMITEC, CEITEC, and the Brazilian Association of the Semiconductor Industry (ABISEMI), reflect growing concerns about the future of the semiconductor sector. These efforts aim to stimulate national semiconductor research and production, seeking to rebalance global technological leadership.

The concept of STEAM in secondary education has been considered in several works in the literature [1-4]. It represents an innovative, interdisciplinary educational approach, founded on the idea that combining different areas can stimulate critical thinking, creativity, and problem-solving in students, which are essential in an increasingly technological and interconnected world. Jeon & Lee [1] developed an educational program based on systems thinking for high school students, highlighting improvements in analytical and problem-solving skills through the promotion of systems thinking. Liu [2] focuses on instructional design in mathematical modeling under the STEAM concept, demonstrating how the integration of interdisciplinary practical activities can foster cooperation and innovation among students, as well as encourage creativity.

In the current context, marked by challenges that demand innovation and continuous adaptation, Brazil faces the need to invest in training specialized labor in semiconductors, particularly sparking the interest of high school students in the fascinating world of microelectronics. This study aims to develop and apply didactic tools to clarify both the manufacturing processes of microchips and the fundamentals of binary logic in transistors, essential for data processing in computers. By using components produced via 3D printing, this work proposes the representation of micrometric and nanometric phases involved in chip manufacturing on an enlarged scale (in centimeters), allowing students a macroscopic and tangible understanding of this complex procedure. The method employed in the didactic proposal, anchored in the principles of maker education and problem-based learning (PBL), seeks to integrate creativity, critical thinking, and technical skills. This practical approach is designed to provide students with direct experience of microchip manufacturing processes, simultaneously promoting innovation and active engagement in learning.

## II. INTRODUCTION TO MICROCHIP MANUFACTURING PROCESSES

The construction of a chip involves numerous, costly technological processes executed in highly controlled environments known as clean rooms. Prior to this, however, complex steps are required which constitute the design and blueprint process of microchips. It begins with the conception and planning of technical specifications for the type of chip to be built, followed by logical design, where the technical team defines the chip's behavior using hardware description languages. This includes the definition of functional blocks such as processors, memory, communication interfaces, and their inter-device interactions. Once simulations demonstrate correct electrical functioning, this model is then transformed into a physical design, detailing the exact location of transistors and interconnections using CAD tools to optimize performance and efficiency. Concurrently, verification occurs to ensure that the design meets all initial specifications. After layout optimization, the design is converted into numerous masks that will be used in the manufacturing process.

Following the design and blueprinting stage, actual manufacturing begins in a clean room environment to prevent any contamination that could affect chip performance. The process starts with the deposition of semiconductor material layers on a silicon substrate, forming the base for integrated circuits. Techniques such as Chemical Vapor Deposition (CVD) are used to create a series of layers that will serve as the foundation for the transistors and other chip structures. This is followed by photolithography, a method that transfers the design mask patterns onto the silicon wafer surface. This technique involves exposing a photosensitive material, called photoresist, to ultraviolet light, which is masked according to the circuit design. The subsequent development reveals the desired pattern, allowing the exposed or protected areas to undergo the next step by using chemical etching. This process selectively removes parts of the material, creating the physical structures of the chip, such as channels for transistors. Ion implantation is used to doping specific areas of the silicon inject critical dopants, altering their electrical properties to form regions that conduct electricity in a controlled manner, essential for transistor functionality. Each step is meticulously controlled and monitored to ensure absolute precision, given the nanometric scale of the structures involved. Finally, metallization is performed to connect the various parts of the circuit. This step involves the deposition of metal layers, typically aluminum or copper, to form the electrical interconnections between the transistors and other chip components, completing the complex manufacturing process. Through these procedures, it is possible to produce microchips that are the heart of modern electronic devices, enabling continual advances in digital technology.

## III. BINARY LOGIC IN TRANSISTORS

To understand the function of the transistor within a microchip, a simple analogy can be utilized, viewing the transistor as a minuscule switch, so small that it is invisible to the naked eye. This switch has the incredible capability to turn on and off billions of times per second, a marvel of modern engineering that lies at the heart of all digital devices. Transistors are fundamentally the building blocks of binary language—the ones and zeros that form the foundation of digital communication. When a transistor is on, it represents a '1'; when it is off, it signifies a '0'. These binary digits are not merely numbers; they are the language through which computers communicate, process information, and make decisions. The magic starts with the transistor's structure, consisting of three parts: the source, the gate, and the drain. The gate can be seen as a control mechanism. When a specific voltage is applied, it allows electronic motion to flow from the source to the drain, turning the switch on, thereby recording a '1'. Removing the voltage stops the flow, turning the switch off, and resulting in a '0'. This ability to switch on and off enables transistors to perform logical operations and process data. The global microelectronics industry tirelessly seeks to shrink transistors to nanoscopic sizes, creating billions of electronic components on a single chip. This miniaturization has been key to increasing computing power and efficiency, leading to the powerful smartphones, laptops, and servers that power our digital world today.

In bit processing, electronic components work together to ensure efficient and precise operation. Transistors act as the primary switching elements representing bits. Resistors ensure that current flows through the transistors at safe, controlled levels. Capacitors help maintain stable voltage and provide quick energy when needed. This interaction ensures that when a bit is processed—for example, when typing the letter 'A'—the transistors within the chip switch between on and off states in a controlled and precise manner, representing the corresponding binary code (01000001 in the case of 'A') through the presence or absence of electron flow. Resistors and capacitors assist in regulating this process, maintaining signal integrity and circuit stability.

Following this contextualization, from manufacturing processes to the functioning of electronic processing that generates bits and, consequently, the operationalization of the chip, arises the challenge of constructing macroscopic and playful pieces that represent the stages of construction and the functioning of the microchip. This task is fundamental to the proposal, as such activities stimulate critical thinking and student interest in learning [5]. Understanding this complex system should spark the interest of high school students, promoting greater engagement during workshops, and facilitating the teaching-learning process proposed by active methodologies [6], with Problem-Based Learning (PBL) being no exception.

## IV. HANDS-ON EDUCATION FOR MICROCHIP MANUFACTURING ENGINEERING

With the rapid advancement of technology and limited classroom time, it becomes increasingly challenging for educators to stay updated and effectively connect theoretical concepts with their practical applications. In response to this challenge, extracurricular workshops offer a valuable alternative. These sessions not only bridge the gap between theory and practice but also ignite students' curiosity, encouraging them to further explore and understand emerging technologies.

Active learning methodologies are increasingly gaining recognition, proving effective not just for student engagement but also for enhancing their learning experiences [7]. Central to this approach is the shift from viewing students as passive recipients of information to engaging them as active participants in their own educational journey. This paradigm shift is particularly pivotal in STEAM education, where hands-

979-8-3315-4064-7/24 $31.00 © 2024 IEEE

on, problem-solving activities encourage students to take ownership of their learning [8]. By actively involving students in practical and theoretical challenges, these methodologies empower them to become key players in their educational development, fostering a deeper understanding of complex concepts through experiential learning [9]. Such active learning strategies are continuously evolving, demonstrating their crucial role in cultivating a proactive, inquiry-based educational environment.

Given the complexities inherent in comprehending the manufacturing and functional mechanisms of transistors, which are key components of microchips, a range of pedagogical strategies have been formulated to assist high school students in mastering these concepts. These methods aim to elucidate the complex processes involved in semiconductor fabrication and transistor functionality, thereby advancing educational engagement in the field of microelectronics.

With this objective in mind, two distinct 3D models were developed for printing, each serving a unique educational purpose. The first model (Figure 1) graphically represents the microchip manufacturing process, as outlined in Section II. This model consists of eight segments, each depicting a critical stage of manufacturing, such as oxidation, photoresist application, and doping. An associated activity organizes students into groups to sequentially assemble a chip, facilitating their understanding of the fundamental concepts.

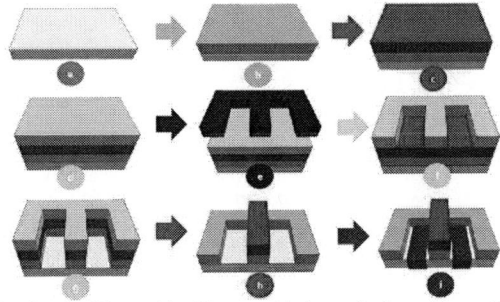

Fig. 1. a) silicon; b) silicon oxidation; c) deposition of electrically conductive polysilicon; d) photoresist deposition; e) photomask deposition; f) and g) result after etching; h) removal of the photoresist and the polysilicon edges; i) final prototype.

The second model (figure 2) features an interactive setup demonstrating binary functionality using circuits and Arduino components.

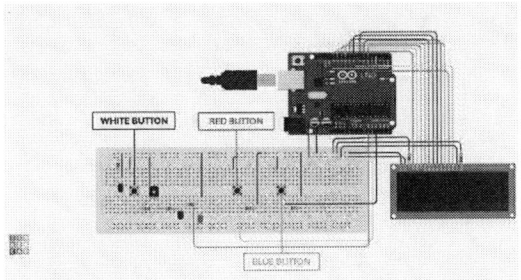

Fig. 2. Schematic representation of the second learning model.

The first red LED represents the transistor's drain in the model, which is always on. The second LED represents the base, activating only the second red and the white LEDs when the white button is pressed, engaging the base of the NPN-type transistor with an operation of ~0.7 V and allowing current to pass. The connected red button is monitored through the analog pin A1, and when activated, it records the value of the white button. The blue button is monitored by the analog port A2, which when activated clears the sequence of assigned information. The LCD is connected in a standard way, with connections as follows: VSS, VCC, V0, RS, RW, E, D0, D1, D2, D3, D4, D5, D6, D7, A, K, with their pin assignments respectively as: GND, 5V, GND, pin 13, pin 12, pin 11, pin 10, pin 9, pin 8, pin 7, pin 6, pin 5, pin 4, pin 3, pin 2, 5V, GND. The resistors seen from left to right in the image are: 467Ω, 5.6kΩ, 100Ω, 100Ω, 1kΩ, 400Ω, 470Ω, 220Ω. Therefore, this model includes pieces representing a single transistor within a microchip, with LEDs indicating when current flows from the drain to the source, triggered by pressing a white button. Students can press a red button to log whether the transistor was 'on' (representing a '1') or 'off' (representing a '0'), which is then displayed on an LCD screen. As inputs accumulate to form an 8-bit binary number, the corresponding character is displayed. Additionally, a blue button allows students to clear the LCD and restart the process. Arduino is tasked with continuously monitoring each button press. Specifically, when the red button is pressed, the Arduino checks the input from the white button to determine the value to be displayed on a particular column of the LCD screen. This process repeats until eight values are registered. Subsequently, these values are compiled into a binary sequence and transformed into a displayable character on the LCD (figure 3).

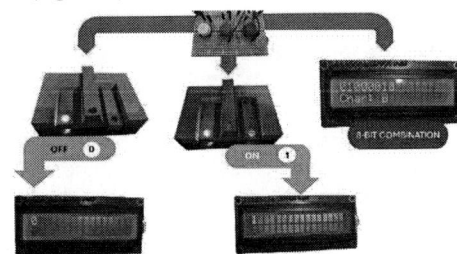

Fig.3. Complete representation of the final model

Figure 3 illustrates the operation of the created model. When the white button is pressed, the Arduino interprets this as a '1' and displays it on the LCD, otherwise, displays '0'. When the red button is pressed, then the Arduino record the value of the white button and go to the next LCD's column. After this action is performed 8 times, an 8-bit combination is formed, the Arduino interprets and displays the corresponding character on the LCD. The blue button functions as a reset, clearing the screen to enable the student to try various 8-bit combinations. This practical setup is designed to provide students with a tangible and visual understanding of electronic communication through problem-solving tasks introduced during the activity, fostering an almost intuitive grasp of the concepts involved [10]. As an example, by providing a straightforward introduction to the logic underlying binary number patterns, students should be empowered to demonstrate various combinations. This approach can also be applied to letters, using a table to illustrate the concept and then showing how this concept extends to technological operation. Using this methodology, students are placed at the center of their learning

979-8-3315-4064-7/24 $31.00 © 2024 IEEE          27

experience, actively determining the values within the system, thereby engaging in active learning and becoming integral to their educational journey. In this context, the role of the educator is to facilitate the learning process by providing guidance and addressing questions during the activity, rather than acting solely as the repository of knowledge.

## V. EVALUATION, SCALEMENT AND POTENTIAL IMPACT

To comprehensively assess the efficacy of the proposed educational approach, a combination of qualitative and quantitative methodologies will be employed. Qualitative evaluation will involve discursive textual analysis (DTA), a methodology that seeks to develop new understandings about phenomena and discourses. The practical exercise of DTA will utilize a data collection instrument through inquiries to students at the end of the workshop application. Quantitative evaluation will be conducted through rigorous statistical analysis of survey and test data, providing measurable insights into the effectiveness of the educational intervention. This mixed-methods approach ensures a holistic understanding of the impact of the workshop on student learning outcomes and engagement. The scalability of 3D printed models is a key strength of this educational approach, allowing for the introduction of more complex content, such as the broader process of transistor manufacturing or various transistor associations that form logic gates. This scalability enables the educational approach to be adapted to different levels of complexity, catering to the diverse needs and abilities of students across a range of educational contexts. The potential impact of the workshop on the career choices of students is expected to be significant and multifaceted. Introducing new electronics and programming concepts to students through practical, hands-on methods such as 3D printed models and binary pattern logic can develop their deeper and more enduring interest in STEM fields. Moreover, by providing students with a clear and practical understanding of how electronic devices function, this educational intervention can spark their curiosity and motivate them to delve further into subjects such as physics and chemistry, potentially inspiring them to pursue higher education in engineering and exact sciences. The objective of the workshop would be fully achieved if, in the near future, these students follow careers in the field of microelectronics. The results of the applied methodologies will be published in a science education journal.

## FINAL CONSIDERATIONS

This study integrated STEAM and Maker Education into a workshop aimed at Brazilian high school students, promoting an understanding of microchip manufacturing. Through hands-on activities with 3D printed models, students engaged with complex concepts such as microchip manufacturing and binary logic. This approach aimed to clarify the intricate stages of manufacturing and reinforce skills such as problem-solving, creativity, and critical thinking, which are present in the current technology-driven scenario. The positive results observed in student engagement and knowledge retention underscore the effectiveness of this educational model. It also highlights the potential of interdisciplinary learning to equip students with the technical and cognitive skills needed for innovation in the constantly evolving technology sector. The models presented can contribute to the education of both high school and higher education students, as they offer a scalable model for educational institutions that aim to modernize their curricula and better prepare students for future technological challenges. The flexibility in constructing models that represent the microchip manufacturing processes and their operating principles allows the workshops to be adapted based on the knowledge level of the students. The application of the workshops can contribute to the awakening of science, as well as the future training of human resources in the field of microelectronics in Brazil.

## ACKNOWLEDGMENT

The authors acknowledge the financial support of INCT-Nano e Microeletrônica para tecnologias habilitadoras – NAMITEC, CNPq Project: 406193/2022-3, and CNPq Process 406311/2023-4.

## REFERENCES

[1] J. Jeon, H. Lee, "The Development and Application of STEAM Education Program based on Systems Thinking for High School Students," *Journal of The Korean Association For Science Education*, vol. 35, p. 1007, Dec. 2015.

[2] L. Liu, "Instructional Design of Mathematical Modeling in High School Under the STEAM Concept," *Journal of Education and Development*, vol. 6, p. 17, May 2022.

[3] M. Zhang and Y. Liu, "A Study of Problem-solving Skills Development Model for Junior High School Students under C-STEAM Concept," *ICEKIM 2023*, Nanjing, People's Republic of China, May 2023

[4] M. H. Land, "Full STEAM Ahead: The Benefits of Integrating the Arts Into STEM," *Procedia Computer Science*, vol. 20, pp. 547-552, 2013.

[5] J. C. Guimarães Junior, J. M. da Silva, A. R. de Sousa, E. H. S. Rodrigues de Melo, and F. da Silva Araujo, "A arte de ensinar e humanizar: práticas pedagógicas inovadoras para uma educação centrada no aluno," *Contribuciones a Las Ciencias Sociales*, vol. 17, no. 1, pp. 1578-1588, Jan. 2024.

[6] P. Leão, C. Coelho, C. Campana, and M. Henriques Viotto, "Flipped classroom goes sideways: reflections on active learning methodologies," *Revista de Gestão*, vol. 30, no. 2, pp. 207-220, 2023.

[7] B. Dogani, "Active learning and effective teaching strategies," *International Journal of Advanced Natural Sciences and Engineering Researches*, vol. 7, pp. 136-142, Apr. 2023.

[8] M. G. Bertrand and I. K. Namukasa, "STEAM education: student learning and transferable skills," *Journal of Research in Innovative Teaching & Learning*, vol. 13, no. 1, pp. 43-56, 2020.

[9] A. Portillo-Blanco, J. Gutiérrez-Berraondo, L. Trombetti, K. Zuza, S. Sirmakessis, A. Moriconi, A. Iturbe-Zabalo, L. Barelli, and S. Pasqua, "Innovative teaching methods in engineering education: the STEAM-Active project," 2023 *32nd Annual Conference of the European Association for Education in Electrical and Information Engineering (EAEEIE)*, pp. 1-5, 2023.

[10] Y. Kubota, A. Takahashi, Y. Hayakawa, Y. Kashiwaba, and K. Yajima, "Analysis of Active Learning Suitability of Subjects in Information and Electronics," *International Journal of Engineering Pedagogy*, vol. 7, no. 3, pp. 19-33, 2017.

[11] C. Jeyamala e A. M. Abirami, "Enhancing Student Learning and Engagement in Freshman Course on Problem Solving using Computers," *Journal of Engineering Education Transformations*, vol. 33, pp. 192-200, jan 2020.

# Reversible Memory Operation of Al/TeO$_2$-ZnO-Au/TeO$_2$-ZnO/p-Si MIS Structures

José Augusto Martins Garcia
Escola Politécnica da USP
Universidade de São Paulo
São Paulo, Brazil.
https://orcid.org/0009-0008-1022-6288

Leonardo Bontempo
Escola Politécnica da USP
Universidade de São Paulo
São Paulo, Brazil.
leonardoobontempo@gmail.com

Daniel Keij Kumada
Escola Politécnica da USP
Universidade de São Paulo
São Paulo, Brazil.
https://orcid.org/0009-0007-4116-2785

Luciana Reyes Pires Kassab
Departamento de Ensino Geral
Faculdade de Tecnologia de São Paulo
São Paulo, Brazil.
https://orcid.org/0000-0002-6795-5712

Sebastião Gomes dos Santos Filho
LSI/PSI/EPUSP
Universidade de São Paulo
São Paulo, Brazil.
https://orcid.org/0000-0002-0324-5703

*Abstract*—**This study investigates the reversible memory operation of Al/TeO$_2$-ZnO-Au/ TeO$_2$-ZnO/p-Si MIS structures using a TeO$_2$-ZnO (TeZnO) matrix in the interface of the p-Si substrate and a thicker TeO$_2$-ZnO-Au (TeZnO-Au) composite (~ 170 nm). From C-V electrical characterization, it was shown a charging and de-charging process of the TeZnO/TeZnO-Au interface. During the forward sweep from -2 V to 2V of the C-V acquisition, electrons are injected into the TeZnO/TeZnO-Au interface which promotes a charging process corresponding to an increase of the flat band voltage from ~0.25 V to ~0.50 V. I-V electrical characterization revealed a voltage-dependent memory effect, with distinct hysteresis behaviors observed at cycles with voltage sweep ranging from -1V to 1V. Also, for cycles ranging from -4 V to 4V, the MIS structure with the TeO$_2$ZnO interlayer exhibited enhanced charge storage and retention capabilities, underscoring its potential as a versatile multibit storage device. Tests with multiple cycles of voltage sweeps demonstrated consistency in the current response, corroborating the relevance and reliability of this memory technology. This memory device emerges as a promising candidate to address the escalating demand for enhanced data storage density and versatility in modern semiconductor memory technologies.**

*Keywords — Memory, TeO$_2$, ZnO, Gold, Composites*

## I. INTRODUCTION

Glasses based on tellurium dioxide (TeO$_2$) offer a unique combination of properties that render them a versatile material for various technological advancements. Their remarkable transparency window (400-5000 nm), high polarizability, and comparatively low melting point in comparison to silicates make them applicable in diverse fields [1]. Moreover, they exhibit low cut-off phonon energy (~800 cm$^{-1}$), high vitreous stability, chemical durability, and a substantial refractive index (~2.0), all of which are essential for ultrafast optical switching. Particularly noteworthy is their exceptionally low OH$^{-1}$ content, rendering them valuable for near-infrared applications [2, 3]. The combination of TeO$_2$ with zinc oxide (ZnO) provides a robust platform for the development of functional materials. TeO$_2$-ZnO composites, whether in bulk or thin-film forms, have demonstrated potential across various domains [1-3]. For instance, these materials have displayed laser action at 1064 nm with differing concentrations of Nd$^{3+}$ ions [4]. Incorporating rare-earth ions into TeO$_2$-ZnO glasses, along with the integration of silver and gold nanoparticles, enhances photoluminescence through plasmonic effects, thus exhibiting promise in photonic devices [1-4]. Furthermore, TeO$_2$-ZnO glasses doped with Eu$^{3+}$ ions has been demonstrated to enhance the efficiency of silicon solar cells [5]. One particularly intriguing property of TeO$_2$-ZnO-Au composites lies in the conduction phenomena of thin insulating films due to the current transport through available energy levels, either by internal charge transfer between the donor-acceptor pairs or by the hopping process due to the charged or noncharged metallic nanoparticles [6]. Previous work has shown that by introducing gold nanoparticles (Au NPs) into TeO$_2$-ZnO-Au (TeZnO-Au) films, bistable electrical behavior can be induced and the leakage current of the device can be influenced by the gate structure and the bias voltage direction [7]. However, the chemical nature of tellurium oxide phases complicates the understanding of these materials. Another study has identified different polymorphs of TeO$_2$ in TeZnO thin films, each possessing distinct characteristics and formation conditions [8]. The advancement of multibit memory technologies has become imperative in addressing the escalating demand for enhanced data storage density, commensurate with the augmented processing capabilities of modern semiconductor technologies like bi-layer memristors, MIS structures, and capacitive multi-bit memory [9-11]. This progress is particularly crucial in the realm of neuromorphic computing and machine learning, where the need for high-density, low-power memory solutions is paramount. [12-14].

This paper investigates the memory phenomenon observed in Al/TeO$_2$-ZnO-Au/TeO$_2$-ZnO/p-Si structures. The aim is to clarify the role of a thin insulating layer positioned between the TeZnO-Au interface and the Si substrate. To achieve this, a different metal-insulator-semiconductor structure (MIS) was proposed. The device was fabricated using a boron-doped silicon wafer (p-Si) as the semiconductor substrate. Two distinct scenarios were evaluated for the insulating layer. In the first scenario, a single layer of TeO$_2$-ZnO-Au (TeZnO-Au) was deposited onto the p-Si substrate to be the gate insulator dielectrics. In the second scenario, a two-layered structure was employed. The first one consisted of a thin layer of TeZnO without Au, which was deposited directly onto the p-Si substrate. The second layer was comprised of TeZnO-Au and was deposited onto the first layer. The purpose of the second scenario was to investigate the influence of the TeZnO interlayer without Au. Aluminum (Al) contacts served as the metallic layer in all instances. The present study represents innovation among multi-bit capacitive memories. We highlight that this effect was also observed in organic materials within pinMOS structures [11] and ferroelectric polymers in double capacitor structures [15].

979-8-3315-4064-7/24 $31.00 © 2024 IEEE

## II. MATERIALS AND METHODS

In this study, boron-doped silicon wafers were employed for semiconductor device applications. Boron-doped wafers were selected, featuring a <100> crystal orientation and resistivity of 1 to 10 $\Omega$.cm. Subsequently, the wafers underwent a RCA cleaning procedure [8], to remove impurities and contaminants from the silicon surface. Finally, a diluted hydrofluoric acid bath in deionized water (1HF(49%):100H$_2$O) was utilized to strip away the native oxide layer. Two Metal-Insulator-Semiconductor (MIS) devices were fabricated using Radio Frequency (RF) magnetron sputtering at a frequency of 14 MHz. To ensure deposition purity, a base pressure of $6.7 \times 10^{-3}$ Pa was employed, and the work pressure was set at $6.7 \times 10^{-1}$ Pa with a controlled flow rate of 18 sccm argon and 0.5 sccm oxygen. The TeZnO target was fabricated from 85% TeO$_2$ and 15% ZnO powders, both 99.999% pure, using an 8-ton uniaxial press and sintering at 515 °C for 10 hours, resulting in a 5 cm diameter and 4 mm thickness target.

Two different Metal-Insulator-Semiconductor (MIS) devices were fabricated. For the first one, after performing the cleaning procedure, a TeZnO-Au composite layer was deposited at room temperature for 75 min. they were using a co-deposition technique, with the first electrode at 50 W RF power and the second at 7 W RF power, targeting a commercial Au target of 99.999% purity to obtain a 160-nm thick layer as illustrated in Fig. 1(a). For the second MIS device, a TeZnO thin layer was first sputtered at 50 W RF power for 7 min. to obtain a ~15nm thick layer without Au. Following this, a TeZnO-Au composite layer was deposited onto this former layer at room temperature for 75 min. using a co-deposition with the first electrode at 50 W RF power and the second at 7 W RF power, targeting a commercial Au target of 99.999% purity to obtain a ~170nm thick layer as illustrated in Fig. 1(b). The deposition of aluminum contacts was achieved through thermal evaporation, resulting in a deposited layer 0.5-$\mu$m thick. A mechanical mask was employed during the evaporation process to define the area of the Al contacts, which was $7.8 \times 10^{-3}$ cm$^2$ after measuring by profilometry. Similarly, the rear contact of the silicon wafer was formed using also aluminum evaporation, ensuring uniformity in thickness comparable to the top contacts.

Following fabrication, two types of MIS capacitors were underwent electrical characterization using the Keithley 2400 Source Meter Unit (SMU), which has an accuracy of 0.012%. This instrument enabled the extraction of parameters from the Current–Voltage (I–V) characteristics of the MIS capacitors. The leakage current was quantified by applying an external voltage ramp to the capacitors and subsequently capturing and analyzing the resulting I–V characteristics. Importantly, all electrical measurements were performed at room temperature. Profilometry analyses were performed to accurately measure the total thickness of the films deposited by sputtering. Additionally, Scanning Electron Microscopy (SEM) images were obtained using the JEOL JCM-6000 benchtop microscope to examine and characterize the interface between the p-Si substrate and the insulating layer.

## III. RESULTS AND DISCUSSION

Using profilometry analysis, the average thickness of the TeZnO-Au layer was (167±12) nm. Conversely, when the TeZnO-Au layer was deposited on the TeZnO thin layer without Au, the total thickness was (182±24) nm. Assuming that the TeZnO-Au layer has the same thickness in both MIS devices, the average thickness of the TeZnO layer was estimated as approximately (15±2) nm. To corroborate the profilometry analysis, SEM images were also acquired. These images further corroborated the results of total thickness obtained by profilometry with a relative error lower than 9%. Also, the thinner layer of TeZnO without Au is not distinguishable in the SEM image, as shown in Fig 2b.

To evaluate the charge storage capacity of the two proposed MIS devices, capacitance-voltage (C-V) measurements were conducted using an HP 4280A C Meter operating at a frequency of 1 MHz. During the C-V characterization, a forward voltage scan ranging from -2V to 2V was performed, followed by an inverse voltage scan from 2V to -2V, with a voltage step of 0.2V. It was observed that the inclusion of an additional TeZnO layer induced the memory effect in the MIS device. This memory effect led to the presence of hysteresis in the capacitance response between the forward and inverse voltage scans of the MIS fabricated with the interfacial TeZnO layer (TeZnO-Au+TeZnO) as shown in Fig. 3.

Current-voltage (I-V) characterizations were conducted for each MIS device to explore the operational characteristics of the memory effect. These I-V characterizations employed a Keithley 2400 SMU, initiating with a forward voltage sweep ranging from -1V to 1V, followed by an inverse voltage sweep from 1V to -1V. Figure 4 shows the I-V results. The obtained I-V data demonstrated that the MIS device fabricated without the TeZnO thin layer did not manifest the memory effect. Conversely, the TeZnO-Au+TeZnO device exhibited a featured hysteresis in current response between the ascending and descending voltage sweeps, as shown in Fig. 4. This observed hysteresis is an indicative of the memory effect inherent to the TeZnO-Au+TeZnO device. Specifically, the forward sweep from -1V to 1V corresponds to the charging state of the memory (Set 1), while the inverse sweep from 1V

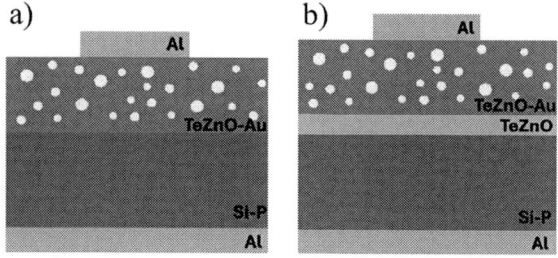

Fig. 1. Schematic illustrations of MIS structure without the TeZnO interfacial layer (a) and the MIS structure with the inclusion of the TeZnO layer (b).

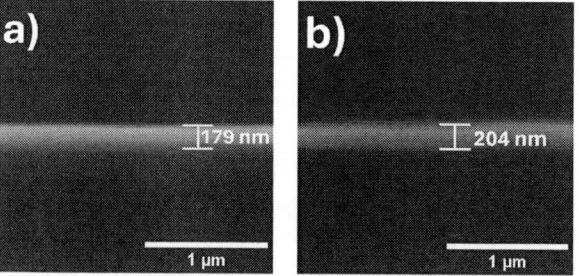

Fig. 2. SEM images of MIS structure without the TeZnO interfacial layer (a) end the MIS with the inclusion of the TeZnO layer (b).

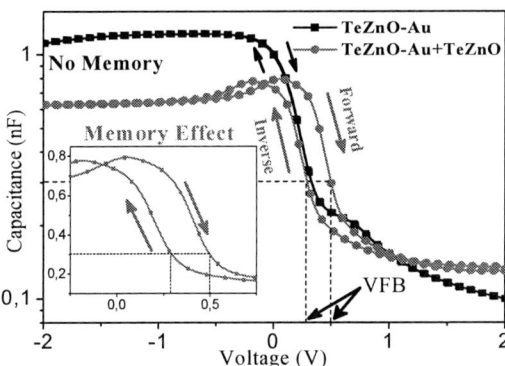

Fig. 3. The C-V characteristics of the TeZnO-Au and TeZnO-Au+TeZnO devices. Inset shows the hysteresis in the capacitance response with expanded scale of the abcissa axis for the the TeZnO-Au+TeZnO and the estimated VFB with $[(C_{OX}/C_{MOS})^2-1](V_G)$ method. [16].

to -1V represents the de-charging state of the memory state (Set -1). Notably, in this device, reading can be performed at 0V, where the device maintains a stored charge at the TeZnO/TeZnO-Au interface.

These findings elucidate the role of the TeZnO as an interface layer between p-Si and TeZnO-Au that activates the observed memory effect shown in Figs. 3 and 4, having the TeZnO-Au dielectrics as a conducting layer for electrons due to Au NPs in the charging and de-charging cycles [13, 14]. For the TeZnO-Au case without the TeZnO layer in Fig. 3, it is noteworthy that the forward and inverse sweeps present a decreasing capacitance in the depletion region at a flat band voltage ($V_{FB}$) around ~0.25V while, for the TeZnO-Au+TeZnO case, the flat band voltage is around ~ 0.50 V in the forward sweep and ~0.25V in the inverse sweep, which is approximately the same value of the forward and inverse sweeps in the case without TeZnO layer. VFB was estimated from the Graphical $[(C_{OX}/C_{MOS})^2-1](V_G)$ method. [16]. These results can be explained by assuming a charging and de-charging process of the TeZnO/TeZnO-Au interface. During the forward sweep, electrons are injected into the TeZnO/TeZnO-Au interface which promotes a charging process corresponding to an increase of the flat band voltage to ~0.50V. On the other hand, during the inverse sweep, electrons are removed from the TeZnO/TeZnO-Au interface which promotes a de-charging process corresponding to a decrease of the flat band voltage to ~ 0.25V. In addition, it is

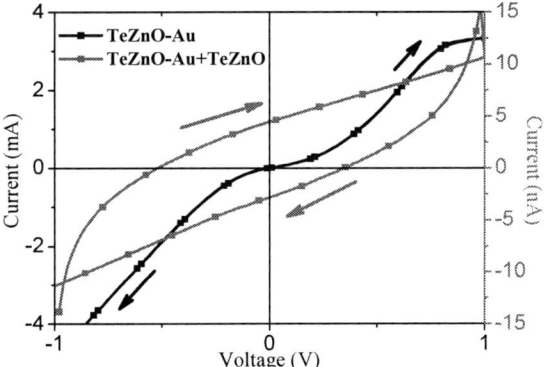

Fig. 4. I-V characteristics of the TeZnO-Au and TeZnO-Au+TeZnO devices. The left current label corresponds to response of the TeZnO-Au device, and the right label corresponds to the response of TeZnO-Au+TeZnO device.

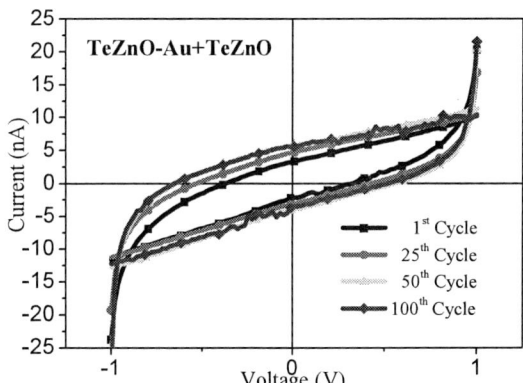

Fig. 5. Current hysteresis obtained after 1, 25, 50 and 100 sweep cycles between -1 and 1V.

important to note that the charging and de-charging processes do not occur for the TeZnO-Au case without the TeZnO layer because the TeZnO/TeZnO-Au interface does not exist and, as a result, the flat band voltage does not change and remains at ~0.25V.

The charging and de-charging mechanisms explained from the C-V characteristics are also manifested in I-V characteristics, as previously described in Fig. 4. The intermediate TeZnO layer, particularly when fabricated as an insulating layer between the TeZnO-Au composite and p-Si layer acts as a passivation layer that controls the charge injection at the TeZnO/TeZnO-Au interface. Therefore, the memory effect observed in the TeZnO-Au +TeZnO device can be attributed to the unique properties of the TeZnO-Au composite in association with the TeZnO insulating layer. The incorporation of Au NPs within the TeZnO matrix introduces bandgap trap states in the structure. This altered property allows the transport of electrons through the TeZnO-Au composite as reported previously [7].

Conducting multiple cycles between -1V and 1V demonstrated consistency in the response curves corresponding to "Set 1" and "Set -1", as depicted in Fig. 5 obtained after 1, 25, 50, and 100 sweep cycles. This underscores the potential of TeZnO-Au+TeZnO as a capacitive memory device. The TeZnO-Au+TeZnO device displays unique hysteresis responses based on the applied bias voltages, as shown by subsequent I-V characterizations. I-V symmetric sweeps between -1V to 1V, -2V to 2V, -3V to 3V, -4V to 4V and -5 to 5 reveal varied current behaviors in the TeZnO-Au+TeZnO memory device, shown in Fig. 6. An increase in bias voltage magnitude amplifies the hysteresis, near-symmetrically, highlighting the device's potential for multi-bit storage in contemporary semiconductor memory technologies.

The different hysteresis curves observed when applying different bias voltages to the TeZnO-Au+TeZnO device can be attributed to the different levels of the charge trapping and de-trapping of electrons at the TeZnO/TeZnO-Au interface. Different bias voltages mean to selectively populate the TeZnO/TeZnO-Au interface with different charge concentrations and, consequently, distinct hysteresis behaviors. However, the memory behavior was significantly altered after applying a bias voltage of -5V, as shown in the inset of Fig. 6. Following this application, the device no longer exhibited the previously observed near-symmetric effect, even

979-8-3315-4064-7/24 $31.00 © 2024 IEEE

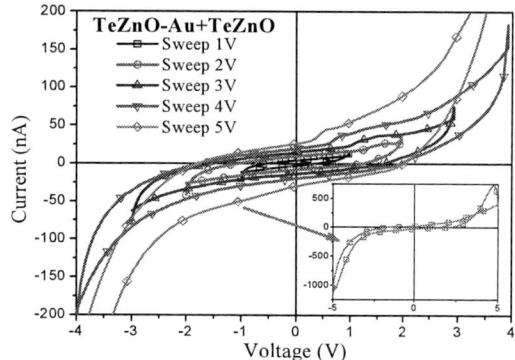

Fig. 6. Distinct current hysteresis behaviors were observed in the results following multiple voltage symmetric sweeps of 1V, 2V 3V, 4V and 5V. With insertion of complete 5V sweep.

during voltage sweeps of 1V, 2V, 3V, or 4V. This suggests a structural modification that impacts the behavior of charges at the interface and alters the current conduction mechanisms. These changes indicate the need for further investigation to fully understand the underlying processes and their implications for device performance. Future studies will focus on elucidating these characteristics in detail. In the TeZnO-Au+TeZnO device, the electrons will be predominantly localized at the TeZnO/TeZnO-Au interface. The TeZnO-Au composite presents conducting properties that are influenced by the charge trapping process that involves capturing and confining charge carriers, particularly electrons, within localized energy states, which can stem from NPs presence in the material [7]. These sites establish discrete potential energy wells or traps within the material's energy bandgap, enabling electron localization and confinement during the conduction of electrons through the TeZnO-Au composite. When all trapping sites are filled, the charging by electrons allows electric current to flow through the material. The applied electrical bias or device operation introduces an external electric field, altering trapped electron energy distributions and elevating them to higher energy states in the conduction band. This energy shift enables trapped electrons to surpass potential barriers imposed by filled trap states, becoming mobile and contributing to the conduction current for the charging and de-charging process of the TeZnO/TeZnO-Au interface. The proposed TeZnO-Au+TeZnO memory presents an alternative charge storage and voltage-dependent memory behaviors that signify its potential as a multibit memory to address the escalating demand for enhanced data storage density and versatility in modern semiconductor technologies.

## IV. CONCLUSIONS

In this study, MIS structures were fabricated with p-Si and two distinct insulator conditions, focusing on the TeZnO-Au composite layer. Our findings elucidated the pivotal role of the TeZnO-Au composite layer in facilitating a pronounced memory effect within the MIS device when a TeZnO interface layer between TeZnO-Au and p-Si is used. This memory effect, evidenced by distinct hysteresis curves in capacitance and current-voltage characteristics, can be attributed to the charge trapping and de-trapping processes at the TeZnO/TeZnO-Au interface. Multiple cycles of voltage sweeps demonstrated consistency in the current response of the TeZnO-Au+TeZnO memory device, corroborating the relevance and reliability of this memory technology. The

observed variability in hysteresis curves across different bias voltages underscores the TeZnO/TeZnO-Au for charge trapping and de-trapping processes, further highlighting the TeZnO-Au+TeZnO architecture's potential as a versatile multibit storage device. The TeZnO-Au+TeZnO memory device emerges as a highly promising candidate in the evolving landscape of semiconductor memory technologies. Studies about possible applications for ReRAM memories will be conducted in the future.

## ACKNOWLEDGMENT

This study was financed in part by the Coordenação de Aperfeiçoamento de Pessoal de Nível Superior – Brasil (CAPES) – Finance Code 001. Acknowledgment is to CNPq (Grant: INCT de Fotônica /465.763/2014 and Grant: 305745/2023-9) for their support in this research endeavor.

## REFERENCES

[1] T. A. de Assumpcão, et al. "Frequency upconversion properties of Tm3+ doped TeO2–ZnO glasses containing silver nanoparticles." J. Alloys Compds, vol. 536, pp. 504-S506, 2012.

[2] V. D. del Cacho, A. L. Siarkowski, N. I. Morimoto, B. H. V. Borges, and L. R. P. Kassab, "Fabrication and characterization of TeO2–ZnO rib waveguides.", ECS Trans., vol. 31, pp. 225–229, 2010.

[3] X. Zhou, Y. Wang, G. Wang, L. Li, K. Zhou, and Q. Li, "Cooperative downconversion and near-infrared luminescence of Tb3+/Yb3+ co-doped tellurite glass.", J. Alloys Compd., vol 579, pp. 27–30, 2013.

[4] L. Moreira, et al, "The effect of excitation intensity variation and silver nanoparticle codoping on nonlinear optical properties of mixed tellurite and zinc oxide glass doped with Nd2O3 studied through ultrafast zscan spectroscopy." Opt. Mater., vol. 79, pp. 397–402, 2018.

[5] J. A. M. Garcia, L. Bontempo, L. A. Gomez-Malagón, and L. R. P. Kassab, "Efficiency boost in Si-based solar cells using tellurite glass cover layer doped with Eu $^{3+}$ and silver nanoparticles.", Opt. Mater., vol. 88, pp. 155–160, 2019.

[6] J. G. Simmons, and R. R. Verderber, "New conduction and reversible memory phenomena in thin insulating films", Proc. R. Soc. Ser. A, vol. 301, pp. 77–102, 1967.

[7] L. Bontempo, S. G. dos Santos Filho, and L. R. P. Kassab. "Conduction and reversible memory phenomena in Au-nanoparticles-incorporated TeO2–ZnO films." Thin Solid Films, vol. 611, pp. 21-26, 2016.

[8] L. Bontempo, S. G. dos Santos Filho, and L. R. P. Kassab. "Process oxygen flow influence on the structural properties of thin films obtained by co-sputtering of (TeO2) x-ZnO and Au onto Si substrates." Nanomaterials, vol. 10, n. 9, pp. 1863, 2020.

[9] S. Stathopoulos, A. Khiat, M. Trapatseli, S. Cortese, A. Serb, I. Valov, and T. Prodromakis, "Multibit memory operation of metal-oxide bi-layer memristors.", Sci. Rep., vol. 7, pp. 17532, 2017.

[10] M. N. Koryazhkina, et al. Bipolar resistive switching in metal-insulator-semiconductor nanostructures based on silicon nitride and silicon oxide. In: Journal of Physics: Conference Series. IOP Publishing. pp. 012028, 2018.

[11] Y. Zheng, et al. Introducing pinMOS memory: A novel, nonvolatile organic memory device. Advanced Functional Materials, vol. 30, n. 4, pp. 1907119, 2020.

[12] J. Kang, et al. Cluster-type analogue memristor by engineering redox dynamics for high-performance neuromorphic computing. Nature Communications, vol. 13, n. 1, pp. 4040, 2022.

[13] Y. Xi, et al. In-memory learning with analog resistive switching memory: A review and perspective. Proceedings of the IEEE, vol. 109, n. 1, pp. 14-42, 2020.

[14] T. Soliman. First demonstration of in-memory computing crossbar using multi-level Cell FeFET. Nature Communications, vol. 14, n. 1, pp. 6348, 2023

[15] W. Y. Kim Free fabrication of a low‐voltage multi‐bit memory device based on a ferroelectric polymer and photosensitive film. Micro & Nano Letters, vol. 14, n. 2, pp. 202-205, 2019.

[16] K. Piskorski, and H. M. Przewlocki.The methods to determine flat-band voltage V FB in semiconductor of a MOS structure. In: The 33rd International Convention MIPRO. IEEE, pp. 37-42, 2010.

# A Simple Method of Fabrication of the Stainless Steel/ Copper Oxide Nanoparticles Hybrid Structure for Sensing Applications

Andrei Alaferdov
*Department of Hardware and Microelectronics*
*Instituto de Pesquisas Eldorado*
Campinas, Brazil
andrei.alaferdov@eldorado.org.br

Carolina Carvalho Previdi Nunes
*Department of Hardware and Microelectronics*
*Instituto de Pesquisas Eldorado*
Campinas, Brazil
carolina.previdi@eldorado.org.br

Matheus Dias Sousa
*Department of Hardware and Microelectronics*
*Instituto de Pesquisas Eldorado*
Campinas, Brazil
matheus.sousa@eldorado.org.br

Igor Fernandes Namba
*Department of Hardware and Microelectronics*
*Instituto de Pesquisas Eldorado*
Campinas, Brazil
igor.namba@eldorado.org.br

Fabio Domingues Caetano
*Department of Hardware and Microelectronics*
*Instituto de Pesquisas Eldorado*
Campinas, Brazil
fabio.caetano@eldorado.org.br

Fernando Idalirio de Lima Leite
*Department of Hardware and Microelectronics*
*Instituto de Pesquisas Eldorado*
Campinas, Brazil
fernando.leite@eldorado.org.br

*Abstract*—The hybrid structure of stainless steel films decorated by copper oxide nanoparticles was fabricated using dc sputtering technique following the thermal oxidation method. The character of oxidation and chemical stability over time were evaluated using energy-dispersive X-ray and X-ray photoelectron spectroscopies. The relatively high amount of the CuO phase in nanoparticles and a low degree of oxidation of stainless steel make this hybrid structure a promising candidate for mono- and disaccharide sensing applications.

*Keywords—hybrid structure, copper nanoparticles, stainless steel, sputtering, thermal oxidation*

## I. INTRODUCTION

The fast-growing sector of sensory systems, which are pervasive in our life, require low-cost, stable and simple to fabricate new materials that provide excellent sensing performance. The nanomaterials are one of the most promising structures for diverse sensing applications [1], specifically metal oxide structures, which start widely to be used as a sensitive element in biosensors [1, 2].

The copper oxide nanostructures, due to their extraordinary optical and catalytic properties together with specific surface morphology, are found in various sensing structures [3], [4], microfluidic [5] and MEMS [6] applications among many other devices. It is well known, that copper oxide II nanostructures – CuO, where Cu is in +2 oxidation state, gain a charge during the electrochemical oxidation processes of mono and disaccharides, i.e. can be used as the sensitive element in glucose sensors [7-9].

The common methods, such a thermal oxidation of copper [10]–[12] or thermal decomposition of metalorganic compounds [7], [13] used for fabrication of the CuO nanoparticles (NPs), usually require high-temperature (about 400 °C) and/or long time (a few hours) treatment procedures. Moreover, the additional step of transferring fabricated NPs to another (conducting) substrate is often necessary for further applications.

Here we developed a cost-effective fabricating process of stainless steel /copper oxide NPs hybrid structure on flexible polymer substrate using dc sputtering and thermal oxidation techniques. The employment of polyimide as durable and biocompatible substrate, and stainless steel (SS) as mechanically stable and resistant to aggressive environments conducting layer allows using this hybrid film as a low-cost alternative to expensive gold or unstable carbon-based structures for sensing mono- and disaccharides structures in the medical, food and agriculture industries among other sensing applications.

## II. MATERIALS AND METHODS

Using two chambers magnetron dc sputtering machine (21-LAT-001, LAT, ROK) equipped with 254 mm diameter copper (99.997 % purity) and 316 stainless steel targets, three types of samples were fabricated on 95 µm thick polyimide substrate (HSF095G, Eleven Electron, ROK): 1.6 µm thick pure SS films, 6 µm thick pure Cu films and 1.6 µm thick SS films decorated by Cu NPs films. The sputtering procedure was carried out at room temperature and pressure of $4\text{-}6\times10^{-4}$ mbar, using argon (99.999 % purity) as worked gas which was introduced in the chamber at a flow rate of 15 sccm for 25 s for bulk films and at 20 sccm for 100 s in the case of nanoparticles. Applying a power of 2 kW and 3 kW for SS and Cu targets, respectively, the deposition was done during 700 s for SS films, 14 min for Cu films and 6 s for Cu NPs with a target-to-substrates distance of 90 mm for Cu and 81 mm for SS target. The fabrication of SS film/Cu NPs structure was performed in only one manufacturing process composed of two steps: first – deposition of SS film in chamber one and second – deposition of Cu NPs in chamber two.

Taking into account the fact that CuO growth process usually is ranged between 300 and 850 ºC (a high-yield CuO growth takes place between 400 °C and 700 °C in the air or oxygen atmosphere) [6], in order to oxidize the NPs of Cu up to CuO, the thermal oxidation of the obtained hybrid structure was performed at 320 ºC during 2 h in the air using reflow oven (TSM, ROK).

The investigation of the fabricated samples morphology was carried out using scanning electron microscope – SEM (JSM-IT200, JEOL, Japan) equipped with silicon-drift energy dispersive spectroscopy – EDS detector. Previous as well as posterior chemical analysis were performed by scanning the area of 100 µm² in at least three different places of the sample. A detailed study of thin film and nanoparticles structure was done employing high resolution SEM – HR SEM (Nova 200 Nanolab, FEI, USA). A detailed chemical investigation of the samples was performed at three different places of each studied sample, employing X-ray photoelectron spectroscopy – XPS technique (K-Alpha, Thermo Fisher Scientific, USA). The X-ray photoelectron spectra were acquired using

979-8-3315-4064-7/24 $31.00 © 2024 IEEE

monochromated, micro-focused, low-power Al Kα X-ray source (hv = 1486.6 eV). The deconvolution of the $Cu2p_{3/2}$ line was performed in the CasaXPS software (version 2.3.14) using GL(30) and GL(90) profiles as fitting functions and the standard Shirley background for achieving better agreement between experimental data and the summary curve of the fitting components. Electrical measurements were performed in the van der Pauw geometry utilizing a digital multimeter (34461A, Keysight, USA). For the evaluation of the chemical stability, one of each type of fabricated samples was kept at normal conditions for different time intervals and then submitted to EDS analysis.

## III. RESULTS AND DISCUSSIONS

### A. Fabrication of hybrid structure by sputtering

SEM investigation of as-fabricated pure SS and Cu films points to the fact that these samples present a uniform and homogenous granular structure (Fig. 1) with grain medium lateral size of about 364 nm and 18.5 nm for stainless steel and copper, respectively (see examples and grain size distributions in Fig. 2).

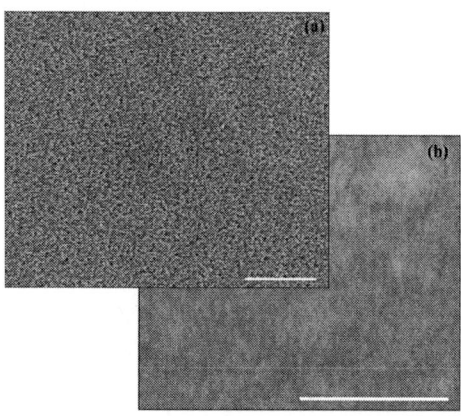

Fig. 1. SEM image of the typical stainless steel (a) and copper (b) films deposited by dc sputtering. Scale bars: 5 μm (a) and 300 nm (b).

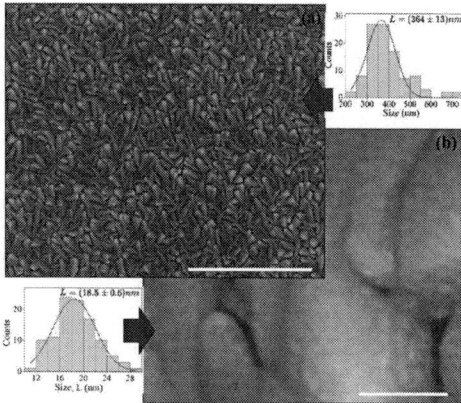

Fig. 2. HR SEM images of stainless steel decorated by copper oxide NPs structure: (a) general view and (b) stainless steel grain decorated by Cu oxide NPs. Scale bars: 2 μm (a) and 100 nm (b). Insets show the size distribution for SS and CuO garins, respectively.

The detailed analysis of HR SEM images indicates that Cu NPs also cover uniformly the superficial SS grain facets and have a size around 20 nm (white spots in Fig. 2 b). Therefore, the selected sputtering conditions allow to obtain a Cu NPs

over granular SS film. Notable, this granular structure provides a bigger effective area where higher amount of Cu NPs can be deposited (against the flat surface). Thus, this morphology can be useful for electrochemical sensing applications, where a bigger area for chemical reaction processes leads to a better sensing performance.

### B. Thermal oxidation

EDS analysis of fabricated samples showed notable growth of oxygen concentration after thermal oxidation process (Table 1). Specifically, the huge increase, more than twentyfold, was detected for Cu films, while only 2.5 times growth was observed in the case of pure SS films. Taking into account this fact, an increase of about 7 times of oxygen amount in hybrid structures (Table 1) can be associated predominantly with Cu NPs oxidation processes.

Moreover, the Cu/O ratio for the treated Cu film (1.4) does not correspond to either pure $Cu_2O$ (2:1) or pure CuO (1:1) case, i.e. bulk Cu after treatment is composed of at least two types of copper oxides. Along with this, the value of this ratio for Cu NPs in hybrid structure (which equals to 0.8, considering the subtraction of 4.1 at % of oxygen, related with the oxidation of SS film) indicates that the nanoclusters are predominantly oxidized up to CuO. The fact that carbon concentration (signature of organic contamination) does not increase after the thermal treatment procedure characterizes this thermal treatment as a high-quality process.

TABLE I.   COMPARISON OF C, O AND Cu CONCENTRATIONS OBTAINED FROM EDS ANALYSIS OF SEVERAL SAMPLES

| Sample | Element (at %) | | |
|---|---|---|---|
|  | O | Cu | C |
| SS film | 1.7±0.5 | 0 | 4.6±0.4 |
| treated SS film | 4.1±1.6 | 0 | 4.2±0.8 |
| Cu film | 1.8±0.2 | 91.5±0.5 | 6.3±0.4 |
| treated Cu film | 38.1±2.8 | 54.0±2.9 | 4.8±0.6 |
| hybrid structure | 1.7±0.5 | 5.8±0.2 | 5.1±0.9 |
| treated hybrid structure | 11.3±0.8 | 5.6±0.1 | 5.3±1.9 |

In order to evaluate the efficiency of the oxidation procedure and to determine the origin of the copper oxide in treated samples, the $Cu2p_{3/2}$ line of XPS spectra was analyzed in pure Cu films (bulk Cu) and SS film/Cu NPs hybrid structure before and after thermal treatment (Fig. 3). As can be seen in Fig. 3, the shapes of the curves for the samples of bulk Cu and Cu NPs are similar. Along with this, the thermal treatment leads to strong chemical shift of this line, and the difference between bulk and nanoparticles curves starts to be well-pronounced.

With the aim to determine the chemical state of the copper before and after thermal treatment, the deconvolution of the $Cu2p_{3/2}$ line was done (see inset in Fig. 3). Four typical components were detected (Table 2): component peak between 932.2 eV and 933.1 eV of binding energy, is usually associated with metallic Cu; near 932.1 eV is assigned to $Cu_2O$; from 933.1 to 934.1 eV corresponds to CuO; and localized in the 934.4 - 935.6 eV region is related to the presence of $Cu(OH)_2$ on the copper surface [14].

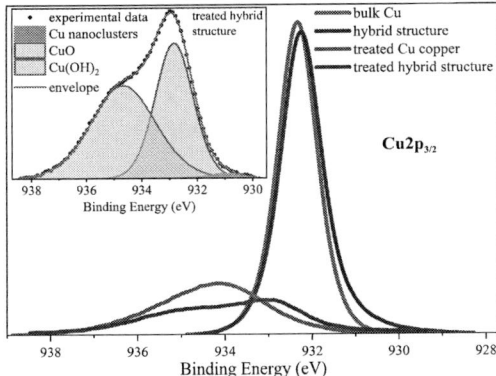

Fig. 3. Comparison of Cu2p₃/₂ XPS lines of four samples. The typical example of Cu2p₃/₂ line deconvolution (sample of treated hybride strture) is shown in inset.

Notably, beyond the typical components found at the deconvolution of Cu2p₃/₂ line, another one, small but well-pronounced, was found near 931 eV. This component can be associated with small clusters of Cu atoms on the surface – nanoparticles of copper [15], [16]. Thus, the emergence of this component, only in hybrid structures (see Table 2), also points to the fact that the sputtering conditions, described in the experimental section, provide the formation of copper nanoparticles in this fabrication process.

As can be seen from Table 2, as-fabricated Cu NPs in hybrid structure are already mostly oxidized up to +1 oxidation state ($Cu_2O$), different from bulk copper samples, when only 1/3 of the superficial layer is composed of $Cu_2O$. The data presented in Table 2 clearly indicate that thermal treatment at 320 °C for 2 h leads to a complete oxidation of Cu NPs to the +2 oxidation state represented by CuO (~40 %) and Cu(OH)₂ (~60 %), that corroborates with conclusions obtained after EDS analysis.

TABLE II.    RESULTS OF Cu2p₃/₂ XPS LINE DECONVOLUTION.

| Sample | Area of component (%) | | | | |
|---|---|---|---|---|---|
| | NP Cu | metallic Cu | Cu₂O | CuO | Cu(OH)₂ |
| bulk Cu | absent | 64.2±7.7 | 33.2±9.4 | 2.7±4.0 | absent |
| hybrid structure | 2.8±1.8 | 33.8±5.0 | 62.0±5.1 | absent | 1.4±2.1 |
| treated bulk Cu | absent | 10.2±3.4 | absent | 14.7±11.1 | 75.1±9.3 |
| treated hybrid structure | 2.6±0.9 | absent | absent | 38.6±2.3 | 58.8±2.5 |

The facts that (i) as-fabricated Cu NPs, in comparison with bulk Cu, are mostly oxidized up to the $Cu_2O$ and (ii) the quantity of CuO* after the thermal treatment is more than 2 times higher for the nanoparticles than for bulk Cu, corroborate with the hypothesis that nanostructures demonstrate an easier ability to oxidize in comparison with bulk samples [10], [12].

*C. Stability*

The chemical stability of obtained films was evaluated performing EDS analysis after several time intervals from the moment the samples were fabricated. The values of concentration† of each element in not treated as well as treated samples were normalized by the value of concentration of as-fabricated not treated sample for better comparison (Fig. 4). For example, normalized concentration of oxygen on the *i* day of treated sample are calculated as

$$nc\,(O_t)_i = {c(O_t)_i}\Big/{c(O_{nt})_{af}}\,100\%\,,$$

where $c(O_t)_i$ is the oxygen concentration on the *i* day of treated sample and $c(O_{nt})_{af}$ is the oxygen concentration of as-fabricated not treated sample.

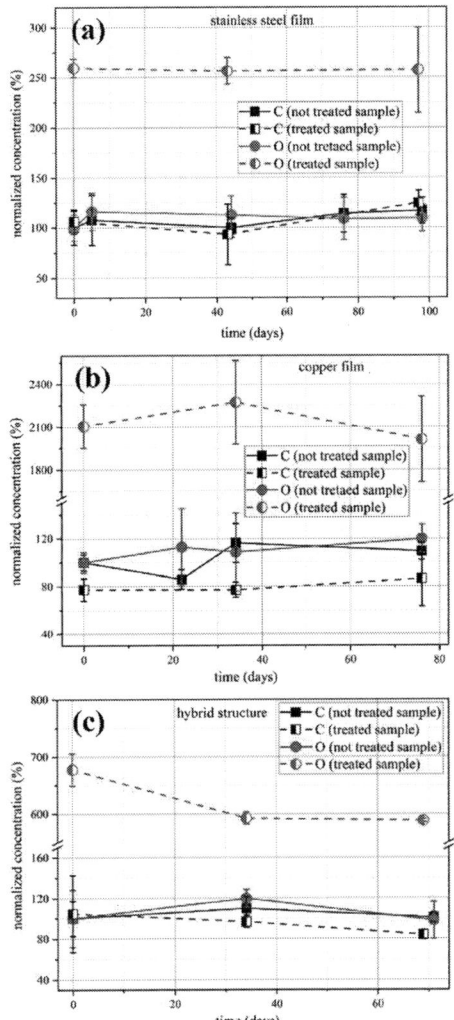

Fig. 4. Evolution of C and O concentration over time for no treated and treated samples for three types of samples: SS film (a), Cu film (b) and hybrid structure (c).

The concentration of oxygen almost does not change within the error bar limits over the studied time in the treated as well as not treated Cu and SS films (Fig. 4 a and b) and in the not treated hybrid structure samples (Fig. 4 c). This can be related to the fact that the surface of these samples are mostly composed of strongly hydrophobic (non-reactive with

---

* The formation of the CuO requires higher energies in comparison with Cu₂O [10], [12].

† Hereinafter we will talk about average values of concertation.

979-8-3315-4064-7/24 $31.00 © 2024 IEEE                35

OH/water molecules) $Cu_2O$ or $Cu(OH)_2$ compounds in bulk Cu and hybrid films (see Table 2) [5], [17], [18] or/and possess strong hydrophobic nanostructured (granulated) morphology [19] in SS or Cu films (Fig. 1). The slight decries of oxygen concentration in treated hybrid sample (Fig. 4 c) can be associated with desorption of OH groups, water or oxygen molecules from CuO hydrophilic regions (Table 2).

Considering the fact that the average values of the elements concentration was calculated using the EDS measurements results, obtained from several places of the sample, the error bar magnitude for the average value can be used as an estimate of the uniformity of the element distribution. It is notable, that the error bar of oxygen concentration increases over time for treated Cu and SS samples, while it is relatively stable in not treated samples. These facts indicate that thermal treatment can modify some regions of this type of sample by manner that it changes its ability to adsorb/desorb OH groups, water or oxygen molecules on the surface which alternates the oxygen concentration, leading to the nonhomogeneous of its chemical composition over time. Along with this, the error bar of oxygen concentration is reducing over time in not treated as well as treated hybrid structures (Fig. 4 c) which point to the improvement of its chemical composition homogeneity over time. Finally, the relatively stable carbon concentration over time in all types of samples points to their low ability to form organic compounds on its surfaces.

Electrical measurements performed in pure SS films showed that thermal treatment does not reduce significantly its electrical conductivity: from 0.55 kS cm$^{-1}$ for as-fabricated sample to 0.36 kS cm$^{-1}$ for treated sample. This issue and the fact that the thermal treatment at 320 °C for 2 h leads to just a soft oxidation of SS samples, which are stable over time (see Table 1 and Fig. 4 a), indicate that this material can act as a reliable conductive network in hybrid structures.

## IV. CONCLUSIONS

In summary, the simple, low-cost fabrication process of stainless steel/copper oxide nanoparticles hybrid structure using dc sputtering and thermal oxidation techniques was demonstrated. Such advantages as the possibility to obtain two-component hybrid structure in just one technological procedure together with using low-temperature thermal oxidation, which provides a relatively high amount of CuO phase (confirmed by EDS and XPS), turns this fabrication process a promising technology for several sensing applications that require monosaccharide and disaccharide structures detection. The stability of the obtained structures was proven by the EDS analysis performed along 70 days.

Tuning the sputtering conditions for Cu NPs, with the aim to obtain a specific morphology of nanostructures which will correspond to a better sensing performance, and optimization of the thermal oxidation conditions (reducing time and temperate of treatment maintaining the high yield of the CuO) will be the scope of the further studying.

## ACKNOWLEDGMENT

This research used facilities of the Brazilian Nanotechnology National Laboratory (LNNano), part of the Brazilian Centre for Research in Energy and Materials (CNPEM), a private non-profit organization under the supervision of the Brazilian Ministry for Science, Technology, and Innovations (MCTI), proposal number: 20240098. The authors would like to thank to Dr Ângela Albuquerque (LNNano) for performing XPS analysis.

## REFERENCES

[1] A. Tripathi and J. Bonilla-Cruz, "Review on Healthcare Biosensing Nanomaterials," *ACS Appl. Nano Mater.*, vol. 6, no. 7, pp. 5042–5074, 2023, doi: 10.1021/acsanm.3c00941.

[2] T. Dutta, T. Noushin, S. Tabassum, and S. K. Mishra, "Road Map of Semiconductor Metal-Oxide-Based Sensors: A Review," *Sensors*, vol. 23, no. 15, 2023, doi: 10.3390/s23156849.

[3] A. Baghdasaryan and T. Bürgi, "Copper nanoclusters: Designed synthesis, structural diversity, and multiplatform applications," *Nanoscale*, vol. 13, no. 13, pp. 6283–6340, 2021, doi: 10.1039/d0nr08489a.

[4] M. Lettieri, P. Palladino, S. Scarano, and M. Minunni, "Copper nanoclusters and their application for innovative fluorescent detection strategies: An overview," *Sensors and Actuators Reports*, vol. 4, no. June, p. 100108, 2022, doi: 10.1016/j.snr.2022.100108.

[5] Y. Wu *et al.*, "Thermal Oxidation Fabricated Copper Oxide Nanotip Arrays with Tunable Wettability and Robust Stability: Implications for Microfluidic Devices and Oil/Water Separation," *ACS Appl. Nano Mater.*, vol. 4, no. 5, pp. 4713–4720, May 2021, doi: 10.1021/acsanm.1c00316.

[6] L. Xiang, J. Guo, C. Wu, M. Cai, X. Zhou, and N. Zhang, "A brief review on the growth mechanism of CuO nanowires via thermal oxidation," *J. Mater. Res.*, vol. 33, no. 16, pp. 2264–2280, Aug. 2018, doi: 10.1557/jmr.2018.215.

[7] S. K. Meher and G. R. Rao, "Archetypal sandwich-structured CuO for high performance non-enzymatic sensing of glucose," *Nanoscale*, vol. 5, no. 5, p. 2089, 2013, doi: 10.1039/c2nr33264g.

[8] J. M. Marioli and T. Kuwana, "Electrochemical characterization of carbohydrate oxidation at copper electrodes," *Electrochim. Acta*, vol. 37, no. 7, pp. 1187–1197, Jun. 1992, doi: 10.1016/0013-4686(92)85055-P.

[9] J. Huang, Y. Zhu, X. Yang, W. Chen, Y. Zhou, and C. Li, "Flexible 3D porous CuO nanowire arrays for enzymeless glucose sensing: in situ engineered versus ex situ piled," *Nanoscale*, vol. 7, no. 2, pp. 559–569, 2015, doi: 10.1039/C4NR05620E.

[10] A. Yabuki and S. Tanaka, "Oxidation behavior of copper nanoparticles at low temperature," *Mater. Res. Bull.*, vol. 46, no. 12, pp. 2323–2327, 2011, doi: 10.1016/j.materresbull.2011.08.043.

[11] J. T. Chen *et al.*, "CuO nanowires synthesized by thermal oxidation route," *J. Alloys Compd.*, vol. 454, no. 1–2, pp. 268–273, 2008, doi: 10.1016/j.jallcom.2006.12.032.

[12] J. Leitner, D. Sedmidubský, M. Lojka, and O. Jankovský, "The Effect of Nanosizing on the Oxidation of Partially Oxidized Copper Nanoparticles," *Materials (Basel).*, vol. 13, no. 12, p. 2878, Jun. 2020, doi: 10.3390/ma13122878.

[13] J. Naktiyok and A. K. Özer, "Synthesis of Copper Oxide (CuO) from Thermal Decomposition of Copper Acetate Monohydrate $(Cu(CH_3COO)_2.H_2O)$," *Ömer Halisdemir Üniversitesi Mühendislik Bilim. Derg.*, vol. 8, no. 2, pp. 1292–1298, Jul. 2019, doi: 10.28948/ngumuh.598177.

[14] National Institute of Standards and Technology, "NIST X-ray Photoelectron Spectroscopy Database (SRD 20), Version 5.0." National Institute of Standards and Technology, Gaithersburg MD, p. 20899, 2000. doi: https://dx.doi.org/10.18434/T4T88K.

[15] J. M. Burkstrand, "Unusual core level spectra of copper on polystyrene," *Surf. Sci.*, vol. 78, no. 3, pp. 513–517, Dec. 1978, doi: 10.1016/0039-6028(78)90229-7.

[16] A. J. Pertsin and Y. M. Pashunin, "An XPS study of the in-situ formation of the polyimide/copper interface," *Appl. Surf. Sci.*, vol. 47, no. 2, pp. 115–125, Feb. 1991, doi: 10.1016/0169-4332(91)90026-G.

[17] Q. Pan, H. Jin, and H. Wang, "Fabrication of superhydrophobic surfaces on interconnected $Cu(OH)_2$ nanowires via solution-immersion," *Nanotechnology*, vol. 18, no. 35, p. 355605, Sep. 2007, doi: 10.1088/0957-4484/18/35/355605.

[18] O. Glemser and H. Sauer, "Handbook of Preparative Inorganic Chemistry," 2nd Ed., G. Brauer, Ed. NY: Academic Press, 1963, p. 1013.

[19] R. Ori, "New insights into hydrophobicity at nanostructured surfaces : Experiments and computational models," vol. 12, pp. 1–21, 2022, doi: 10.1177/1847980421106231.

# Effect of interface traps on the different conduction mechanisms of MISHEMT from 200 K to 450 K

Welder F. Perina[1], Joao A. Martino[1] and Paula G. D. Agopian[1,2]
[1]LSI/PSI/USP, University of Sao Paulo, Sao Paulo, Brazil
[2]UNESP, Sao Paulo States University, Sao Joao da Boa Vista, Brazil
email:welder.perina@usp.br

*Abstract*— In this work, the effect of the interface traps on the different conduction mechanisms of a Metal Insulator Semiconductor High Electron Mobility Transistor (MISHEMT) operating in multiple temperatures (from 200 K to 450 K) is evaluated through High Electron Mobility Transistor (HEMT) and Metal Insulator Semiconductor Field Effect Transistor (MISFET) numerical simulations. The interface traps of the MISFET poses barely any effect on the threshold voltage. For the HEMT, however, the behavior with traps enabled is basically the opposite when the traps are disabled. Combining the opposite behaviors of the threshold voltage curves of the MISFET and HEMT as a function of temperature, gives a possible explanation of the threshold voltage rebound effect observed in MISHEMT.

*Keywords*— *MISHEMT, multiple conductions, temperature, GaN, threshold voltage*

## I. INTRODUCTION

The High Electron Mobility Transistor (HEMT) can used in simple designs for power switch applications, RF and microwave circuits [1, 2], and has been commercially used in the field of power electronics and high frequency operation since 1983 [3]. It also operates in harsh environments [4]. This device is based on the two-dimensional electron gas (2DEG) which forms a region of high mobility and high density of carriers at the interface of the heterostructure, usually AlGaN/GaN, being the main mechanism of the HEMT [5].

However, at scaled dimensions the HEMT presents a high gate leakage current and current collapse due to electrons capture by traps at the passivation layer. By introducing an insulator between the gate metal and the semiconductor, a better immunity against the gate leakage current while mitigating the current collapse, giving birth to the Metal Insulator Semiconductor High Electron Mobility Transistor [6, 7, 8]. For applications requiring high frequency or high power operations, the MISHEMT is considered one of the most promising candidates in the future [8, 9].

While many studies strive to acquire a more stabilized threshold voltage ($V_{TH}$) for normally-off operation, through the research of new materials and process steps [1-2, 5-11], another important point is to understand the working principle of the device. These devices present multiple conductions, composed of HEMT and MISFET, which impacts the device's behavior [12-16]. When operating under different temperatures the device presents a rebound effect on the threshold voltage [15, 16]. This work focuses on studying the influence of interface traps on threshold voltage of

simulated HEMT and MISFET in order to understand the rebound effect in the threshold voltage of a MISHEMT.

## II. DEVICE CHARACTERISTICS

The rebound effect was observed on a MISHEMT [13, 15] fabricated at imec – Belgium – with channel width of 5 µm, gate length of 600 nm, 2 nm gate insulator of $Si_3N_4$, 15 nm barrier layer of AlGaN, 1 nm spacer of AlN, and 2.5 µm GaN grown on a silicon platform, as presented in figure 1A. The simulated MISFET, figure 1B, and simulated HEMT, figure 1C, were designed with the same materials and dimensions.

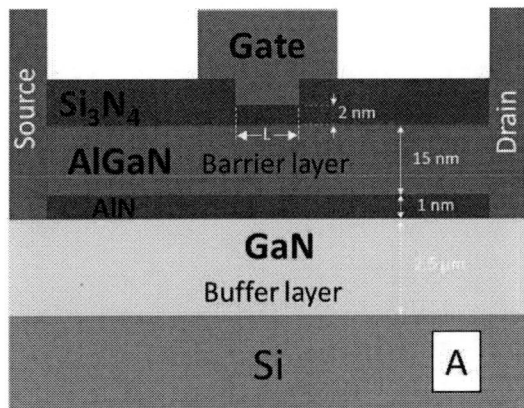

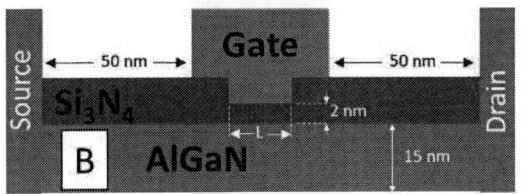

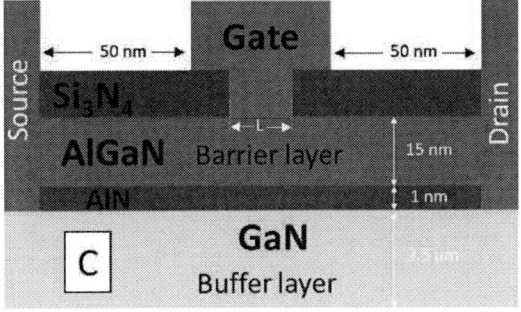

979-8-3315-4064-7/24 $31.00 © 2024 IEEE

Fig. 1. Cross section structure of the MISHEMT (A), simulated MISFET (B) and simulated HEMT (C) used in this work.

The MISFET was evaluated using Sentaurus Synopsys TCAD simulations. The device has a channel width of 1 μm, gate length of 600 nm, 2 nm gate insulator of Si3N4, a 15 nm barrier layer of AlGaN. The distance between gate and drain (LGD), and between gate and source (LGS) were kept at 50 nmsdfsdfsdf.

The HEMT was evaluated using Sentaurus Synopsys TCAD simulations. The device has a channel width of 1 μm, gate length of 600 nm, 5 nm of Si3N4 for passivation, 15 nm layer of AlGaN, 1 nm spacer of AlN, and 2.5 μm GaN. The distance between gate and drain (LGD), and between gate and source (LGS) were kept at 50 nm.

## III. RESULTS AND ANALYSIS

Figure 2 presents the drain current as a function of gate voltage for a MISFET with traps, disabled (top) and enabled (bottom), at the Si3N4 interface, at different temperatures. For increasing temperature, is possible to see the decrease on the current for high $V_{GS}$, due to mobility degradation, and the shift of the $V_{TH}$ to the left due to the reduction of the Fermi potential. The acceptor-type trap was used thus inducing a capture of electrons reducing the current drive. Also, there is a slight shift in the threshold voltage, which will be thoroughly approached in figure 3. The second derivative was the method used for the VTH extraction [17, 18].

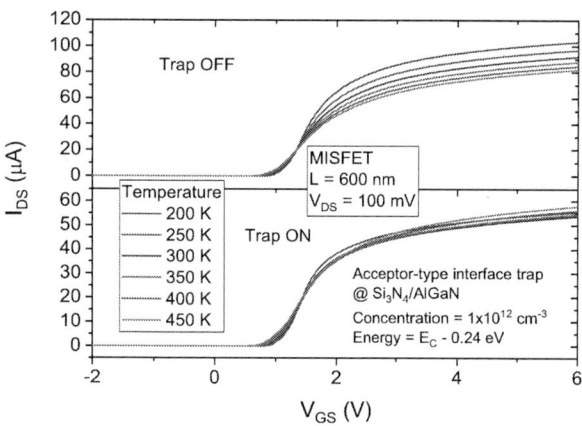

Fig. 2. Drain current as a function of gate voltage for different temperatures, for a MISFET with interface traps at Si3N4/AlGaN disabled (top) and enabled (bottom).

The $V_{TH}$ behavior of the MISFET as a function of temperature follows the usual trend of decreasing with increasing temperature. The presence of the traps does not change the behavior significantly, possibly because the charges accumulated at the traps are not enough to affect the high mobility channel.

Figure 4 presents the drain current as a function of gate voltage for a HEMT at different temperatures, with traps enabled (bottom) and disabled (top). The energy which usually the traps are found are in the range of $E_C - 0.5$ eV [18, 19] with a density concentration above $10^{12}$ cm$^{-3}$ [20, 21]. When the traps are disabled, it is possible to observe an increase of the drain current for increasing temperature, at

high $V_{GS}$. However, when the traps are considered, the drain current decreases for increasing temperature due to the mobility degradation.

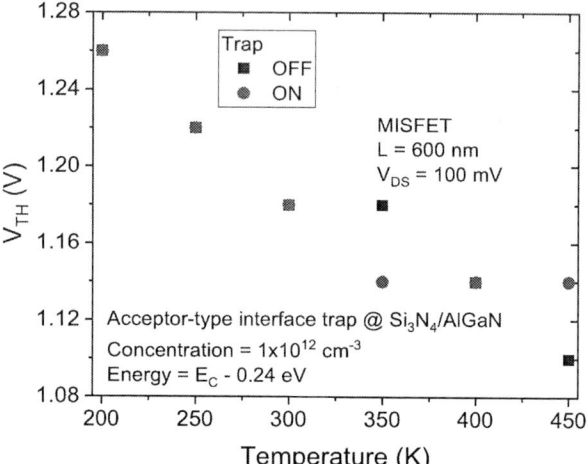

Fig. 3. Threshold voltage as a function of temperature for a MISFET with interface traps at Si3N4/AlGaN disabled (top) and enabled (bottom).

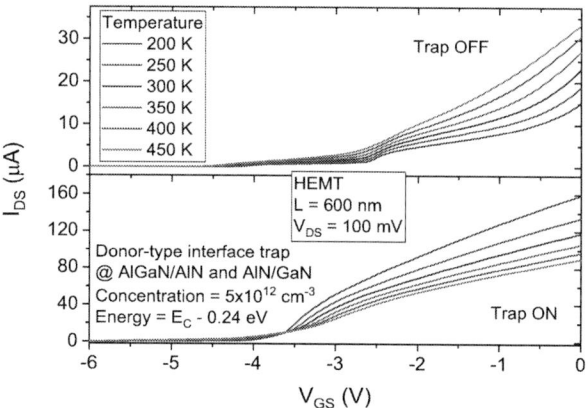

Fig. 4. Drain current as a function of gate voltage for different temperatures, for a HEMT with interface traps at the Si3N4/AlGaN, AlGaN/AlN and AlN, disabled (top) and enabled (bottom).

The threshold voltage as a function of temperature of the HEMT is presented in figure 5, for traps disabled and enabled. For increasing temperature, it is possible to see a slight decrease of the $V_{TH}$ when the traps are disabled due to depletion depth reduction. However, when traps are enabled, the behavior changes since the increasing temperature increase the energy and numbers of ionized donor-like traps, increasing the capture of electrons thus reducing the positive charge resulting in the increase of $V_{TH}$ [22, 23].

Figure 6 presents the threshold as a function of temperature of the HEMT and MISFET. For increasing temperature, the MISFET fermi potential reduces thus decreasing the $V_{TH}$, the behavior of the HEMT however is the opposite, since the traps at the 2DEG interfaces (AlGaN/AlN and AlN/GaN) plays a major role and are thermally assisted.

979-8-3315-4064-7/24 $31.00 © 2024 IEEE

With further improvement, the devices should have approximate values of $V_{TH}$ while maintaining the same behavior. Combining both the trend of MOS and HEMT $V_{TH}$ curves, should give the approximate behavior of the MISHEMT rebound effect, where there was a decrease of VTH temperature from 200 K to 350 K and a great increase from 350 K to 450 K [15,16].

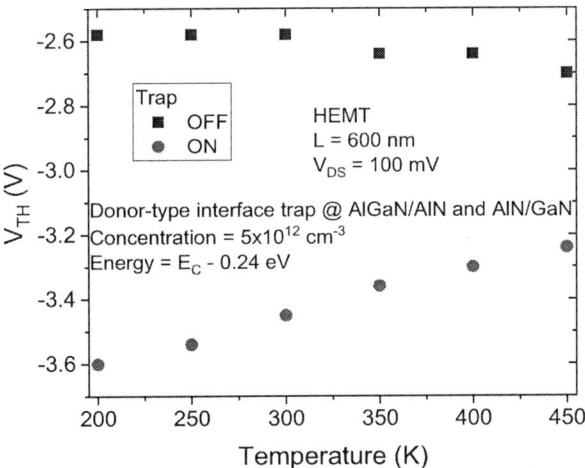

Fig. 5. Threshold voltage as a function of temperature for a HEMT with interface traps at the Si3N4/AlGaN, AlGaN/AlN and AlN, disabled (top) and enabled (bottom).

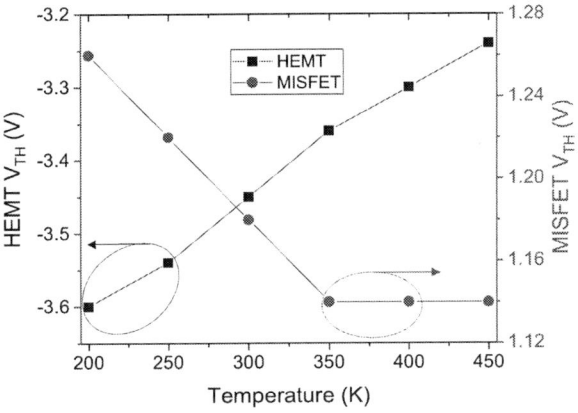

Fig. 6. Threshold voltage as a function of temperature of MISFET and HEMT with traps enabled.

## IV. CONCLUSIONS

In this work the effect of interface traps on different conduction mechanism measured by the threshold voltage of simulated HEMT and MISFET was studied in a large range of temperature (from 200 K to 450k) in order to explain these effects on MISHEMTs. Through the threshold voltages curves of MISFET, it was possible to observe that the interface traps barely affect the device's behavior. The HEMT, however presented a completely different behavior when the traps are considered. When comparing the results of the threshold voltages of both simulated devices, with traps enabled, is possible to infer that, if the conduction mechanisms was adjusted properly for experimental MISHEMT, the MOS conduction (MIS channel) should

dominate the behavior until 350 K, while for higher temperatures the HEMT (2DEG) should be the dominating mechanism, resulting in the threshold voltage rebound effect.

### ACKNOWLEDGMENT

The authors acknowledge CNPq, São Paulo Research Foundation - FAPESP (under grant #2020/04867-2) and Coordenação de Aperfeiçoamento de Pessoal de Nível Superior - Brasil (CAPES) - Finance Code 001 for the financial support.

### REFERENCES

[1]  W. -M. Wu et al., "ESD HBM Discharge Model in RF GaN-on-Si (MIS)HEMTs," in IEEE Transactions on Electron Devices, vol. 69, no. 4, pp. 2180-2187, April 2022.

[2]  J. He, et al. "Normally-off AlGaN/GaN MIS-HEMTS with Low RON and Vth hysteresis by functioning in-situ SiNx in regrowth process." IEEE Electron Device Letters, v. 43, n. 4, p. 529–532, 2022.

[3]  T. Mimura. "Special contribution invention of high electron mobility transistor (HEMT) and contributions to information and communications field." Fujitsu Scientific and Technical Journal, v. 54, n. 5, p. 3–8, 2018.

[4]  J. Ghosh, S. Das,S. Mukherjee,S. Ganguly,A. Laha. "A study of electrical characteristics of Gd2O3/GaN and Gd2O3/AlGaN/GaN MOS heterostructures." Microelectronic Engineering, v. 216, February, p. 111097, 2019.

[5]  L. He, et al. "A review of selective area grown recess structure for insulated-gate E-mode GaN transistors". Japanese Journal of Applied Physics, v. 59, 2019.

[6]  A. M. Nahhas, "Review of AlGaN/GaN HEMTs based devices." American Journal of Nanomaterials, v. 7, n. 4, p. 10–21, 2019.

[7]  U. Peralagu, et al, "CMOS-compatible GaN-based devices on 200mm-Si for RF applications: Integration and Performance." Technical Digest - International Electron Devices Meeting (IEDM), v. 2019- December, p. 398–401, 2019.

[8]  M. Whiteside, S. Arulkumaran, G. I. Ng, "Demonstration of vertically-ordered h-BN/AlGaN/GaN metal-insulator-semiconductor high-electron-mobility transistors on Si substrate." Materials Science and Engineering: B, v. 270, p. 115224, 2021.

[9]  A. Grill, et al, "Electrostatic coupling and identification of single-defects in GaN/AlGaN Fin-MIS-HEMTs", Solid-State Electronics, v. 156, p. 41–47,January 2019.

[10] L. He, et al, "Threshold voltage engineering in Al2O3/AlGaN/GaN MISHEMTs with thin barrier layer: MIS-gate charge control and high threshold voltage achievement". In 33rd International Symposium On Power Semiconductor Devices And ICs (ISPSD), p. 339–342, 2021.

[11] H. Y. Lee, T. W. Chang, E. Y. Chang, N. Rorsman,C. T. Lee. "Fabrication and characterization of GaN-based fin-channel array Metal-Oxide-Semiconductor High-Electron Mobility Transistors with Recessed-Gate and GaO gate insulator layer." IEEE Journal of the Electron Devices Society, v. 9, n. March, p. 393–399, 2021.

[12] P. G. D. Agopian, et al, "Gate dielectric material influence on DC behavior of MO(I)SHEMT devices operating up to 150 °C", Solid-State Electronics, v. 185, p. 108091, 2021.

[13] W. F. Perina, J. A. Martino, and P. G. D. Agopian, "Experimental analysis of mishemt multiple conductions from 200K to 450K", SBMicro 2022 - 36th Symposium on Microelectronics Technology and Devices, p. 1-4, 2022

[14] B. G. Canales and P. G. D. Agopian, "MISHEMT's multiple conduction channels influence on its DC parameters", Journal of Integrated Circuits (JICS), v. 18, n . 4, p. 1-5, 2023.

[15] W. F. Perina, J. A. Martino, E. Simoen, U. Peralagu, N. Collaert and P. G. D. Agopian, "Study of the effect of multiple conductions on threshold voltage in a MIS-HEMT from 450 K down to 200 K", 2023 37th Symposium on Microelectronics Technology and Devices (SBMicro), Rio de Janeiro, Brazil, 2023, pp. 1-3.

[16] W. F. Perina, J. A. Martino, E. Simoen, U. Peralagu, N. Collaert and P. G. D. Agopian, "Experimental study of MISHEMT from 450 K down to 200 K for analog applications", Solid-State Electronics, v. 208, 108742, 2023.

[17] G. Pananakakis, G. Ghibaudo and S. Cristoloveanu, "Threshold voltage in FD-SOI MOSFETSs", Solid-State Electronics, 2024.

[18] A. Ortiz-Conde, F. J. Garcia Sanchez, J. J. Liou, A. Cerdeira, M. Estrada and Y. Yue, "A review of recent MOSFET threshold voltage extraction methods", Microelectron. Reliab., vol. 42, 583, 2002.

[19] J. Yang, et al., "Electron tunneling spectroscopy study of electrically active traps in AlGaN/GaN high electron mobility transistors", Applied Physics Letters, v. 103, 223507, 2013.

[20] M. Meneghini, et al., "GaN-based power devices: Physics, reliability, and perspectives", Journal of Applied Physics, v. 130, 181101, 2021.

[21] S. Yang, et al., "Investigation of SiNx and AlN Passivation for AlGaN/GaN High-Electron-Mobility Transistors: Role of Interface Traps and Polarization Charges", Journal of the Electron Devices Society, v. 8, 2020.

[22] H. Y. Wong, Nelson Braga, R. V. Mickevicius, F. Gao, and T. Palacios, "Study of AlGaN/GaN HEMT degradation through TCAD simulations", 2014 International Conference on Simulation of Semiconductor Processes and Devices (SISPAD), p.97-100, 2014.

[23] M. A. Alim, A. A. Rezazadeh and C. Gaquiere, "Anomaly and threshold voltage shifts in GaN and GaAs HEMTs over temperature", 2017 12th European Microwave Integrated Circuits Conference (EuMIC), Nuremberg, Germany, pp. 33-36, 2017.

[24] Huang, Huolin, Feiyu Li, Zhonghao Sun, and Yaqing Cao. "Model Development for Threshold Voltage Stability Dependent on High Temperature Operations in Wide-Bandgap GaN-Based HEMT Power Devices", Micromachines 9, no. 12: 658, 2018.

# Influence of the Temperature on the Operational Transconductance Amplifier designed with triple gate SOI FinFETs

Henrique Hilkner[1], Paula Ghedini Der Agopian[1,2], *Senior member IEEE*,
and Joao Antonio Martino[1], *Senior member IEEE*

[1]LSI/PSI/USP, University of Sao Paulo, Sao Paulo, Brazil
[2]UNESP, Sao Paulo State University, Sao Joao da Boa Vista, Brazil
e-mail: henriquehilkner@usp.br

## ABSTRACT

*This paper presents the temperature influence on the Operational Transconductance Amplifier (OTA) designed with triple gate Silicon-On-Insulator (SOI) FinFETs. The FinFET electrical characteristics were obtained experimentally at room temperature, and the results were used to model it using Lookup table in Verilog-A. Based on the transistors experimental data, the OTA was simulated at room temperature and also for low and high temperature (from 180 to 600 K) using simulation device characteristics, with and without thermal compensation (TC). The increases in temperature resulted in the decreases of the voltage gain, gain-bandwidth (GBW) due to the mobility degradation, and in the increases in phase margin. For both circuits, from 180 to 600 K, the values of the parameters variations and the resulting curves were very similar: the results demonstrated an average reduction per 100 K of 6.8% in voltage gain at low frequencies, 17.1% in GBW, 14.5% in the transistor efficiency $g_m/I_D$, and a slight average increase of 1.9% in phase margins. The little differences between the OTA circuits with and without TC bias circuit do not require a complex thermal compensated bias circuit.*

*Keywords: OTA circuit, Transistor, Analog amplifier, SOI FinFET, Thermal analysis, Temperature.*

## I. INTRODUCTION

The reduction of power, delay and the increase in circuit density make scaling extremely attractive for digital systems. The two principal reasons for the dominance of Complementary Metal-Oxide Semiconductor (CMOS) technology in today's semiconductor industry are the zero static power dissipation of CMOS logic and the scalability of Metal-Oxide Semiconductor Field-Effect Transistors (MOSFETs) [1].

As the dimensions of transistors are shrunk, the close proximity between the source and the drain reduces the ability of the gate electrode to control the potential distribution and the flow of current in the channel region, and undesirable effects, called the "short-channel effects" start plaguing MOSFETs. In a continuous effort to increase current drive and better control of short channel effects, silicon-on-insulator (SOI) MOS transistors have evolved from classical, planar, single-gate devices into three-dimensional devices with a multi-gate structure [2]. Among different device architectures, FinFET is a potential candidate, since it can control the channel from all the three sides of the gate [3].

The Operational Transconductance Amplifier (OTA) topology finds application in many analog and digital systems [1]. Some applications require electronic circuits capable of operating at temperatures up to 300 °C. SOI devices and circuits present three advantages in this field over their bulk counterparts: the absence of thermally-activated latch-up, reduced leakage currents and, in thin-film, fully depleted devices, a smaller variation of threshold voltage with temperature. On the other hand, SOI transistors are thermally insulated from the substrate by the buried insulator. As a result, removal of excess heat generated within the device is less efficient than in bulk, which may cause an elevation of the device temperature, depending on the bias condition. As SOI MOSFETs present several properties that allow them to operate in harsh environments [4], this paper aims to study the temperature influences on the behavior of OTA circuit designed with triple-gate SOI FinFETs, based on experimental data at room temperature, simulation and extrapolation to high temperature 3D transistors simulation, and the OTA circuit simulation using lookup table models and Verilog-A codes.

## II. DEVICE CHARACTERISTICS

The studied devices are triple gate SOI FinFETs, n-type and p-type, fabricated in Imec/Belgium, with the characteristics shown in tables I and II. Figure 1 shows the schematic view of a triple gate SOI FinFET.

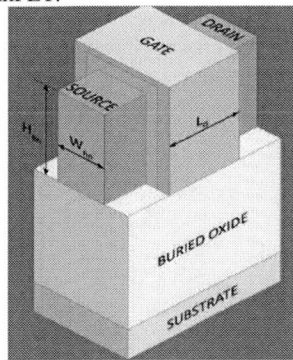

**Figure 1.** Schematic view of triple gate SOI FinFET.

To analyze the components, the tridimensional models were created using Sentaurus Workbench, a 3D simulator software, with the same dimensions of the real transistors. Important values, like mobility, drain-source current, work function, threshold voltage were calibrated and compared to experimental data measured at 300 K.

Table I presents the geometrical parameters of the transistors.

TABLE I – DIMENSIONS CHARACTERISTICS

| Parameter | Value (nm) |
|---|---|
| Channel Width ($W_{fin}$) | 20 |
| Channel Length ($L_G$) | 150 |
| Equivalent Oxide Thickness | 1.5 |
| Fin Height ($H_{fin}$) | 65 |
| Buried Oxide Thickness | 145 |

Table II presents the doping and material parameters of SOI triple-gate FinFETs.

TABLE II – DOPING AND MATERIALS DEVICES CHARACTERISTICS

| Region | n-type | p-type | Doping (atom/cm³) |
|---|---|---|---|
| Channel | Boron | Boron | $1.10^{15}$ |
| Source | Arsenic | Boron | $8.10^{19}$ |
| Drain | Arsenic | Boron | $8.10^{19}$ |

| Region | Material |
|---|---|
| Gate Oxide | HfSiON on SiO$_2$ |
| Buried Oxide | SiO$_2$ |
| Contact Metal | Si-poli on TiN |

## III. LOOKUP TABLE METHOD AND VERILOG-A MODEL VALIDATION

Before designing the OTA simulation circuit, it was necessary to simulate transistors that would respond the closest as possible to the experimental data measured from real devices. A 150 nm channel length n-type and p-type SOI FinFETs were chosen in order to avoid any short-channel effects. The experimental data from the transistors were measured at room temperature (300 K), with $0 \leq |V_{GS}| \leq 1.5$, and $|V_{DS}| = 50$ mV (triode), 0.7 V and 1.0 V (saturation).

Once the experimental data measurements were finished, the simulated devices were created in the Sentaurus Workbench software, using the real dimensions, and calibrated, so it was possible to match important curves and parameters, like drain-source currents, work functions, threshold voltages and mobility values. For a more accurate result, the simulations were set to take into account the mobility dependences on lateral and vertical electric fields. The simulated values were compared to the experimental ones.

Once the curves were fit, the devices were validated, and it was able to create the lookup tables, which were generated using $0 \leq |V_{GS}| \leq 1.5$ with 10 mV steps, $0 \leq |V_{DS}| \leq 1.5$ with 50 mV steps, all curves were simulated for the temperatures from 180 K up to 600 K. The lookup tables were associated to Verilog-A language on Cadence Virtuoso software, in order to create the models [9].

The gate capacitances values were measured at room temperature and since the temperature influence on CV curve is negligible it was considered for all analyzes.

Figures 2 and 3 show the obtained comparisons among the experimental and simulation curves at 300 K, and the simulated curves from 180 K up to 600 K, for $V_{DS} = 0.7$ V (n-type) and $V_{DS} = -0.7$ V (p-type) SOI FinFET. The lookup table model was derived from the Technology Computer-Aided Design (TCAD) outcomes, and so the simulated curves from Sentaurus and Cadence yield identical.

## IV. OTA CIRCUIT WITHOUT AND WITH THERMAL COMPENSATION

After creating the transistors model using the lookup table approach and the Verilog-A language, the circuits presented in figure 4 were simulated, using a $g_m/I_D \cong 8$ V$^{-1}$ as a condition to guarantee the transistors in strong inversion, which resulted in a polarization current $I_{BIAS}$ of 6 μA.

Two different bias circuits are considered in the OTA simulation: the first simulation is biased using an inverter with feedback circuit made with transistors, while the second has an idealized current source representing a more complex current source topology (no temperature dependent).

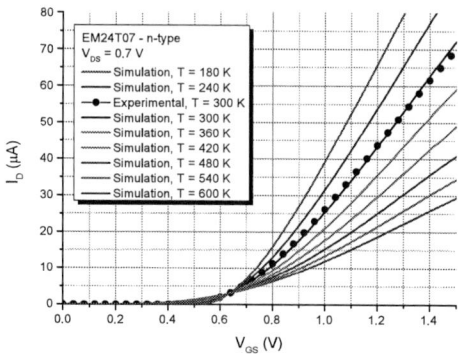

**Figure 2.** Experimental and simulated drain current as a function of gate voltage for n-type triple gate SOI FinFET with $V_{DS} = 0.7$ V.

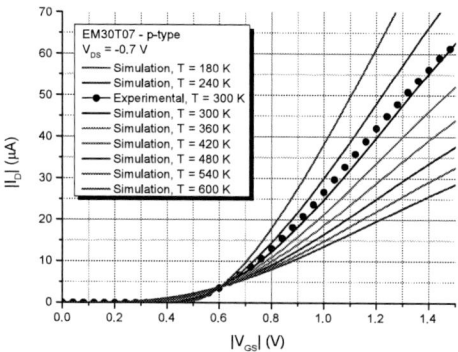

**Figure 3.** Experimental and simulated drain current as a function of gate voltage for p-type triple gate SOI FinFET with $|V_{DS}| = 0.7$ V.

The studied OTA circuit is basically designed in three parts. The first part is the Bias circuit, as described before. The second part is the first amplification stage, composed of a differential circuit amplifier with active load. The third part is a common source amplifier also with active load, responsible for the second amplification stage. The capacitor $C_C$ is used to improve the circuit's stability, while $C_L$ simulates the capacitance of a supposed next stage of amplification [5].

The OTA voltage gain ($A_{vt}$) can be estimated by the multiplication of the gains of the first stage ($A_{v1}$) and the second stage ($A_{v2}$), as shown in eq. (1), (2) and (3) [1].

$$|A_{v1}| = \frac{g_{m2}}{g_{d2} + g_{d4}} = g_{m2} \times R_{out1} \qquad (1)$$

$$|A_{v2}| = \frac{g_{m6}}{g_{d6} + g_{d7}} = g_{m6} \times R_{out} \qquad (2)$$

$$|A_{vt}| = |A_{v1}| \times |A_{v2}| \qquad (3)$$

Which $g_{m2}$ is the transconductance of M2 transistor; $g_{d2,4}$ are the output conductances of M2 and M4 transistors; $R_{out1}$ is the output resistance of the first stage; $g_{m6}$ is the transconductance of M6 transistor; $g_{d6,7}$ are the output conductances of the M6 and M7 transistors; and $R_{out2}$ is the output resistance of the second stage. Figure 4 shows both simulated circuits.

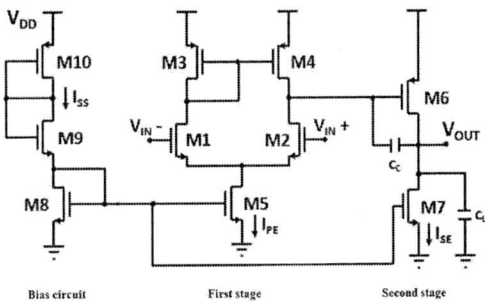

(A)

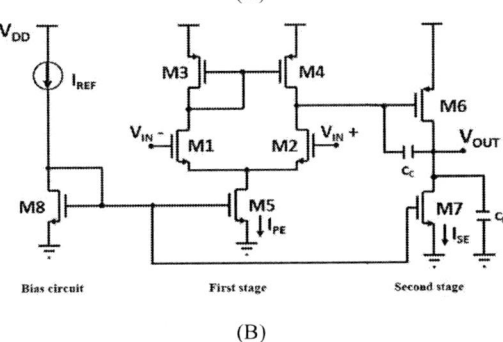

(B)

**Figure 4.** OTA circuit with bias circuit without thermal compensation (A) and with thermal compensation (B).

## V. RESULTS AND ANALYSIS

Table III shows the number of fins in parallel that is used in each SOI FinFET designed using CADENCE simulator in order to have a $g_m/I_D \cong 8$ $V^{-1}$ for M1 and M2 at room temperature. The $g_m/I_D \cong 8$ $V^{-1}$ condition was chosen to guarantee the transistor' strong inversion at room temperature. The number of fins were defined based on the ratio from n and p transistor's experimental drain currents, as well as to achieve a higher gain in the second stage of amplification in order to improve the circuit stability. The supply voltage $V_{DD}$ was projected to obtain $V_{DS} \cong 0.7$ V on each transistor. The $I_{REF}$ from the current source was set to 6 µA, approximately the same current from the bias circuit at 300 K.

**TABLE III – FinFETs Designed Devices**

| Transistors | Number of Fins ($N_{fins}$) | Transistors | Number of Fins ($N_{fins}$) |
|---|---|---|---|
| M1, M2 | 5 | M6 | 42 |
| M3, M4 | 6 | M7 | 35 |
| M5 | 10 | M8, M9, M10 | 1 |

Table IV shows the parameters of the OTA circuits used in the simulations in both topologies.

**TABLE IV – OTA Circuit parameters**

| Parameters | Value |
|---|---|
| $V_{DD}$ | 2.1 V |
| $V_{CM}$ | 1.4 V |
| $g_m / I_D$ | 8 $V^{-1}$ |
| $C_L$ | 100 fF |
| $C_C$ | 46 fF |

Figures 5 and 6 show the total voltage gains and phase margins from both OTA circuits without and with thermal compensation, respectively.

Tables V and VI present the OTA performances obtained.

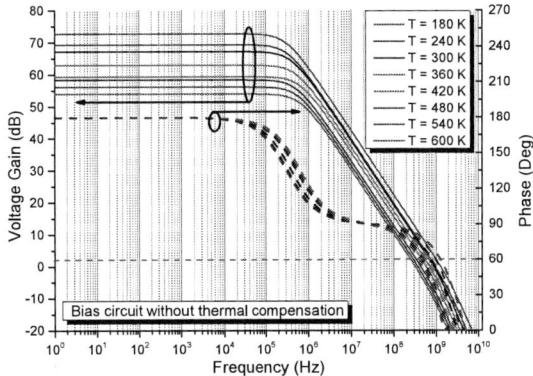

**Figure 5.** Voltage gain and phase margin of OTA circuit without thermal compensation.

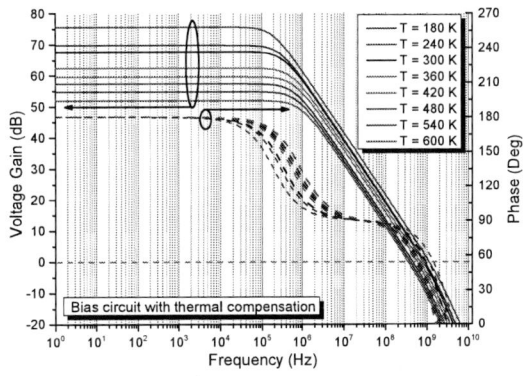

**Figure 6.** Voltage gain and phase margin of OTA circuit with thermal compensation.

**TABLE V – No thermal compensated OTA's figures of merit**

| Results | 180K | 240K | 300K | 360K | 420K | 480K | 540K | 600K |
|---|---|---|---|---|---|---|---|---|
| $I_{SS}$ (µA) | 8.0 | 7.0 | 6.2 | 5.7 | 5.4 | 5.1 | 4.9 | 4.7 |
| P (µW) | 787.8 | 687.0 | 616.8 | 567.4 | 530.5 | 502.9 | 481.3 | 464.6 |
| $A_V$ (dB) | 72.8 | 69.4 | 67.2 | 63.0 | 59.4 | 58.4 | 56.2 | 54.0 |
| GBW (MHz) | 1250 | 929 | 755 | 608 | 504 | 423 | 361 | 313 |
| $g_m/I_D$ ($V^{-1}$) (M1, M2) | 10.0 | 9.5 | 7.7 | 6.8 | 6.1 | 5.5 | 4.9 | 4.5 |
| PM (deg) | 60.0 | 61.0 | 60.2 | 61.7 | 62.9 | 64.2 | 64.6 | 65.7 |

**TABLE VI – Thermal compensated OTA's figures of merit**

| Results | 180K | 240K | 300K | 360K | 420K | 480K | 540K | 600K |
|---|---|---|---|---|---|---|---|---|
| $I_{SS}$ (µA) | 6.0 | 6.0 | 6.0 | 6.0 | 6.0 | 6.0 | 6.0 | 6.0 |
| P (µW) | 594.9 | 593.8 | 593.0 | 591.8 | 590.8 | 589.8 | 588.2 | 586.1 |
| $A_V$ (dB) | 75.7 | 69.7 | 67.7 | 62.4 | 59.6 | 57.3 | 54.8 | 51.9 |
| GBW (MHz) | 1116 | 872 | 724 | 606 | 515 | 455 | 398 | 352 |
| $g_m/I_D$ ($V^{-1}$) (M1, M2) | 11.9 | 9.5 | 7.9 | 6.7 | 5.7 | 5.0 | 4.5 | 4.0 |
| PM (deg) | 61.1 | 61.2 | 61.4 | 61.8 | 62.9 | 64.1 | 64.7 | 65.3 |

Based on tables V and VI, it is possible to observe the polarization currents $I_{SS}$ and $I_{BIAS}$ values of both circuits. For the circuit without thermal compensation, there was a reduction in $I_{SS}$ values as the temperature increased due to the mobility degradation of MOS transistors which decreases the drain current [7], and also confirmed by the curves obtained in figures 2 and 3. The circuit with thermal compensation is able to guarantee constant bias current, not being influenced by temperature. As a consequence, the values of power dissipated in the circuit vary for the first case, and remain constant for the second case.

Based on figures 5 and 6, it is possible to notice that, for both circuits, the voltage gains at low frequency suffered reductions as the temperature increased. The values are presented in tables V and VI, and can be explained since the circuit gain $A_V$ is proportional to $g_m/g_D$ and the degradation of mobility results in $A_{Vt}$ reduction.

The reductions in the circuit's gain curves also resulted in reductions in the gain-bandwidth (GBW) values, which, as the tables show, also suffered reductions due to the mobility degradation with the temperature increase. Since the GBW values are proportional to the $g_m$, it is expected that there will be a greater reduction of these values in the circuit without thermal compensation, in which there is a reduction in the current values as the temperature increases. This effect is confirmed by the values presented in the tables, where it is possible to observe greater variations in the circuit without thermal compensation in relation to the circuit with thermal compensation, resulting in lower GBW values for higher temperature values.

It is possible to notice that there were no significant changes in the phase margins for both circuits. In both cases, the phase margins increased as the temperature increased, but without major differences that directly affected the stability of the circuit. The positioning of the poles of the circuit demonstrates that, as designed, the $C_C$ capacitor is correctly performing its function, since the gain is reduced by 20 dB/dec until the circuit's gain is unitary (0 dB) and the phase margins are above 60 degrees [1].

For ambient temperatures, it is possible to notice that the efficiency of the M1 and M2 nFinFETs remained close to the design value, mainly for the circuit with thermal compensation, which guarantees that the value chosen for the bias current is correct, and all the transistors are in the same inversion mode region.

For both forms of polarization, the circuits behaved similarly, which demonstrates that there were no major differences between the ways of polarizing the OTA. In both cases, the voltage gains, gain-bandwidths and efficiencies $g_m/I_D$ reduced as the temperature increased, as well as the phase margins increased slightly.

The table below presents the differences in percentage values of the results obtained in tables V and VI.

From Table VII, it is possible to observe that, for the same range of temperature variation, the differences in average terms were very close in both circuits. Also, the little differences between gain values and phase margins suggest that the construction of an OTA circuit with SOI FinFETs transistors is quite stable with temperature variation. Table VIII presents the variation in values in percentage terms, for every 100 K temperature increment.

Since SOI and FinFET devices present a small variation of threshold voltage with the temperature $dV_{TH}/dT$ [2,4], it contributes to the great stabilities that both circuits demonstrated, even biasing the circuit with a not ideal current source. From the point of view of using the circuit under temperature variations, the results suggest that polarization using an ideal current source may not have a great influence on the results, when compared to polarization without thermal compensation.

TABLE VII – VARIATIONS DUE TO 100 K TEMPERATURE INCREMENT WITH AND WITHOUT THERMAL COMPENSATION

| Variation 180-600K | Av dB/100K | GBW MHz/100K | Efficiency $(g_m/I_D)$/100K | PM ° / 100K |
|---|---|---|---|---|
| No TC | -4.46 | -222.9 | -1.32 | 1.36 |
| With TC | -5.65 | -182.0 | -1.88 | -0.91 |

TABLE VIII – PERCENTAGE VARIATIONS DUE TO 100 K TEMPERATURE INCREMENT WITH AND WITHOUT THERMAL COMPENSATION

| Percentage variations | Av [%]/100K | GBW [%]/100K | Efficiency [%]/100K | PM [%]/100K |
|---|---|---|---|---|
| No TC | -6.1 | -17.8 | -13.3 | 2.3 |
| With TC | -7.5 | -16.3 | -15.8 | 1.5 |
| Average | -6.8 | -17.1 | -14.5 | 1.9 |

## VI. CONCLUSION

The impact of temperature from 180 to 600 K on the 2-stage OTA circuit designed with triple-gate SOI FinFETs was evaluated. OTA circuits with and without thermal stability was studied. The obtained results showed that the temperature increase results in a decrease of voltage gain and gain-bandwidth due to the mobility degradation.

For both circuits, biased with and without thermal compensation, the values of the parameters variations and the resulting curves were very similar: the results demonstrated an average reduction per 100 K of 6.8% in voltage gain at low frequencies, 17.1% in GBW, 14.5% for $g_m/I_D$, and a slight average increase of 1.9% in phase margins.

The similarities from the obtained curves demonstrate that, in terms of temperature variation, biasing the OTA circuit with bias thermal compensation was not necessary.

### ACKNOWLEDGMENTS

The authors acknowledge CNPq, São Paulo Research Foundation - FAPESP (under grant #2020/04867-2) and Coordenação de Aperfeiçoamento de Pessoal de Nível Superior - Brasil (CAPES) - Finance Code 001 for the financial support, and Imec for supplying the studied devices.

### REFERENCES

[1] RAZAVI, Behzad. Design of Analog CMOS Integrated Circuits. [S.l: s.n.], 2017.

[2] COLINGE, J.-P. et al. FinFETs and Other Multi-Gate Transistors. Boston, MA: Springer US, 2008.

[3] BHATTACHARYA, D.; JHA, N. K. Finfets: from devices to architectures. Advancesin Electronics, Hindawi, v. 2014, 2014.

[4] COLINGE, J.-P. Silicon-on-Insulator Technology: Materials to VLSI. Boston, MA: Springer US, 2004.

[5] CARUSONE, T. C.; JOHNS, D.; MARTIN, K. Analog integrated circuit design. New York: Wiley. Hannane Gholamnataj was born in Babolsar, Iran, on September, v. 16, p. 1984, 2012.

[6] Wu, Y.-C., Jhan Y.-R. 3D TCAD Simulation for CMOS Nanoeletronic Devices. © Springer Nature Singapore Pte Ltd. 2018.

[7] SEDRA, A. S.; SMITH, K. C. Microelectronic circuits. [S.l.]: Pearson Prentice Hall, 2007.

[8] Sentaurus Device User Guide, Synopsys, Inc., 2011.

[9] Cadence Design. Cadence ® Verilog ® -A Language Reference. [S.l: s.n.], 2006

[10] DE SOUZA, B. R. Study of the radiation effect of a transconductance operational amplifier implemented with SOI FinFETs. Master Thesis. University of Sao Paulo, Brazil 2021.

# Impact of Ionizing Radiation and Temperature on the Performance of pMOSFETs with Different Layouts

Guilherme Inácio Grandesi
Electrical Engineering
Department
Centro Universitário FEI
São Bernardo do Campo, Brazil
ggrandesi@gmail.com

Paulo R. Garcia Jr.
Electrical Engineering
Department
Centro Universitário FEI
São Bernardo do Campo, Brazil
uniepajunior@fei.edu.br

Alexis V. Boas
Electrical Engineering
Department
Centro Universitário FEI
São Bernardo do Campo, Brazil
alexiscvboas@fei.edu.br

Renato Giacomini
Electrical Engineering
Department
Centro Universitário FEI
São Bernardo do Campo, Brazil
renato@fei.edu.br

L. E. Seixas
Electrical Engineering
Department
CTI/MCTI
Campinas, Brazil
luis.seixas@cti.gov.br

Marcilei A. Guazzelli
Electrical Engineering
Department
Centro Universitário FEI
São Bernardo do Campo, Brazil
marcilei@fei.edu.br

*Abstract*—**This study investigates whether there is a difference in behavior between different layouts of MOSFET transistors, ELT and Rectangular, in terms of their functionalities at different temperatures, reflecting the effects of exposure to ionizing radiation and the bias mode as it was irradiated. The analysis focuses on the subthreshold slope and $I_{ON}/I_{OFF}$ ratio, among other parameters, to evaluate device robustness. This study highlights the importance of layout design in enhancing the radiation resilience of semiconductor devices.**

*Keywords—MOSFET, ELT, ionizing radiation, Total Ionizing Dose, temperature effects, power transistor*

## I. INTRODUCTION

In electronic devices such as MOSFETs, the effects of ionizing radiation impact their operational characteristics and can cause degradation or even functionality failure [1,2,3]. Total Ionizing Dose (TID) is cumulative and can affect devices depending on their constructive characteristics.

Ionizing radiation is known for its ability to excite and ionize atoms of the matter it interacts with. Total Ionizing Dose (TID) prompts the creation of electron-hole pairs mainly in oxide regions – those with more structural defects and impurities – leading to an accumulation of positive charges or trapped holes (due to their lower mobility) in areas such as gate oxide and the Si/SiO₂ interface on MOSFETs [3,4], impacting the electric field and functionalities of the device.

Power transistors need to withstand higher voltages at the gate and typically have thicker oxide layers, which can accentuate the effects caused by radiation through the increase in oxide-trap charge, which directly impacts the threshold voltage variation [5]. Also, these trapped charges in the oxide cause an "OFF" state leakage current to flow [5].

In aerospace applications, electronic systems comprising numerous transistors are highly exposed to ionizing radiation, leading to Total Ionizing Dose (TID) accumulation in these devices. This research aims to evaluate pMOSFET power transistors with ELT (Enclosed Layout Transistor) and Rectangular layouts, having identical channel length and width dimensions (W = L = 0.6 μm), as well as gate oxide thickness ($t_{ox}$ = 42 nm). The devices were designed at the Information Technology Center (CTI) using SOI CMOS technology and manufactured by CEITEC – Brazil. The PPTLEXT06SOID4 chip has five pMOSFET power transistors: two ELTs and three conventional rectangular ones.

In the active region of the rectangular (or conventional) MOSFET, leakage current generation can occur due to construction imperfections originated from the manufacturing process (LOCal Oxidation of Silicon, LOCOS) [4,6]. As the oxide grows, it can diffuse into the silicon at the edges of the gate layer, as shown in Figure 1. Oxide-trap charges can reduce threshold voltage for these parasitic transistors [5], allowing current to flow between the drain and source even when the main transistor is turned off, increasing the leakage current.

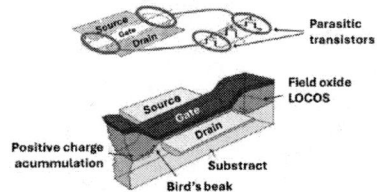

*Figure 1 - Parasitic transistors and trapped charges*

The ELT geometry aims to mitigate the effects of ionizing radiation by reducing construction imperfections at the oxide interfaces with the transistor's active region, which arise from the manufacturing process explained earlier. In this transistor, the drain or source terminal is surrounded by the transistor's active region under the gate [7].

The increase in ambient temperature, induced by a heat treatment process, causes atomic motion (vibration) [8] and could potentially enable the recombination of trapped charges (TID) within the gate oxide and at the SiO₂/Si interfaces. As temperature increases, the intrinsic carrier concentration, $n_i$, rises, reducing the Fermi potential, $\Phi_F$ [8,9]. Consequently, the threshold voltage (Equation 1), $V_{th}$, decreases, implying that the transistor requires less voltage to conduct current between the source and drain [8,9,10].

$$V_{th} = \Phi_{ms} - \frac{Q_{ox}}{C_{ox}} + \frac{2qN_{it}\Phi_F}{C_{ox}} + 2\Phi_F + \frac{\sqrt{4q\varepsilon_{Si}Na\Phi_F}}{C_{ox}} \quad (1)$$

where $V_{th}$ is the threshold voltage, $\Phi_{ms}$ the metal-semiconductor potential difference, $Q_{ox}$ the total charge in the oxide layer, $C_{ox}$ the capacitance of the oxide layer, $N_{it}$ the interface trap state density, $\Phi_F$ the Fermi potential, q the elementary charge of the electron, $\varepsilon_{Si}$ the permittivity of silicon, and $N_a$ the acceptor dopant concentration.

A parameter that increases directly with temperature is the subthreshold slope, S, shown in Equation 2. This slope rising tends to lead to an increase in leakage current ($I_{DS\_leak}$), unlike the saturation current, $I_{DSsat}$, which tends to decrease (Eq. 3) as mobility reduces with temperature [8,9,10].

979-8-3315-4064-7/24 $31.00 © 2024 IEEE

$$S = \frac{1}{\frac{\partial \log I_{DS}}{\partial V_{GS}}} = \ln 10 \frac{kT}{q} \left(1 + \frac{C_{Si} + C_{it}}{C_{ox}}\right) \quad (2)$$

$$I_{Dsat} = \mu_n C_{ox} \frac{W}{L} \frac{(V_{GS} - V_{th})^2}{2} \quad (3)$$

where S is the subthreshold slope, $I_{DS}$ the drain-source current, $V_{GS}$ gate-source voltage, k the Boltzmann constant, T the temperature, q the elementary charge of the electron, $C_{Si}$ and $C_{it}$ the capacitances of the silicon and the interface traps, respectively, $C_{ox}$ the oxide capacitance, $I_{Dsat}$ the drain saturation current, $\mu_n$ the electron mobility, W the channel width and L the channel width.

The inverse trends of "degradation of $I_{DS}$ and mobility" and "degradation of $V_{th}$" converge at a gate bias point where there is no variation in the drain current, known as the Zero Temperature Coefficient (ZTC) point, illustrated in Fig 2 [11].

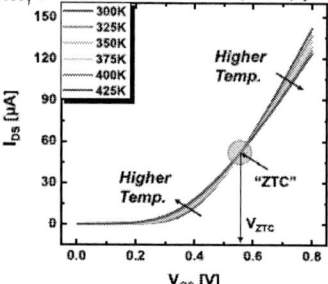

Figure 2 – example ZTC point [11]

For gate voltage $V_{GS}$ smaller than the $V_{ZTC}$ point, the dominant decrease in threshold voltage is responsible for increase drain current with temperature. On the other hand, in cases where gate voltage $V_{GS}$ is higher than the $V_{ZTC}$ point, mobility degradation dominates and is responsible for the decrease in drain current as temperature rises [12].

Analyzing post-radiation devices' response to temperature variations is a comparative tool to assess the resilience of various transistor layouts, focusing on parameters like subthreshold slope and $I_{ON}/I_{OFF}$ ratio. The subthreshold slope (S), defined in equation 2, indicates the voltage required to alter output current by 1 decade [2,10]. A smaller S is preferable, indicating the transistor's ability to switch current levels with lower applied voltage.

The $I_{ON}/I_{OFF}$ ratio is related to the gain and power consumption of the transistor [2]. A higher ratio indicates that the off-state current is much lower than the on-current, and energy consumption is reduced for most digital applications.

## II.  MATERIAL AND METHODOLOGY

To ensure accuracy in comparing the transistor layouts while respecting the experiment's limitations (2 devices), one transistor from each geometry was chosen from the five with the closest original characteristic curves.

The Devices Under Study (DUT) were irradiated up to 600 krad with 10 keV x-ray beam and subjected to a temperature test to compare whether, among different device layouts, TID damage could impact the temperature response differently depending on whether the device was polarized during irradiation, i.e., with an applied external electric field.

Results in the next section show the curves concerning temperature, ranging from 223 K to 353 K, steps of 10 K. The terms to be presented, "biased mode" and "unbiased mode", mean that bias conditions during radiation were different, the former having an electric field applied by an external voltage ($V_{GS} = -5$ V; $V_{DS} = -10$ mV) and the latter not ($V_{GS} = V_{DS} = 0$ V).

The processes carried out from the device irradiation to the temperature test are schematized in Figure 3. The temperature test was conducted in a closed-chamber cryostat, allowing the separate analysis of different layout transistors (one of each) on the same chip for the extraction of their characteristic curves ($I_D$ x $V_G$).

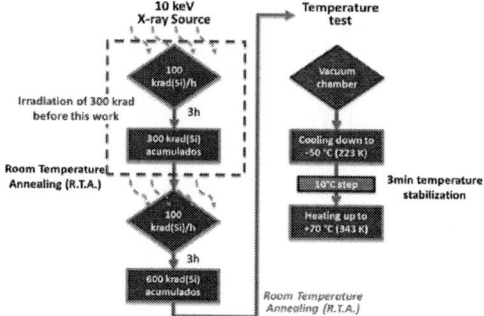

Figure 3 – Radiation and Heat treatment processes

All temperature-dependent measurements were conducted with the drain biased with $V_{DS} = -10$ mV, the gate voltage $V_{GS}$ was varied from 0 to -6 V, and the drain current $I_D$ (output current) was measured.

The next section will discuss two similar drain-source currents: $I_{OFF}$ and $I_{DS\_leak}$. $I_{OFF}$ refers to the drain-source current $I_{DS}$ when the gate-source voltage $V_{GS} = 0$ V, indicating that the transistor is turned off. Conversely, $I_{DS\_leak}$ represents $I_{DS}$ within a constant region before conduction, observed in this work over a $V_{GS}$ range from 0 to -1.5 V, with the transistor also turned off. The $I_{DS}$ current for the transistor in the on-state, denoted as $I_{ON}$, was measured at $V_{GS} = -5$ V.

## III.  RESULTS

Attempting to compare the behavior of post-irradiated devices under temperature tests, the difference in the $I_D$ x $V_G$ and gm x $V_G$ curves of 1) the virgin device, 2) the device at room temperature before the temperature test, and 3) the device during the temperature test at 303 K are shown. The curves of the unbiased mode are represented in Fig. 4. Biased mode is not shown because $gm_{máx}$ and $I_{DS}$ values decreased, representing poor effects to examine in this first analysis.

In the unbiased mode, the thermal treatment showed a greater effect on the parameters than in the other. In the unbiased mode, $gm_{máx}$ increased more in ELT than in the rectangular one. In the ELT, $gm_{máx}$ is more precisely defined, indicating better channel control. This reflects the effectiveness of variations in the gate voltage $V_G$ in controlling the channel current $I_D$.

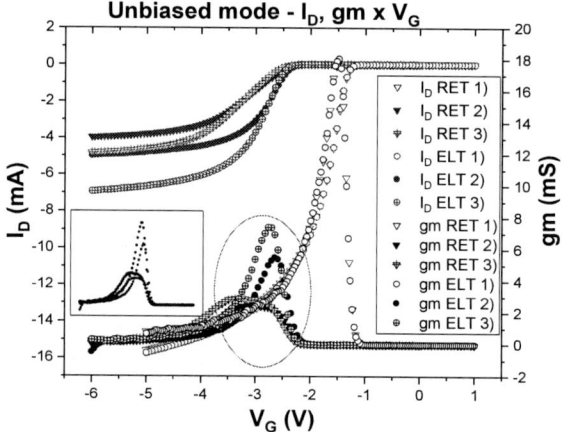

Figure 4 – Unbiased mode – $I_D$ x $V_G$ (left) and gm x $V_G$ (right)

Plotting the $I_D$ x $V_G$ curve as a function of temperature, observed in Fig. 5, we note ZTC appearing in all instances.

979-8-3315-4064-7/24 $31.00 © 2024 IEEE

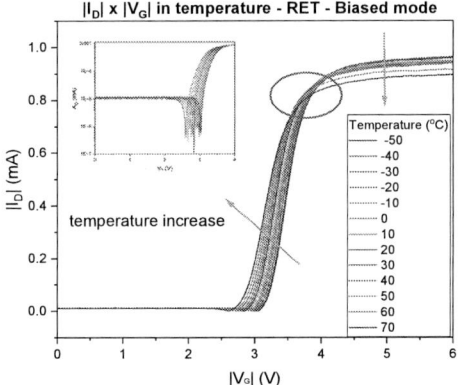

*a) RET in Biased mode*

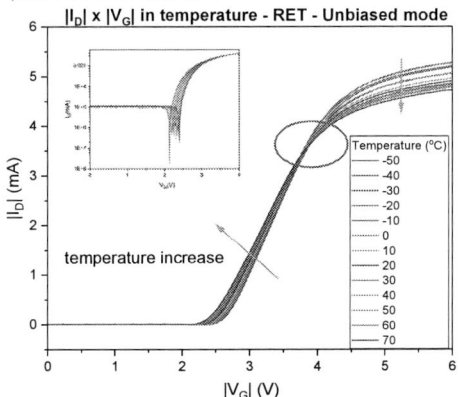

*b) RET in Unbiased mode*

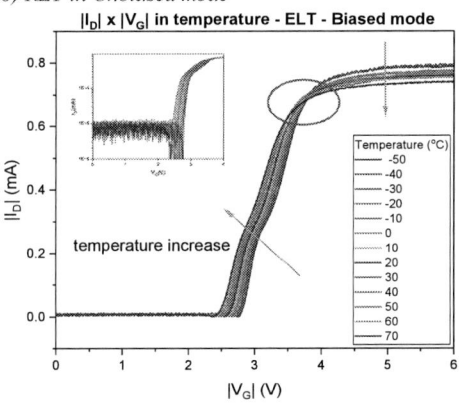

*c) ELT in Biased mode*

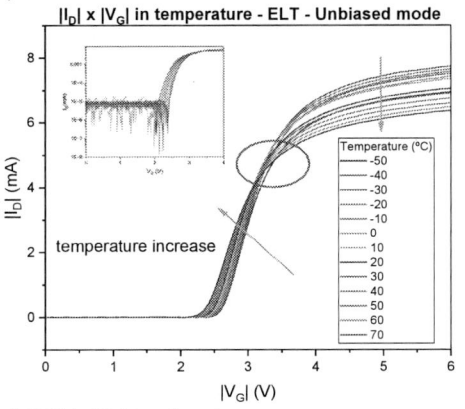

*d) ELT in Unbiased mode*
*Figure 5 – $I_D$ x $V_G$ for all temperatures*

A mathematical analysis of the curves was conducted, calculating the barycenter of all the intersections of the different curves as a function of temperature, along with the standard deviation, to assess if any of the scenarios was more definitive in determining the temperature-independent bias point. This analysis is presented in Table 1.

*Table 1 – Determination of ZTC coordinates for $V_G$ (V) and $I_D$ (A) and respective standard deviation*

| Mode | Baricenter coordinate (average value of ZTC) | Standard deviation |
|---|---|---|
| **ELT biased** | ($V_G$=3.955; $I_D$=0.704E-3) | (X=0.334;Y=3.818E-5) |
| **RET biased** | ($V_G$=3.945; $I_D$=0.848E-3) | (X=0.236;Y=4.568E-5) |
| **ELT unbiased** | ($V_G$=3.424; $I_D$=5.15E-3) | (X=0.250;Y=52,21E-5) |
| **RET unbiased** | ($V_G$=3.807;$I_D$=3.47E-3) | (X=0.139;Y=30.85E-5) |

According to the curves shown in Fig. 5, in every instance, the device transitions from the cutoff zone to the subthreshold region is more quickly with an increase in temperature. This indicates a reduction in the subthreshold voltage, meaning the device starts to conduct with a lower voltage applied to the gate. This effect is better illustrated in Fig. 6 below, showcasing this parameter's reduction relative to temperature.

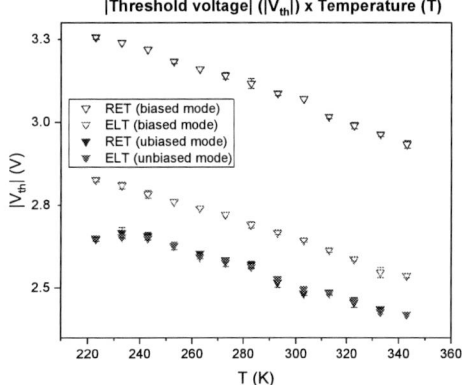

*Figure 6 – $V_{th}$ x $T$ – all modes and layouts*

There is almost no variation in the leakage current as a function of temperature, shown in Fig. 7. The ELT demonstrates lower leakage current values, as expected, since it eliminates the bird's beak regions.

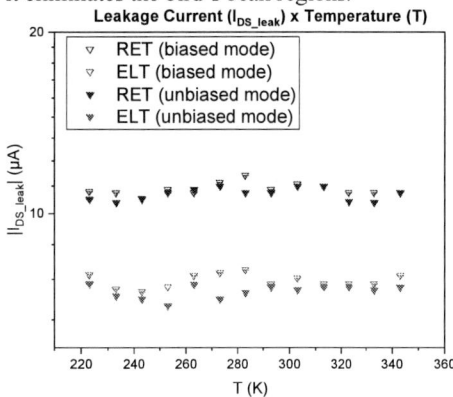

*Figure 7 – $I_{DS\_leak}$ x $T$ – all modes and temperatures*

For the analysis of the subthreshold slope, the range of interest concerns temperatures above 300 K, where it is possible to observe the effects more clearly since the generation of pairs is higher [4], typically increasing current. By examining the plotted curves below and their corresponding linear adjustments, it becomes evident that only the rectangular pMOSFET exhibited distinct behavior, displaying a more significant variation when subjected to bias during irradiation (see Figures 8 and 9).

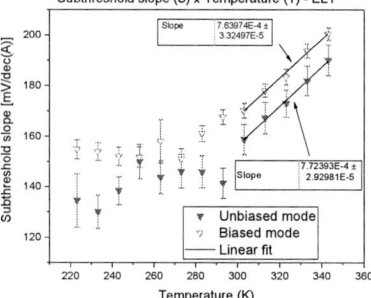

*Figure 8 - Subthreshold slope versus temperature in the ELT*

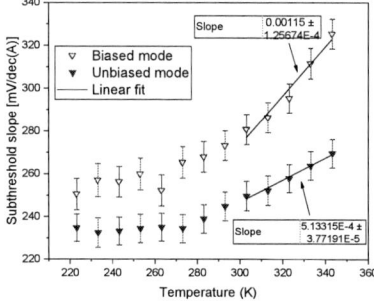

*Figure 9 - Subthreshold slope versus temperature in the RET*

Conversely, for the ELT, irrespective of the presence of an applied electric field during irradiation (due to bias on biased mode), the response to temperature effects remained consistent, as indicated by the parallel lines in Figure 8.

When weakly trapped, charges generated during radiation exposure are more influenced by the electric field, which was much more pronounced in the rectangular layout than the ELT. This indicates that the rectangular layout is more susceptible to the electric field and radiation, as both devices received the same radiation dose. Imperfections in oxide growth in layouts where the isolation between source/drain and gate is worse are susceptible to retaining more charges, a vulnerability that ELT tends to be more resilient [4,7].

The behavior of the $I_{ON}$ and $I_{OFF}$ currents in both types of transistors remained consistent and within the standard deviation as temperature increased. Nevertheless, the primary effect observed is that the presence of a field (bias) during irradiation made the devices more vulnerable to the effects of TID, as the $I_{ON}$ current is nearly one decade lower in the biased mode compared to the unbiased one, shown in Fig. 10.

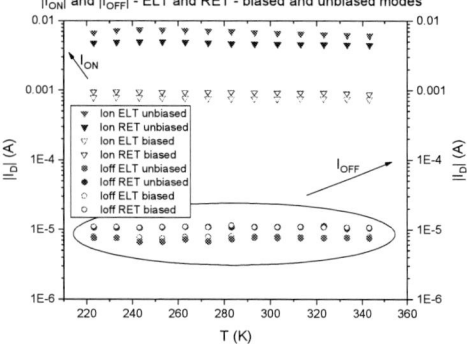

*Figure 10 - $I_{ON}$ and $I_{OFF}$ x T, all modes and layouts*

Furthermore, as expected for this geometry, the ELT exhibits lower $I_{OFF}$ compared to the other. The enclosed

geometry offers an advantage by avoiding the "bird's beak" phenomenon inherent in the rectangular layout.

## IV. CONCLUSIONS

Regardless of the bias mode during radiation, the threshold voltage and leakage current were equally affected by temperature. As expected, $V_{th}$ decreases with increasing temperature because the impacts on $n_i$ and $\Phi_F$ cause a reduction of this parameter.

Biasing during radiation adversely affects the subthreshold slope response to temperature. Only the ELT layout shape demonstrated insensitivity to bias during radiation exposure when utilizing this parameter to assess radiation resilience. One possible explanation is that the rectangular layout accumulated a higher concentration of mobile charges than the ELT, exhibiting a greater impact on the applied field. The enclosed ELT layout exhibited greater resilience against the harmful effects of radiation.

Under the conditions of this study, the ELT transistors were expected to have significantly lower $I_{OFF}$ current. However, this difference is less pronounced in power transistors due to their larger dimensions, which support high voltages and result in lower current density, which can reduce the sensitivity to variations in $I_{OFF}$ current. Nonetheless, the analysis of Fig. 7 shows that the ELT design results in lower leakage current ($I_{DS\_leak}$) due to its geometry, as expected [4,7].

## ACKNOWLEDGMENT

The authors acknowledge financial support from the funding agencies. FAPESP. Brazil 2022/09131-0, 2018/25225-9, 2020/04867-2, 2019/07764-1; CITAR: Proc. 01.12.0224.00; INCT_FNA, Proc. 464898/2014-5; CNPq: 408800/2021-6, 301576/2022-0, 30360/2020-9.

## REFERENCES

[1] OLDHAM, T.R. "Total Ionizing Dose Effects in MOS oxides and Devices". IEEE Transactions on Nuclear Science v.50, p.483, 2003.

[2] REZENDE, S. R.. Materiais e Dispositivos Eletrônicos. 3.ed. São Paulo: Editora Livraria da Física, 2014.

[3] A. C. V. Bôas et al. "Ionizing radiation hardness tests of GaN HEMTs for harsh environments." Microelectronics Reliability, Vol. 116, 2021, https://doi.org/10.1016/j.microrel.2020.114000.d.

[4] L. E. Seixas et al., "Improving MOSFETs' TID Tolerance Through Diamond Layout Style," in IEEE Transactions on Device and Materials Reliability,2017

[5] J. R. Schwank et al., "Radiation Effects in MOS Oxides," in IEEE Transactions on Nuclear Science, vol. 55, no. 4, pp. 1833-1853, Aug. 2008, doi: 10.1109/TNS.2008.2001040.

[6] P. E. Allen and D. R. Holberg, "CMOS Analog Circuit Design," 2nd Edition, Oxford University Press, New York, 2004.

[7] De Lima, J. ; Gimenez, S. ; Cirne, K. . MODELING AND CHARACTERIZATION OF OVERLAPPING CIRCULAR-GATE MOSFET AND ITS APPLICATION TO POWER DEVICES. IEEE Transactions on Power Electronics , v. 27, p. 1622-1631, 2012.

[8] CAPARROZ, L. F. V.,Efeito da Radiação em Transistores 3D em Baixas Temperaturas, Escola Politécnica da Universidade de São Paulo, São Paulo, 2017.

[9] Caparroz, Luis & Bordallo, Caio & Martino, J.A. & Simoen, Eddy & Claeys, Cor & Agopian, Paula. (2018). Analysis of proton irradiated n- and p-type strained FinFETs at low temperatures down to 100 K. Semiconductor Science and Technology. 33. 10.1088/1361-6641/aabab3.

[10] J.A. Martino, M.A. Pavanello, P.B. Verdonck.Caracterização Elétrica de Tecnologia e Dispositivos MOS. São Paulo: Thomson, 2003.

[11] W. Chen, M. Zheng, Y. Lyu and L. Cai, "Determining the Zero-Temperature-Coefficient Point From Device Simulation to Circuit for Improving Temperature Variation Immunity," in IEEE Transactions on Electron Devices, vol. 70, no. 3, pp. 864-870, March 2023, doi: 10.1109/TED.2023.3234897.

[12] Y. Q. Aguiar, A. L. Zimpeck, C. Meinhardt and R. A. L. Reis, "Temperature dependence and ZTC bias point evaluation of sub 20nm bulk multigate devices," 2017 24th IEEE International Conference on Electronics, Circuits and Systems (ICECS), Batumi, Georgia, 2017, pp. 270-273, doi: 10.1109/ICECS.2017.8291999.

# A frugal integrated circuit packaging for non-planar surface-conformant electronics applications

Leonardo Shimizu Yojo
*Instituto SENAI de Inovação em Microeletrônica (ISI-ME)*
Manaus, Brazil
leonardo.yojo@am.senai.br

Fávero Guilherme Santos
*Instituto SENAI de Inovação em Microeletrônica (ISI-ME)*
Manaus, Brazil
favero.santos@am.senai.br

Louise Patron Etcheverry
*Instituto SENAI de Inovação em Microeletrônica (ISI-ME)*
Manaus, Brazil
louise.etcheverry@am.senai.br

Carlos R. P. dos Santos Junior
*Instituto SENAI de Inovação em Microeletrônica (ISI-ME)*
Manaus, Brazil
carlos.pereira@am.senai.br

Fagnaldo Braga Pontes
*Instituto SENAI de Inovação em Microeletrônica (ISI-ME)*
Manaus, Brazil
fagnaldo.pontes@am.senai.br

Elvio C. Dutra e Silva Jr.
*Instituto SENAI de Inovação em Microeletrônica (ISI-ME)*
Manaus, Brazil
elvio.dutra@am.senai.br

*Abstract—* **This work presents a proposal for semiconductor packaging for application in flexible electronics, with simplified manufacturing, aiming to create a low-cost alternative to meet the global demand for semiconductors. It consists of two layers: a base substrate and an interconnection substrate, which can be manufactured from FR-4 laminates, polyimide, polyamide, PET, PEN, and LCP. A proof of concept was fabricated to demonstrate its feasibility. After packaging assembly, a test silicon die was positioned in the center, and connections were made using gold wire bonding. The correct electrical connection between the die pads and the packaging electrodes was verified by measuring electrical continuity in daisy-chain structures.**

**Keywords— Flexible electronics, integrated circuit packaging, polyimide.**

## I. INTRODUCTION

Semiconductor packaging plays a crucial role in protecting integrated circuits or semiconductor components against physical damage, moisture, dust, among other factors, while ensuring their electrical connections and mechanical support or, in other words, it ensures the reliability and proper operation of semiconductor devices. In a typical classification there are two main types of technologies in semiconductor packaging: Through-Hole Technology and Surface-Mount Technology [1]. Each type has their distinct characteristics that can be exploited to meet different applications and requirements. One possibility, for example, resides on the use of flexible substrates to allow the conformity of circuits on non-flat uneven surfaces [2], [3]. Considering the methods of fabrication, semiconductor substrates packaging can either be subtractive - starting from a laminated film and moving to the definition of tracks through a dry-film lithography; or additive, in which metal tracks are defined from the exposed regions of the substrate previously covered by a photosensitive film where the metal layers are grown [1]. Generally, the manufacturing of these substrates still involves additional processing steps, such as drilling and metallization of the drilled holes for the electrical connection between planes.

In this context this work proposes the design of a semiconductor packaging utilizing flexible substrate and simplified processing, aiming at cost reduction, compared to other packaging [4]. For less complex devices, it is feasible to streamline substrate manufacturing and carry out processing in fewer stages. The following will outline the concept of the flexible packaging proposed in this study. Samples were fabricated as proof of concept using a die with built-in daisy chain structure for electrical continuity evaluation.

## II. FLEXIBLE PACKAGING

The concept behind the proposed design in this type of packaging is to provide a simplified manufacturing alternative for flexible electronics. To achieve this, the solution was devised to minimize the number of required processes while still meeting the needs and requirements of semiconductor packaging. The following will describe the proposal for the low-cost, flexible packaging. Figure 1 provides an overview of the processes developed for its fabrication, as well as the assembly steps and tests conducted in this study.

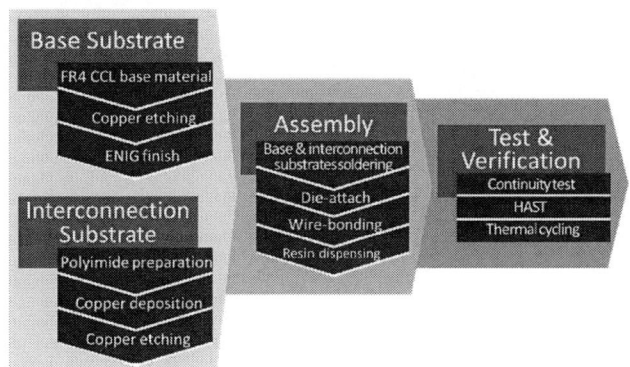

Fig. 1. Workflow of this work.

This packaging is composed of two parts, as depicted in Figure 2, where one can observe schematic diagrams of both sides and their respective lateral views. Firstly, the base substrate serves as the foundation upon which a silicon die is centrally positioned. The base substrate features metal tracks dispersed around its center, which facilitates the electrical connections between the base substrate and the die pads. In the depicted packaging, wire bonding is employed for this purpose, although alternative methods such as flip chip may also be considered. To ensure an optimal surface for the soldering of gold wires, the metal traces should have ENIG

979-8-3315-4064-7/24 $31.00 © 2024 IEEE

plating finishing. The second component is the interconnection substrate, crafted from flexible base material. It must be slightly larger in size than the base substrate and the interconnection substrate has a hollow interior. Its sides may be defined with corresponding metal traces mirroring those on the base substrate or it may act as a fan-out redistribution layer (RDL) if needed.

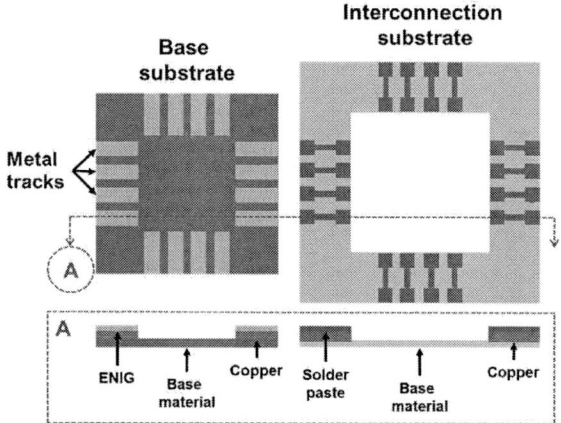

Fig. 2. Schematic representation of the base substrate and the interconnection substrate, top views and lateral views.

Figure 3 illustrates the assembly process of the packaging. The die is fixed at the center of the base substrate with die-attach adhesive (1); the face with the metal structures of interconnection substrate is positioned atop the base substrate (2), ensuring the adequate contact. The electrical connection between the two components is achieved through soldering paste application.

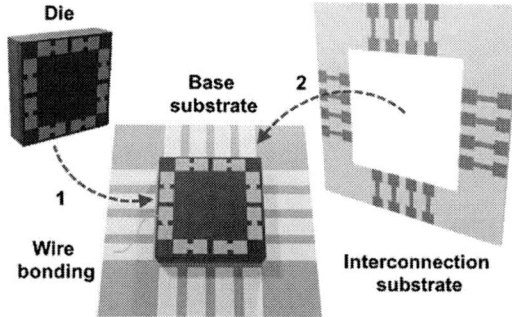

Fig. 3. Packaging assembly.

Following the packaging assembly, electrical connections between the silicon die and the base substrate are established through gold wire bonding. To safeguard the die, a resin molding step or epoxy overflow coating may be applied. A flexible material can be used for this purpose to ensure that the whole packaging is malleable [5]. It is important to observe that the proposed process in this work may also be feasible for non-flexible substrates. For example, it can be easily replicated on LTCC (Low Temperature Co-Fired Ceramics) and glass, commonly used in aerospace applications, high-frequency electronic systems, and microfluidic systems [1].

The soldering of the finalized packaging onto the printed circuit board can be achieved either through solder balls positioned on the interconnection substrate traces or directly through solder paste printing. Figure 4 illustrates an example of the schematic diagram of a completed packaging, depicting both the top view and the lateral view. The final size of the packaging is dependent on the die size, on the quantity of pads, and on the pitch.

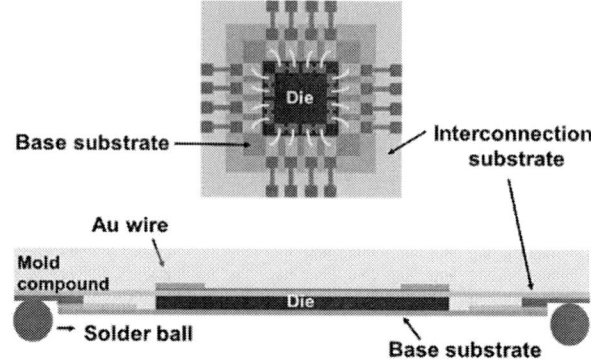

Fig. 4. Final packaging schematic representation, top view and lateral view.

## III. FABRICATION

To demonstrate the feasibility of the proposed packaging, samples were produced as proof of concept. The fabrication process of these samples is detailed below.

The base substrates were prepared from a prefabricated FR4 copper clad laminate designed for integrated circuit packaging. The traces feature ENIG finishing to facilitate gold wire bonding. The choice of this material was made to enable the wire bonding step after the die attach on the base substrate [6], [7].The central spacing of the samples used for the proof-of-concept are square shaped with 0.5 cm side, which are surrounded by the traces for electrical contact. The final thickness of the base substrates is 320 μm.

The interconnection substrates were fabricated using 125 μm-thick polyimide as the support material. This choice was made due to polyimide's high heat stability, chemical solvents resistance, good mechanical and electrical properties, and suitability for flexible packaging manufacturing. The recipe used to grown the copper pads on the polyimide substrate presented in this work are the result of the development made in [8]. Initially, the polyimide underwent argon plasma surface treatment. This procedure was adopted based on observations from [8], which revealed that plasma enhances adhesion between polyimide and the deposited metal. Plasma treatment reduced the delamination rate of deposited films by promoting functional groups on the polyimide film, thus increasing surface energy. Figure 5 illustrates a comparison of the contact angle test results on the polyimide surface before and after argon plasma treatment. The surface became more hydrophilic, with the water droplet angle decreasing from approximately 77.6° to 16.7°.

Subsequently, copper deposition was carried out to form the traces on the interconnection substrate. Approximately 2 μm of copper was sputter-deposited onto the polyimide, which had already been prepared with the hollow center cut off. Other deposition methods can be adopted to provide a low cost process [9]. The size of the cutout in the center is slightly larger than the die size to accommodate it after the packaging assembly. Following this, a photolithography

979-8-3315-4064-7/24 $31.00 © 2024 IEEE

process was employed to define the copper tracks on the interconnection substrate.

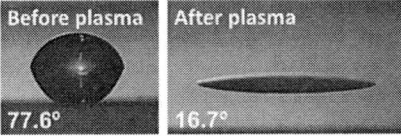

Fig. 5. Droplet contact angle measurement on polyimide before (left) and after (right) plasma treatment.

With both parts prepared, the packaging assembly process was performed as follows: solder paste was applied to the regions on the metal tracks of the base substrate intended for interconnection with the interconnection substrate. The assembly was then placed in a reflow oven to complete the bonding between the two sides. A square silicon test die measuring 5 mm on each side was affixed at the center of the base substrate, as depicted in Figure 4. This die features multiple pairs of short-circuited electrodes, enabling the creation of daisy chain structures for electrical continuity testing. Electrical connections between the die pads and the base substrate were established using 25 μm-thick gold ball-edge wire bonding. Prior to the wire bonding, the surface of the pads were plasma cleaned to increase bondability [10]. The die and the regions of the packaging with wire bonding were safeguarded through the dispensing of epoxy resin. Figure 6 displays images of the fabricated device. Details of the wire bonding process are visible in Figures 6a and 6b, while Figure 6c depicts the complete packaging. In this work, the samples were not soldered to printed circuit board (PCB) and the solder balls deposition was not performed.

Fig. 6. Photographs of the fabricated packaging. a. Au balls of the wire bonding in detail. b. wire bonding on the lateral of the die. c. the complete packaging.

IV. DISCUSSION

Following the fabrication of samples with the test dies, electrical continuity tests were conducted using daisy-chain structures. Since the electrodes of the dies are short-circuited, it is expected to measure a low-resistance path between the packaging terminals. This serves as a straightforward method to verify that the connections were adequately made and that the packaging was assembled correctly. If the continuity measurement results in an open circuit, it can be inferred that there was no connection either in the solder between the base and interconnection substrates or in the wire bonding between the die and the base substrate pads. Figure 7 provides detailed illustrations of the test die and the current path in this type of structure. The resistance associated with the packaging ($R_{pack}$) encompasses all components related to PCB contacts, the deposited metal traces, the interlayer contacts, and the wire bonding. Meanwhile, $R_{die}$ represents the electrical resistance between the two connected electrodes of the die.

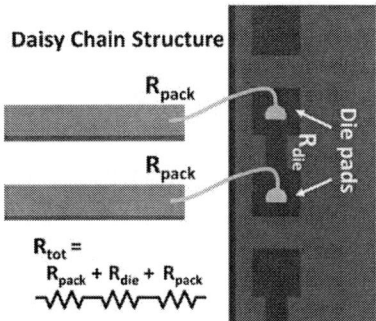

Fig. 7. Daisy chain structure.

Electrical continuity measurements were conducted using a handheld multimeter. All samples tested exhibited continuity, thus confirming the feasibility of the proposed packaging. The average total resistance ($R_{tot}$) was recorded at 0.49 $\Omega$. This value could be further improved through optimization of the packaging fabrication process. Visual inspection revealed oxidation of the copper layer on the interconnection substrate, which affected its contact with the PCB, after the soldering step.

The reliability of the packaging with the epoxy covering was analyzed through unbiased highly accelerated stress test (HAST) and thermal cycling test.

The HAST was performed at temperature of 110 °C and relative humidity of 85% during 96 hours in a noncondensing chamber, with the goal of evaluating the endurance of the device against corrosion under high humidity environment. The electrical continuity was checked every 24 hours of experiment. All the tested contacts remained connected after 48 hours of test, but only 50% endured until the end of the experiment.

The thermal cycling test was performed in a dual chamber equipment with a cold chamber at the temperature of -40 C and the hot chamber at 110 °C. The aim of this test was to determine the ability of the device to withstand the mechanical stress induced by the temperature variation. The electrical continuity was verified after each cycle. It was observed a failure rate of 50% after 2 cycles and the total failure of the continuity in all tested samples after 5 cycles.

It was found that the solder between the base and the interconnection substrates broke after the reliability tests, which ceased the electrical continuity of the daisy-chain structures. The failed devices underwent through the solder reflow step once again and it was observed that the electrical continuity was reestablished. Thus, it was confirmed that the join between both substrates made by the solder represents the weak point in the reliability of the packaging. Further test and improvement shall be made to guarantee the applicability of the packaging such as the verification of molding material fill and the selection of the bond material between the two substrates.

## V. CONCLUSION

This work proposed a low-cost semiconductor packaging approach. Among its distinctive features, we can highlight the capability to perform packaging with flexible materials, enabling application on non-planar surfaces. Additionally, another significant appeal is its simplified fabrication process with a reduced number of process steps, eliminating the need of via manufacturing. The viability of the packaging was verified through the fabrication of proof-of-concept samples, which were produced using test silicon dies with daisy-chain structures. Electrical continuity was measured, demonstrating that the electrical connections were successfully established. Reliability tests showed that there is still room for optimization of the packaging fabrication process, particularly regarding the metallization of the traces composing the interconnection substrate and the joining process of the base and interconnection substrates. Finally, and above all, the presented proposal has emerged as a strong package candidate not only for circuits but also for electronics systems whose technical requisites are related to low-cost and to mechanical conformity on non-planar surfaces.

## REFERENCES

[1] W. J. Greig, "Integrated circuit packaging, assembly and interconnections," *Integrated Circuit Packaging, Assembly and Interconnections*, pp. 1–296, 2007, doi: 10.1007/0-387-33913-2/COVER.

[2] F. D. Egitto, R. N. Das, F. Marconi, B. Wilson, and V. R. Markovich, "Development of electronic substrates for medical device applications," *IMAPS International Conference and Exhibition on Device Packaging*, no. September 2012, 2012, doi: 10.4071/2012dpc-wa23.

[3] S. Gupta, W. T. Navaraj, L. Lorenzelli, and R. Dahiya, "Ultra-thin chips for high-performance flexible electronics," *npj Flexible Electronics*, vol. 2, no. 1, 2018, doi: 10.1038/s41528-018-0021-5.

[4] W. Christiaens, E. Bosman, and J. Vanfleteren, "UTCP: A novel polyimide-based ultra-thin chip packaging technology," *IEEE Transactions on Components and Packaging Technologies*, vol. 33, no. 4, pp. 754–760, 2010, doi: 10.1109/TCAPT.2010.2060198.

[5] B. Zhang, Q. Dong, C. E. Korman, Z. Li, and M. E. Zaghloul, "Flexible packaging of solid-state integrated circuit chips with elastomeric microfluidics," *Sci Rep*, vol. 3, pp. 1–8, 2013, doi: 10.1038/srep01098.

[6] N. B. Jaafar and R. Damalerio, "Challenges of wirebonding with polyimide flexible printed circuit board (FPCB)," *2017 IEEE 19th Electronics Packaging Technology Conference, EPTC 2017*, vol. 2018-Febru, no. 65, pp. 1–5, 2017, doi: 10.1109/EPTC.2017.8277435.

[7] Y. Hin, C. Æ. J. Kim, Æ. D. Liu, P. C. K. Liu, and Æ. Y. Ming, "Comparative performance of gold wire bonding on rigid and flexible substrates," *J Mater Sci: Mater Electron*, vol. 17, pp. 597–606, 2006, doi: 10.1007/s10854-006-0005-4.

[8] L. S. Yojo, F. G. Santos, F. B. Pontes, C. R. P. Dos Santos, W. Hasenkamp, and E. C. Dutra E Silva, "Reliability Aspects and Study of Copper Seed Deposition on Polyimide via Sputtering," *2023 37th Symposium on Microelectronics Technology and Devices, SBMicro 2023*, no. 2, pp. 1–4, 2023, doi: 10.1109/SBMicro60499.2023.10302524.

[9] Y. Wu, Y. Huang, M. Chen, Y. Lin, and S. Fu, "Failure Analysis of Cu Electroplating Process with Polyimide Substrate Fabricated for Flexible Packaging," *2012 7th International Microsystems, Packaging, Assembly and Circuits Technology Conference (IMPACT)*, pp. 94–97, 2012, doi: 10.1109/IMPACT.2012.6420278.

[10] Y. H. Chan *et al.*, "Improvements in Au Wire Bondability of Rigid and Flexible Substrates using Plasma Cleaning," *Packaging Technology*, no. April 2004, pp. 0–5, 2005, doi: 10.1109/ICEPT.2005.1564609.

979-8-3315-4064-7/24 $31.00 © 2024 IEEE

# Development of a double pulse test plataform for switching loss investigation in emerging SiC MOSFET technology

Denison Rodrigo Ferreira Silva
*Faculdade de Engenharia*
*Elétrica e de Computação*
*Universidade Estadual de*
*Campinas*
Campinas, Brazil
d253593@dac.unicamp.br

Joel Felipe Guerreiro
*Instituto de Pesquisas Eldorado*
Campinas, Brazil
joel.guerreiro@eldorado.org.br

Lucas B. Spejo
*Faculdade de Engenharia*
*Elétrica e de Computação*
*Universidade Estadual de*
*Campinas*
Campinas, Brazil
lucas.spejo@gmail.com

Marcos V. Puydinger dos Santos
*Faculdade de Engenharia*
*Elétrica e de Computação*
*Universidade Estadual de*
*Campinas*
Campinas, Brazil
mpuyding@unicamp.br

*Abstract*—The emerging technology of silicon carbide (SiC) MOSFETs has been gaining prominence in power electronics due to its superior performance compared to traditional Silicon (Si) technology. This technology stands out for operating at higher switching frequencies, which is beneficial for improving the power density of the converter by reducing weight or size. However, switching losses can become significant with the increase in switching frequency. In this work, we developed a Double Pulse Test (DPT) platform, enabling the evaluation of SiC MOSFET switching losses, providing essential insights for optimizing design and ensuring a better efficiency. The results obtained in conjunction with LTSpice simulation, demonstrate the challenges of performing DPT in characterizing SiC MOSFETs switching losses.

*Keywords*—*Double Pulse Test, Power Electronics, Switching Losses, SiC-MOSFET.*

## I. INTRODUCTION

In a global context where sustainability has become an undeniable priority, energy efficiency emerges as a crucial component in the search for environmental and economic solutions. Within the field of power electronics, where the demand for more efficient and powerful devices is incessant, the emergence of Silicon Carbide Metal-Oxide-Semiconductor Field-Effect Transistor (SiC MOSFET) has been revolutionary. Unfortunately, converters based on Si technology are reaching their theoretical limits and are less efficient than desired in high-power applications [1]. SiC is a wide-bandgap semiconductor with superior material properties compared to Si. It presents low on-state resistance and the ability to operate at higher temperatures, and, as can be seen in Fig. 1, it can handle high power and switching frequencies. Using power converters with high switching frequencies is another way to reduce the size of the converter, as the dimensions of passive components can be smaller, leading to better power density of converters.

However, with the increase in switching frequency, switching losses become a crucial contributor to the total converter losses [2]. Thus, the double pulse test (DPT) is a widely used method that plays a key role in evaluating the dynamic performance of power semiconductor devices, providing valuable information for the design and development of more efficient and reliable power electronic circuits [5].

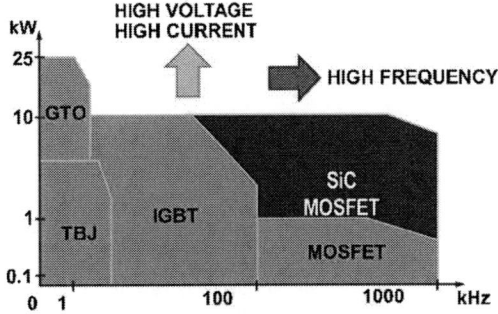

Fig. 1. Power vs Frequency Performance for different electronic power switch technologies. Adapted from [1].

Therefore, in this work we present a DPT prototype specifically designed by our group to evaluate the switching losses of the emerging SiC MOSFET technology. The purpose of the DPT is ultimately to test the MOSFET inside the power convert layout itself. In this sense, all the parasitic components can be considered into the switching loss calculation Moreover, this test is crucial for assessing the switching parameters of the MOSFET under various conditions, including distinct temperatures, currents and gate resistances. With the test results, it is possible to track and model the appropriate behavior of the converter in order to achieve lower switching losses.

## II. DOUBLE PULSE TEST (DPT) CIRCUIT

The circuit for the DPT is shown in Fig. 2. The main idea behind is to test the switching behavior of the circuit during both the turn on and turn off cycles. The parasitic capacitances of the devices are the main responsible for the current-voltage delay during device operation, which causes a non-negligible power loss during switching. Therefore, this test is frequently employed to estimate the transistor switching power loss during a cycle to estimate the efficiency and heat budget. The DPT comprises a power supply ($V_{CC}$), a capacitor (C) to provide current and stabilize the supply voltage during the test, and an inductor (L). In a DPT, it is advantageous to switch an inductive load as it enables forced transistor switching. Consequently, in the worst-case scenario of activation and deactivation transients at a particular current, they can be thoroughly analyzed at a specific drain current [1].

Additionally, the circuit consists of a diode and a SiC MOSFET as the device under test (DUT). The gate drive signal pulses control the MOSFET activation by a voltage applied to the gate relative to the source ($V_{GS}$). Meanwhile, the diode serves as a freewheeling function [2]. Fig. 3(a) illustrates the two pulses applied to the MOSFET, hence the name double pulse test. The time T1 is typically selected in the order of microseconds, while T2 is chosen to be shorter than T1 and can be considered as ½ to ⅔ of T1. This is done to maintain a constant current load during the on and off transitions. On the other hand, time T3 is shorter than T1 although sufficiently long to ensure a new transient measurement. T3 can be taken as half of T1 [5].

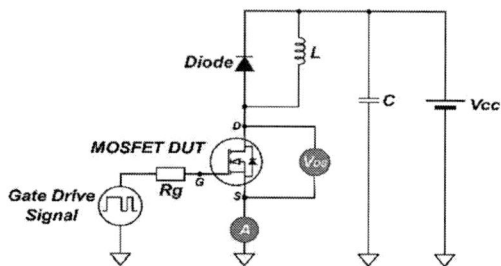

Fig. 2. Basic circuit of the double pulse test.

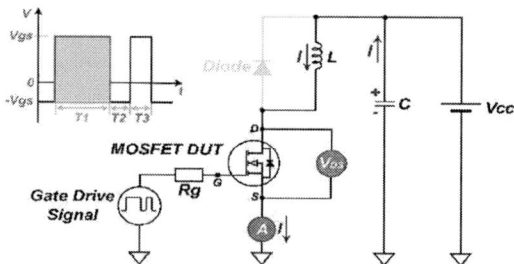

Fig. 4. Operation of the DPT circuit during the first pulse, interval T1.

The first activation pulse will start conducting the MOSFET, but the current is not yet established in the inductor. Therefore, calculating losses at this moment results in an inaccurate evaluation. Hence, it is necessary to load a sufficient current in the inductor, turn it off, and then turn it on again to calculate losses more accurately [5]. Thus, the main measurement region for the DPT purposes is during the transition from T1 to T3, where we will analyze the switching losses in both turn off and turn on regimes, as indicated in Fig. 3(b).

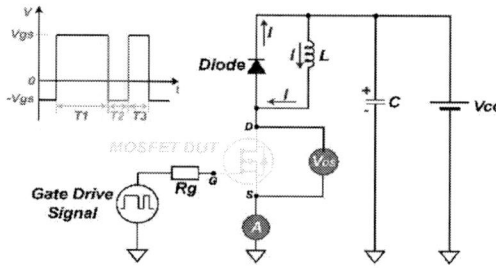

Fig. 5. Operation of the DPT circuit during interval T2.

## III. DESIGN OF THE DPT PLATFORM

Fig. 6 illustrates the sequence used to generate the pulses to activate the MOSFET. The microcontroller used to generate the pulses was the F28379D from Texas Instruments, with an amplitude of 3.3V. This signal passes through a gate driver for amplification and to adjust the voltage and current levels of the command signal. In this test, a bipolar gate drive was used, ranging from -2V to 20V for controlling the MOSFET. Table 1 shows the components used in the test setup.

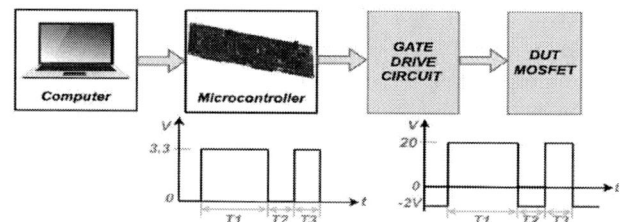

Fig. 6. Steps for generating the pulses to drive the MOSFET.

Fig. 7 illustrates the setup used to perform the tests. Two MOSFETs were used on the board, configured in a half-bridge setup. One MOSFET serves as the test device (low side), and the other (high side) acts as a freewheeling diode due to the presence of the internal body diode. They are positioned underneath the board, with the gate drive fixed on top.

Fig. 3. (a)Double pulses for MOSFET activation, (b) drain-source voltage ($V_{DS}$), and (c) switching current in the MOSFET.

The first pulse at T1 aims to charge the inductor, and, during the transition to the second pulse, the switching losses of the MOSFET are analyzed. Note that during pulse 1, there will be a flow of current passing through the inductor and the MOSFET, as shown in Fig. 4. The MOSFET behaves like a closed switch, while the diode is blocked. At this moment, the MOSFET drain-source voltage ($V_{DS}$) is close to zero, as shown in Fig. 3(b), and its current is the same as that of the inductor, as presented in Fig. 3(c). In the interval T2 between pulses 1 and 2, the ($V_{GS}$) goes to zero (or negative value), turning it into an open switch, as seen in Fig. 5. Consequently, there will be no current flowing through the MOSFET, and ($V_{DS}$) will be equal to $V_{CC}$. It is worth mentioning that at this moment, the current present in the circuit dissipates in the diode. Finally, during pulse 2, in the interval T3, the MOSFET begins conducting again, similarly the state shown in the Fig. 4 during pulse 1.

979-8-3315-4064-7/24 $31.00 © 2024 IEEE 54

TABLE I.    COMPONENTS USED IN THE DPT PLATFORM

| Component | Value/ Ratings |
|---|---|
| Capacitor | EPCOS B43845, electrolytic 1000μF, 250V |
| Inductor | Toroidal Ferrite Core, 840μH, 30A |
| SiC MOSFET | Rohm SCT3060, 600V, 39A |
| DC Source ($V_{CC}$) | Hp 6030A, 200V, 17A |
| Probe ($V_{DS}$) and ($V_{GS}$) | Yokogawa 700924, 100Mhz differential probe 1.4kV |
| Shunt resistor | Powertekuk SDN-10, 100mΩ, 2000MHz. |

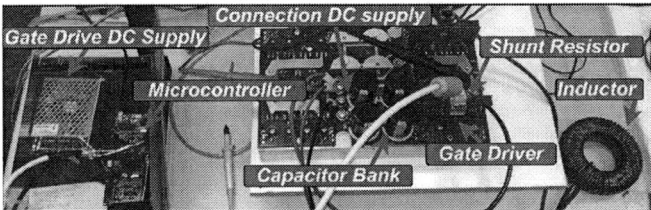

Fig. 7.   Assembly of the DPT on a power board.

### A. Sizing of the Inductance

In the DPT platform, the inductor is used to replicate the conditions of the circuit in a converter design [1]. The value of the inductance can be designed using (2), which arises from (1). Note that the value of inductance, L, depends on the source voltage $V_{CC}$, the time of the first pulse, T1, and the current, I, under test.

$$V_L = L \frac{di}{dt} \quad (1)$$

$$L = \frac{Vcc.T1}{I} \quad (2)$$

### B. Sizing of the Capacitance

From the beginning to the end of the first pulse, as shown in Fig. 3(a), the energy of the capacitor undergoes a variation ($\Delta E_C$), corresponding to the drop in voltage across the $V_{CC}$ bus [3]. To ensure that this variation is as minimal as possible, the value of the capacitor should be carefully chosen. Considering that at the beginning of the first pulse the capacitor is discharged, the stored electrostatic energy can be obtained using (3), and the magnetic energy in the inductor by (4).

$$E_C = \frac{1}{2}.C.V_{CC}^2 \quad (3)$$

$$E_L = \frac{1}{2}.L.I^2 \quad (4)$$

Considering the variation in $V_{CC}$ during T1, (3) can be modified into (5). Thus, to ensure that the energy transfer from the capacitor to the inductor is as constant as possible, by equating (4) and (5), one can find a minimum value for the capacitance (C), as in (6). This guarantees that there will be a smaller voltage drop in $V_{CC}$ between the interval of pulse 1 and pulse 2.

$$\Delta E_C = \frac{1}{2}C.V_{cc}^2 - \frac{1}{2}C.(V_{cc} - \Delta V_{cc})^2 \quad (5)$$

$$C_{min} \geq \frac{L.I^2}{2V_{cc}.\Delta V_{cc} - \Delta V_{cc}^2} \quad (6)$$

In this project, we used $V_{CC}$ = 200 V and I = 10 A, and a voltage variation of 3% in $\Delta V_{cc}$ to obtain a smaller voltage variation during pulse intervals [4]. Consequently, this resulted in C=1000 μF, and L = 840 μH. With these values, T1 = 42μs was found, and intervals T2 and T3 were adopted to be half of T1 to maintain a constant current load during the on and off transitions [5].

### C. Calculation of the Switching Loss

Fig. 8 (a) shows the typical parasitic capacitances in the SiC MOSFET. The main ones are the parasitic capacitances $C_{GS}$, $C_{GD}$, and $C_{DS}$, which represent, respectively, gate-source capacitance, gate-drain capacitance, and drain-source capacitance. These capacitances slow down the switching speed of the device, extending the turn-on and turn-off times, thus limiting the switching frequency [1].

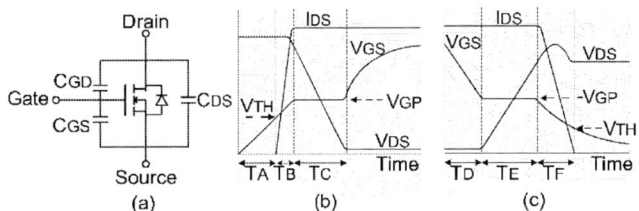

Fig. 8.   (a) Representation of parasitic capacitances $C_{GS}$, $C_{GD}$, and $C_{DS}$ in the MOSFET symbol and equivalent transition during MOSFET (b) turn-on and (c) turn-off periods. Adapted from [4].

Figs. 8 (b) and (c) depict the behavior of the MOSFET DUT during transients during turn-on and turn-off. One can observe that during $T_A$, the MOSFET receives a suitable signal at $V_{GS}$ capable of driving it into conduction. However, the MOSFET does not turn on instantaneously. First, it reaches its threshold voltage ($V_{TH}$), and, hence, $V_{DS}$ starts to decrease linearly. At the beginning of the $T_C$ interval, $V_{GS}$ is constant at $V_{GP}$ (the gate plateau voltage), and all the gate current discharges the capacitance $C_{GD}$. After $T_C$, $C_{GD}$ discharge, and the $V_{DS}$ voltage reaches its minimum value, yielding the full charging of $C_{GS}$, after which the MOSFET can be considered as a closed switch.

Thus, during the turn-off phase of the MOSFET, as shown in Fig. 8 (c), $V_{GS}$ tends to zero. But the MOSFET does not immediately enter the cutoff state. At $T_D$, the capacitance $C_{GS}$ is being discharged, remaining constant at $V_{GP}$ in the interval $T_E$, while $V_{DS}$ starts to rise linearly, and the capacitance $C_{GD}$ begins to charge. It is only when $V_{GS}$ reaches Vth that the current $I_{DS}$ goes vanishes, and the MOSFET enters cutoff.

Therefore, the two most important capacitances to consider in the MOSFET switching are $C_{GS}$ and $C_{GD}$, as they undergo a delay in charging and discharging [4]. With this, one can calculate the MOSFET power loss in these intervals and find the switching losses during turn on and turn off. The energy dissipated in the MOSFET in these intervals is given by (7), enabling the calculation of switching losses in Joules [J] [5]. The integration interval (t) corresponds to the turn-on and turn-off time, as illustrated in Fig. 8.

$$E = \int_0^t p(t)dt = \int_0^t (Vds(t).I(t))dt \quad (7)$$

979-8-3315-4064-7/24 $31.00 © 2024 IEEE

## IV. EXPERIMENTAL RESULTS

Fig. 9 (a) depicts the experimental result of MOSFET switching. The switching curves demonstrate that our DPT prototype works by switching on and off the MOSFET, as expected. It should be noted that the $V_{GS}$ signal originates from the gate driver. During the conduction of the MOSFET, in the T1 and T3 intervals, the $V_{DS}$ voltage is close to zero. In the remaining intervals of $V_{GS}$, the $V_{DS}$ voltage is approximately 200 V, which is the DC voltage used to power the power circuit.

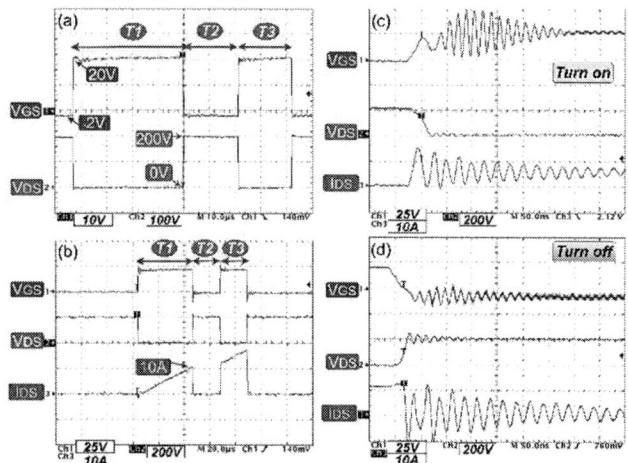

Fig. 9. (a) Experimental result of MOSFET switching without the shunt resistor, (b) MOSFET switching with the shunt resistor, (c) zoom in on turn on, (d) zoom in on turn off.

Preliminary measurements upon inserting the shunt resistor for current reading on the MOSFET, Figs. 9 (b), (c), and (d), revealed a relatively high noise level due to capacitive couplings [7]. However, this issue is being addressed for a future work. Subsequently, a Rogowski coil probe is intended to be acquired to perform the current reading on the MOSFET. Therefore, to estimate the desired switching loss results using (7) during the turn-on and turn-off periods, a Spice simulation of a circuit similar to the one depicted in Figure 2 was conducted using LTSpice software with MOSFET model SCT3060 [6].

The simulation results are illustrated in Figs. 10 (a) and (b). They show the behavior of $V_{GS}$, $V_{DS}$, as well as $I_{DS}$ in the MOSFET during the DPT switching. The shaded area represents the values of $V_{DS}$ and $I_{DS}$ and the time interval over which the switching loss was calculated. In this sense, given the aforementioned limitations with the shunt, the energy dissipated in the SiC MOSFET DUT during switching was estimated via simulation. The energy consumed during the turn on and turn off periods – which consists of the $V_{DS}$–$I_{DS}$ product over the switching time – were found to be 45µJ and 16 µJ, respectively. Thus, the total loss during the MOSFET switching is approximately 61 µJ per cycle.

The simulation was performed by switching the SiC MOSFET at room temperature with the current and voltage as 10 A and 200 V, respectively. The simulation was taken with the same current and voltage parameters as the experimental ones. Therefore, reducing the switching losses in the MOSFET implies reducing the thermal energy dissipation. In addition, the importance of DPT is to assess this MOSFET switching loss in operation for different values of current, voltage, temperature,

and switching frequency. In this sense, it is possible to optimize the converter to obtain lower dissipated thermal energy in the MOSFET, thus achieving better efficiency.

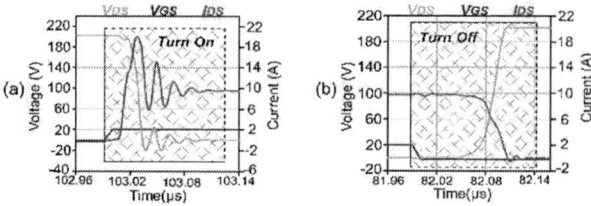

Fig. 10. (a) Turn On, activation of the MOSFET (b) Turn Off, deactivation of the MOSFET.

## V. CONCLUSION

In this work, we present a prototype for performing DPT on SiC MOSFET transistors. The idea behind DPT is to evaluate the transistor switching parameters under various conditions, including temperature and power, thus enabling the optimization of the converter to achieve better efficiency. During the test, we obtained the MOSFET switching with curves of $V_{GS}$ and $V_{DS}$. However, when using the shunt resistor to measure current $I_{DS}$ in the SiC MOSFET, we encountered problems of high oscillations due to capacitive couplings. This issue is being investigated for conducting further tests. As an alternative, we conducted a Spice simulation that assisted us in estimating the switching loss of the SiC MOSFET SCT3060 when switched at 10 A, 200 V, and at room temperature. The estimated switching loss result was 61µJ. Therefore, with this result from the DPT, a horizon of possibilities opens up for studying transistor switching and tracking its switching causes.

### ACKNOWLEDGMENT

The authors would like to thank LCEE lab at FEEC/UNICAMP for the experimental facilities. The authors are also grateful to supported by Fundep research development Foundation Route 2030/Line V #27192.03.02/2022.02-00. This study was financed in part by the Coordenação de Aperfeiçoamento de Pessoal de Nivel Superior – Brasil (CAPES) – Finance Code 001.

### REFERENCES

[1] B. N. Torsæter, S. Tiwari, R. Lund and O. -M. Midtgård, "Experimental evaluation of switching characteristics, switching losses and snubber design for a full SiC half-bridge power module," 2016 IEEE 7th International Symposium on Power Electronics for Distributed Generation Systems (PEDG), Vancouver, BC, Canada, 2016, pp. 1-8, doi: 10.1109/PEDG.2016.7527071.

[2] A. Ghosh, C. N. M. Ho, J. Prendergast and Y. Xu, "Conceptual Design and Demonstration of an Automatic System for Extracting Switching Loss and Creating Data Library of Power Semiconductors," in IEEE Open Journal of Power Electronics, vol. 1, pp. 431-444, 2020, doi: 10.1109/OJPEL.2020.3026896.

[3] Rohde-Schwarz, Appl. Note GFM347, pp.8-26.

[4] Vishay Siliconix, Appl. Note AN608A, pp.2-4.

[5] M. I. Masoud, W. Issa and W. Yates, "A Tutorial on Double Pulse Test of Silicon and Silicon Carbide MOSFETs," 2023 IEEE Workshop on Electrical Machines Design, Control and Diagnosis (WEMDCD), Newcastle upon Tyne, United Kingdom, 2023, pp. 1-6, doi: 10.1109/WEMDCD55819.2023.10110895.

[6] Linear Technology Corporation, LTspice XVII Manual, Analog Devices Inc., 2021.

[7] Z. Zhang, et al., "Methodology for Wide Band-Gap Device Dynamic Characterization," IEEE Trans Power Electron, vol. 32, no. 12, pp. 9307–9318, Dec. 2017, doi: 10.1109/TPEL.2017.2655491.

979-8-3315-4064-7/24 $31.00 © 2024 IEEE

# A Semi-Automatic Tool for the Extraction of RF-Plasmas Parameters by Langmuir Probe

Rodrigo Cicareli
*Electrical Engineering Dept*
*Federal University at São Carlos*
São Carlos, Brazil
rodrigo.cicareli@gmail.com

Giuseppe A. Cirino
*Electrical Engineering Dept*
*Federal University at São Carlos*
São Carlos, Brazil
gcirino@ufscar.br

*Abstract*—**MEMS/NEMS and Integrated circuits technology are direct related to the advance of plasma assisted processes. Cold plasmas can be electrically characterized by the employment of Langmuir probes due to its versatility and low-cost. One can measure the following electrical parameters of plasma: electron temperature, plasma density, floating potential and plasma potential. This work reports a numerical tool, for a semi-automatic analysis of a characteristic curve, resulting from a Langmuir probe measurements on RF plasmas (13.56 MHz). This automation consists of data acquisition and signal conditioning via LabView® software, followed by numerical processing via Matlab® software. The results show the effectiveness of the tool for data analisys and obtained results.**

*Index Terms*—**RF Plasmas, Langmuir probe, parameter extraction, cold-plasmas, plasma etching, plasma deposition**

## I. Introduction

There are several sectors of human activities which involve the application of cold plasmas in their processes, mainly in the semiconductor processing industry, such as thin film deposition and dry etching. From a process control point-of-view, important process variables are plasma density and average electrons and ions energy, among others. Cold plasmas can be electrically characterized by the employment of Langmuir probes due to its versatility and low-cost. It consists of inserting a small electrode (probe) immersed in the neutral region of the plasma. This probe is excited by applying a linearly variable bias voltage, $V_{pr}$. As a response, a bipolar current $I_{pr}$ will flow through the plasma and the external circuit, as shown in figure 1(a). This current is proportional to the potential drop across $R_L$. Figure 1(b) shows schematically the probe circuitry and plasma chamber setup in order to obtain the characteristic curve on a PC screen.

Some limitations of this technique are the relatively difficult data interpretation, and the fact that it is an intrusive method (plasma disturbances may occur and in some cases distort the measurement). The probe construction, data acquisition and data treatment need to be made very carefully in order to avoid some artefacts on the processed data. In this sense, a semi-automatic tool would facilitate the job to extract electric

parameters from a plasma plant. A typical characteristic I-V curve looks like a diode curve due to the difference between ions and electrons mobility.

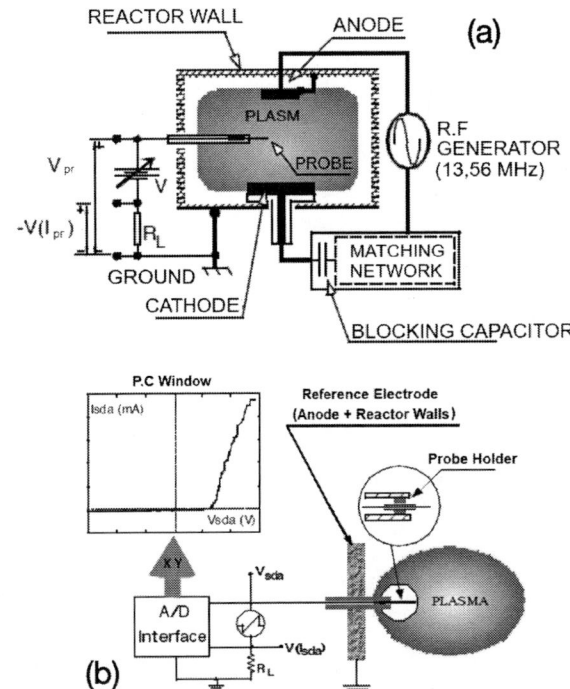

Fig. 1: (a) Schematic view if a RIE chamber, with an electrostatic probe immersed in the plasma bulk; (b) Experimental setup used in measurements with simple probes.

By the interpretation of this characteristic curve one obtains the following plasma parameters: electrons number density, $n_e$; positive ions number density, $n_i$; floating potential, $V_f$; plasma potential, $V_p$; average electron energy, $(kT_e/q)$, where $k = 1.38 \times 10^{-23}$ J/K is the Boltzmann constant and $q = 1.6 \times 10^{-19}$ C is the elementary charge magnitude.

This work reports a semi-automatic numerical tool, for the analysis of a characteristic curve, resulting from a Langmuir probe measurements on RF plasmas (13.56 MHz). This

979-8-3315-4064-7/24 $31.00 © 2024 IEEE

automation consists of collecting data from a single probe through an analog-to-digital converter and thus treating the characteristic I-V curve. It also provides agility and precision, with the appropriate degree of human intervention in the determination of plasma parameters. Interactive interfaces are used to simplify the entire data handling process.

## II. THEORY FOR PLASMA PARAMETERS EXTRACTION

The study discussed here is based on plasma having a Maxwell-Boltzmann energy distribution, together with several assumptions, such as [1,2,3]: plasma consists only of electrons and positive ions; the plasma is isotropic and is in a steady state of charge generation and loss; the probe sheath is collisionless. Since the particles species are not in thermal equilibrium, typically the average electron temperature $T_e$ is two orders of magnitude higher than the positive ions temperature $T_i$, which, in its turn, is approximately twice the neutral species temperature (300K) [4,5,6]. Therefore, the approximation $(T_i/T_e) = 0$ holds.

Once the characteristic curve is obtained and storage as a spreadsheet, the mathematical procedures can now be started to obtain all the desired parameters, as follows.

**Floating Potential**, $V_f$. Its the probe bias at which the net current collected equals zero.

**Plasma Potential**, $V_p$. Analyzing the inflection point within the exponential part of the characteristic curve. As this point normally does not occurs abruptly, so its difficult to visualize it. Two common approaches are to plot the first and second derivatives of the probe curve, and take the probe bias at which a peak (1st derivative) or a zero (2nd derivative) occurs.

**Ion density**, $n_i$. Two alternative ways were employed:

- *OML Theory*, $n_{i\_OML}$. This method employs the orbital-motion-limited (OML) model. In this model the ion current scales as $(V_{pr})^{1/2}$, being more suitable for thick, non-collisional probe sheaths. The ion density can be obtained by [2,3]:

$$n_{i\_OML} = \left(\frac{\pi^2 m_i}{2e^3 A_{pr}^2}\right)^{1/2}\left(\frac{\partial I_i^2}{\partial(V_{pr} - V_p)}\right)^{1/2} \quad (1)$$

Where $m_i$ is the positive ion mass, $A_{pr}$ is the probe area and $V_{pr}$ is the potential applied to the probe. Since the ion current scales as $(V_{pr})^{1/2}$, a plot of $(I_i)^2$ versus $V_{pr}$ results in a linear function. Therefore, $n_{i\_OML}$ is proportional to the square-root of the slope of this linear function, as evidenced by Eq 1,

- *Bohm Theory*, $n_{i\_B}$. From this model, the ion density number is obtained by considering the Bohm velocity at the pre-sheath and sheath interface around the probe surface $v_B = (kT_e/m_i)^{1/2}$.

$$n_{i\_B} = \left[\frac{I_i(V_f - 15V)}{0.61e A_{pr}}\right]\left(\frac{kT_e}{m_i}\right)^{-1/2} \quad (2)$$

where $I_i(V_f - 15V)$ is the ion current evaluated at 15V more negative than $V_f$, which can be directly read from the filtered characteristic curve.

**Electron Temperature**, $T_e$. Within the transition region of the I-V curve (between $V_f$ and $V_p$), the probe current is given mostly by electrons, and presents an exponential relationship with respect to $V_{pr}$. Therefore, by plotting $ln(I_e)$ versus $V_{pr}$ on a semi-log scale, one obtains a linear relationship. $T_e$ can be extracted from the angular coefficient of this linear region of the curve, $\alpha = (e/kT_e)$:

$$ln[I_e(V_{pr})] = \left(\frac{e}{kT_e}\right)V_{pr} + K \quad (3)$$

where $K = ln[en_e A_{pr}(kT_e/2pim_e)^{1/2}] - (eV_p/kT_e)$ is a constant, independent on $V_{pr}$.

However, before apply Eq.3, the following procedure must be performed. The current due to electrons only is obtained by subtracting the original probe current from the ion-only current: $I_e = I_{pr} - I_i$. The former is determined by considering the ion saturation region. The original (filtered) probe current is fitted to a quadratic function, by employing the least-mean-square method, so $I_e$ can be determined. Figure 6a shows ( in blue) the fitted function extending to the potential $V_p$.

**Electron density**, $n_e$. When the voltage applied to the probe is equal to the plasma potential $(V_{pr} = V_p)$, the current collected is equal to the electron saturation current, $I_{e0}$, available directly from the characteristic curve, from which the density $n_e$ can be extracted:

$$n_e = \left(\frac{I_{e0}}{eA_{pr}}\right)\left(\frac{2\pi m_e}{kT_e}\right)^{1/2} \quad (4)$$

Theoretically, for $V_{pr}$ larger than $V_p$, the probe current should be kept at the value $I_{e0}$, with a relatively abrupt change in the slope of the characteristic curve. However, one gets a slightly change in the probe current. This occurs due to some factors, such as the influence of the probe's polarization on the thickness of the probe sheath, reflection or secondary emission of electrons from the surface of the probe and draining of many electrons by the probe, causing severe disturbance of the plasma. Therefore is relatively difficult to identify $I_{e0}$.

## III. SEMI-AUTOMATIC TOOL FOR PLASMA PARAMETERS EXTRACTION

The tool is divided in two environments. In order to acquire and digitize the I-V probe information, a digital acquisition system under LabView® environment was developed. In order to extract the plasma parameters, a graphical interface was developed under Matlab Appdesigner®. Figure 2 shows the whole architecture of the system, with LabView® and Matlab® sections.

At the top, one has the LabView® graphical interface used to acquire the characteristic curve from the plasma plant, digitized by an A/D converter and sent to the a personal computer. There are three tabs to navigate. The first one shows the original (raw) characteristic curve obtained; the second one shows the results of the applied filters, aiming noise reduction; the third one shows the first and second derivatives

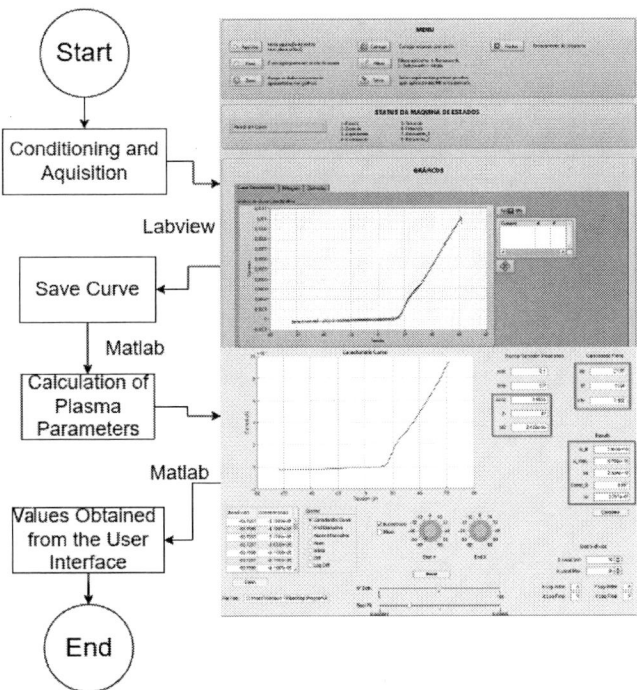

Fig. 2: Simplified schematics of the integration between two environments for plasma parameters extraction: LabView® and Matlab®.

of the (filtered) curve. In the menu, the following options are provided:

- **Acquire:** Starts data acquisition;
- **Load:** Uses an existing curve as input;
- **Stop:** Stops processing;
- **Filter:** Apply the desired filters and insert them in graphs on the "Filtering" tab;
- **Reset:** Cleans the probe channels for a new test;
- **Save:** Saves the characteristic curve obtained in an excel file containing the points in two columns (first for voltage and second for current);
- **Close:** Closes the program.

With the characteristic curve on screen, the user saves the curve as txt extension file, so it can be opened on Matlab®.

At the bottom part of Fig.2, one has the Matlab® graphical interface used to process the characteristic curve. The user has several options for querying and dealing with the curve.

By importing the file generated in Labview®, the user can identify, the spreadsheet with the collected points and the plot of the characteristic curve (inset graphs on figure 2).

Through the "Open" button, the user selects the path of the input file for display in the main graph (Voltage x Current). The input is a text file containing two columns where voltage and current points are presented.

Some fields in the "Plasma Generator Parameters" area are already filled in, such as probe radius, $r_{pr}$, probe length, $l_{pr}$,

and atomic number of the positive ion, $Zi$ (for argon, $Zi = 40$).

With the characteristic curve in hand, it is then possible to filter in the way that is most convenient (Butterworth or Midpoint), by selecting the corresponding field. If one chooses a midpoint operation, a slider command at the bottom called "N Dots" can be dragged to the desired number of points (ranging from 1 to 100), from which data will be averaged out. Similarly, by choosing Butterworth filter, the curve is averaged out by employing a second-order filter, as shown in figure 3. After filtering, the tool returns on screen the value of $V_f$, that is the voltage point where the current equals zero.

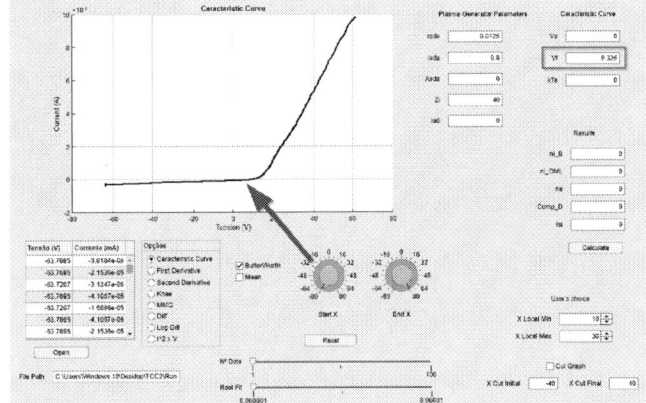

Fig. 3: Characteristic curve filtering by Butterworth and $V_f$ extraction.

Figure 4 exemplifies how the tool works for the first derivative method to obtain $V_p$. The user has the option of choosing the maximum point search interval. The graph then displays the chosen region (two vertical bars), as well as the maximum point found. The two rotary knobs have the function of adjusting the graph scales for better viewing. The value extracted is then displayed in the $V_p$ field.

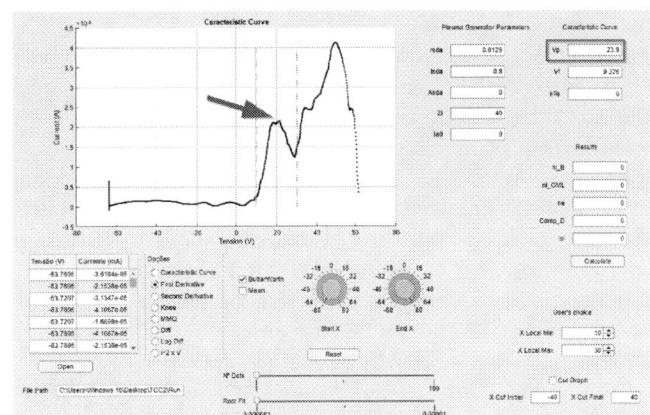

Fig. 4: First derivative and extraction of $V_p$. The two vertical bars indicate the range chosen by the user.

Next plasma parameter to be extracted is the ion density

979-8-3315-4064-7/24 $31.00 © 2024 IEEE

$n_{i\_OML}$. To this end, as indicated by Eq.1 one has to plot $(I_i)^2$ versus $V_{pr}$, resulting in a linear function, from which the angular coefficient must be determined. The user then must choose two points and include them in the highlighted fields. It is then enough for the user to store these values in the fields *X Fit Initial, Y Fit Initial, X Fit Final, Y Fit Final* and press the *Calculate* button. The program will then internally perform a linear fit and the slope of this line is used for obtaining $ni_{OML}$ at the end of the entire procedure. Figure 5 shows the raw data with the fitted region.

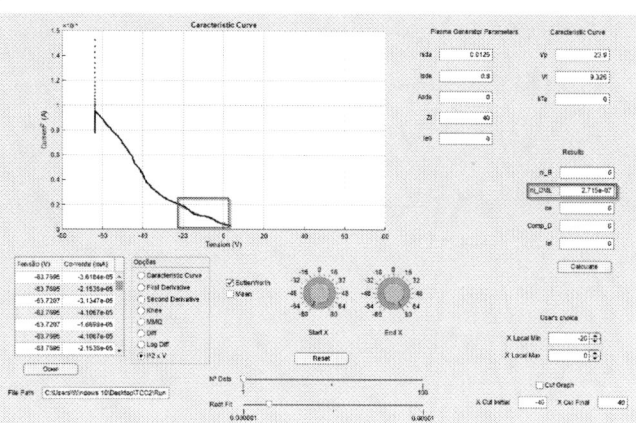

Fig. 5: Filtered characteristic curve (black) and adjusted ion saturation current (red).

The extraction of $n_{i\_B}$ is done by employing Eq.2, after take $I_i(V_f - 15V)$ directly from the characteristic curve. The user must press the button *Calculate* and the field $ni\_B$ is filled out with the resulting value.

The extraction of $T_e$ is done by employing the procedure summarized by Eq.3. Initially $I_e = I_{pr} - I_i$ is obtained, as shown in figure 6a, depicted in blue.

This fitting extends until $V_p$. The *Diff* button finds the difference between the original curve and the one adjusted by a quadratic function. Figure 6b shows a semi-logarithmic $\ln(I_e)$ versus $V_{pr}$.

Here again, the user must choose an interval between $V_f$ and $V_p$, to perform the linear regression and determine its slope. The user then enters manual values for this range in the same fields used to obtain $n_{i\_OML}$ previously, figure 6b.

At the end of the process the screen shows the final result, with all fields properly filled out. All the three values of plasma density must be consistent among each other: $n_{i\_OML} \approx n_{i\_B} \approx n_e$. Concerning the electron temperature, it is important to note that more reliable results is obtained by employing RF chokes (tunned either for the 1st and 2nd harmonics) and compensating electrode [1].

## IV. CONCLUSIONS

In this work, a semi-automatic tool was developed aiming the analysis of a characteristic curve, resulting from a Langmuir probe measurements on RF plasmas (13.56 MHz). The aquisition of the characteristic curve was performed with the

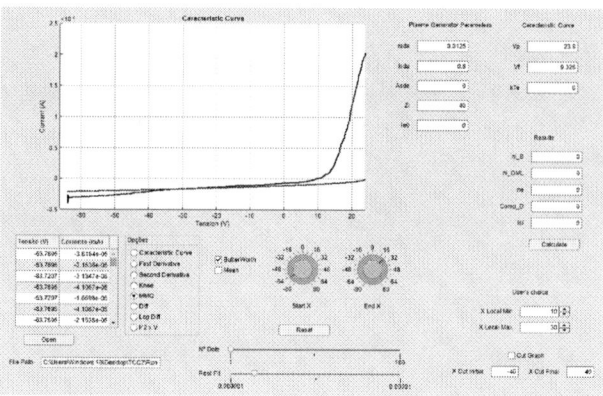

(a)

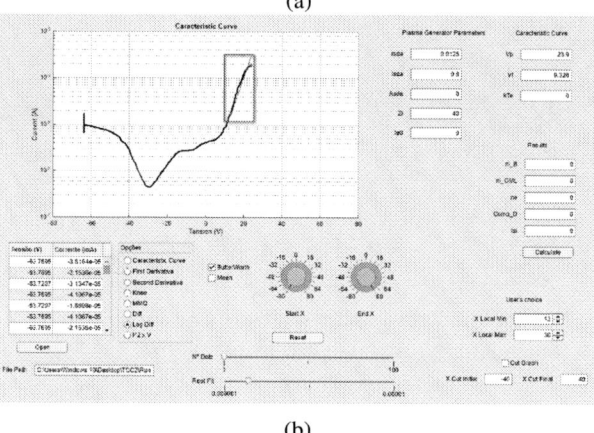

(b)

Fig. 6: (a) Filtered characteristic curve (black) and adjusted ion saturation current (blue); (b) Linear fit within the region chosen by the user.

aid of Labview® USB data acquisition board. The extraction of plasma parameters was performed by a tool developed in Matlab's Appdesigner®. There is some degree of user interactions through graphics and buttons. The authors believe that this automation, employing interactive interfaces, enables greater agility, accuracy and reproducibility in the extraction of plasma parameters.

## REFERENCES

[1] LANGMUIR, I.; MOTT-SMITH, H.M. "The Theory of Colectors in Gaseous Discharges", Physical Review, v.28, p.727-63, 1926.

[2] LANGMUIR, I.; MOTT-SMITH, H.M. "Studies of Electric Discharges in Gases at Low Pressures", General Electric Review, v.27, n.7, part I, p.449-538, 1924.

[3] BOHM, D.; BURHOP, E.H.S.; MASSEY, H.S.W. In The Characteristics of Electrical Discharge in Magnetic Fields, cap.2, Ed.A. Guthrie e R.K. Waterling, USA, 1949.

[4] ALLEN, J.E.; BOYD, R.L.F.; REYNOLDS, P. "The Collection of Positive Ions by a Probe immersed in a Plasma", Proceedings of the Physical Society B, n.70, p.297-304, 1957.

[5] BERNSTEIN, I.B.; RABINOWITZ, N. "Theory of Electrostatic Probes in a Low-Density Plasma", The Physics of Fluids, v.2, n.2, p.112-21, 1959.

[6] LAFRAMBOISE, J. "Theory of Cilindrical and Swpherical Langmuir Probes in a Collisionless Plasma at Rest", Rarefied Gas Dynamics, v.2, p.22-44, 1966.

# Green Synthesis of Anatase Titanium Dioxide Nanoparticles from Joannesia Princeps Extract for Enhanced Photovoltaic Performance

1st Felipe S. C. Portes
*Instituto de Recursos Naturais*
*Universidade Federal de Itajubá*
Itajubá, Brazil
d2022004298@unifei.edu.br

2nd Adhimar F. Oliveira
*Instituto de Física e Química*
*Universidade Federal de Itajubá*
Itajubá, Brazil
adhimarflavio@unifei.edu.br

3rd Maria E. L. González
*Instituto de Física e Química*
*Universidade Federal de Itajubá*
Itajubá, Brazil
mariae@unifei.edu.br

*Abstract*—**Energy demand is growing worldwide due to rapid population growth and technological advancements. An alternative is the conversion of solar radiation into electrical energy through photovoltaic devices. Considering such demand, this work presents the results of the synthesis and characterization of titanium dioxide nanoparticles (NPsTiO$_2$) produced by green synthesis, aiming at their application in thin film solar cells. Green synthesis is an emerging and prosperous process that does not generate environmental contamination and is cost-effective. The green synthesis of TiO$_2$ nanoparticles from plant extracts generates a reliable material for various applications, especially for optoelectronic devices. The synthesis was carried out using the extract of Joannesia Princeps fruit. The obtained NPsTiO$_2$ were characterized through ultraviolet-visible spectroscopy (UV-vis), thermogravimetric analysis (TGA), and X-ray diffraction (XRD). XRD pattern shows that the TiO$_2$ prepared by green synthesis was obtained in the anatase crystalline phase. The UV-vis spectrum of TiO$_2$ exhibited a semiconductor band of TiO$_2$ at 280 nm. These results indicate that the obtained nanoparticles are a promising green alternative for low-cost photovoltaic device construction.**

*Index Terms*—**Titanium dioxides, Green synthesis, Photovoltaic devices**

## I. INTRODUCTION

Currently, the field of photovoltaic technology is making significant advances through the use of wafer-based cells, including traditional crystalline silicon or gallium arsenide, as well as commercial thin-film cells such as cadmium telluride, amorphous silicon, and copper indium gallium selenide. Emerging thin-film technologies, including perovskites, organic materials, and quantum dots, are also gaining prominence [1]. In particular, titanium dioxide (TiO$_2$) stands out due to its exceptional physicochemical properties, which are driving its increasing demand for the development of devices with promising technological and environmental applications [2, 3, 4]. The unique chemical characteristics of TiO$_2$ have facilitated the development of dye-sensitized solar cells [5], photocatalysts for the degradation of pollutants in water and air [6], gas sensors, and photoluminescent materials [7]. Additionally, TiO$_2$ holds considerable promise in the field of transparent conductive oxide technology [8].

The rutile phase of TiO$_2$ remains stable at high temperatures, while anatase and brookite are commonly found in both natural and synthetic nanoscale samples. During heating and thickening processes, anatase can convert to brookite and subsequently to rutile, or brookite can transform into anatase and then into rutile. These transformation pathways reveal an energy balance significantly influenced by particle size. The surface enthalpies of these three polymorphs are sufficiently distinct to permit changes in thermodynamic stability even under conditions that inhibit thickening [9, 10].

The anatase phase of TiO$_2$ is frequently employed in photochemical applications within solar cells [11]. Key properties such as the bandgap, high refractive index, and high dielectric constant make TiO$_2$, especially in its anatase form, highly suitable for constructing photoelectrodes in dye-sensitized solar cells (DSSCs). The high refractive index enhances UV radiation absorption through efficient light scattering, while the high dielectric constant reduces the recombination of photoexcited electrons by providing electrostatic protection.

The increase global population, together with technological advances demand for sustainable and efficient energy solutions [16]. This presents a pressing challenge: the need to develop sustainable and efficient energy sources. Conventional energy generation methods have limitations in meeting growing demand in an environmentally responsible manner. In this context, photovoltaic generation has been experiencing a significant surge in prominence, by the photovoltaic effect in semiconductors, with TiO$_2$ as an interesting semiconductor [17], represents a significant stride towards clean and renewable energy sources. Currently, The most prevalent production of the semiconductors NPS method involves the use of harsh and toxic chemicals, such as xylene [18]. Exposure to these chemicals can cause severe harm to human health and the environment. Given this scenario, Green synthesis offers a promising solution to environmental and health challenges [19]. By employing natural extracts and milder reaction conditions, green synthesis minimizes the generation of toxic waste and reduces the environmental impact associated with traditional production. About the extract he induces the oxidation and stabilization of metal ions; however, the exact mechanism underlying this process remains elusive, with various theories proposed to explain this behavior [20].

The quest for abundant and free solar energy drives research into new materials for optoelectronic devices. Dye-sensitized solar cells are of particular interest due to their high potential for electricity generation, combining low-cost materials with advanced technology. However, their efficiency remains relatively low [12, 13]. The exploration of new materials is crucial in DSSC research, with semiconductors such as $TiO_2$ and zinc oxide (ZnO) frequently used. Variations in photovoltaic parameters are linked to several key factors: the bandgap, which varies among oxides, particle size, and recombination rates [14, 15]. Various synthesis methods are documented in the literature, including sol-gel, co-precipitation, and polymer precursor methods, among others [9]. Among these methods, it is worth mentioning that the green synthesis method allows the production of $TiO_2$ without using environmentally harmful reagents. Considering this concern, this work presents the outcomes of the preparation and characterization of $TiO_2$ produced via green synthesis, based on the aqueous extraction of Joannesia Princeps's fruit.

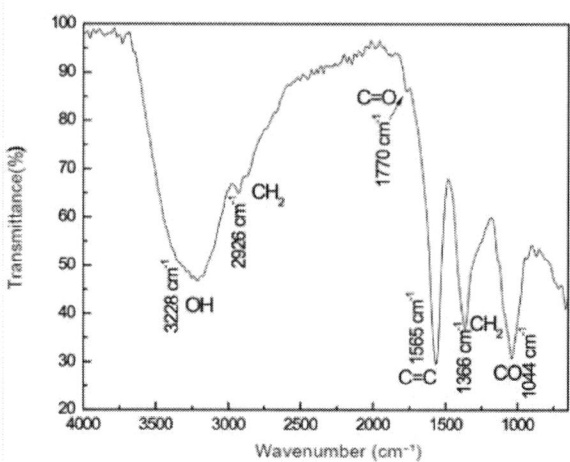

Fig. 1. Ftir of extract.

## II. EXPERIMENTAL METODOLOGY

To prepare the aqueous extract, 63.658g of Joannesia princeps fruit were previously hand-cut and dried at 70°C for 24 hours. The obtained extract was added to a beaker containing 20 mL of distilled water and subjected to an ultrasonic bath for 30 minutes at 50°C. Filtration was performed using a paper filter to separate the extract from the solid residue. This extract was characterized using Fourier-transform infrared spectroscopy (FTIR) in the range of 600-4000 cm$^{-1}$ with a resolution of 4 cm$^{-1}$, utilizing an attenuated total reflectance (ATR) accessory on an IRTracer-100 instrument, and thermogravimetric analysis (TGA) on a TGA-50 instrument, both from SHIMADZU trademark. In addition, UV-vis analysis was carried out by a Cary 50 Bio instrument, from the Varian brand, for characterization and comparison with the titanium dioxide nanoparticles (NPsTiO$_2$) obtained through green synthesis.

The green synthesis of NPsTiO$_2$ was executed by using 70 mL of the previously obtained extract at a concentration of 50 g/L, and 6 mL of TiCl3 (15%, Riedel-de-Haën) was added dropwise while the solution was heated to 50°C and stirred for 2 hours. After this period, the solution was stored in the refrigerator for 24 hours for sedimentation. In the subsequent process, the precipitate was washed with distilled water and centrifuged to remove impurities from the synthesis.

The NPsTiO$_2$ were characterized through TGA measurement. Subsequently, the NPsTiO$_2$ were calcined at a temperature of 500°C for 4 hours in an oven. The calcined NPsTiO$_2$ UV-vis, and X-ray diffraction (XRD) using the X'Expert PRO equipment from Pan Analytical, in the 2θ range of 5° to 90°.

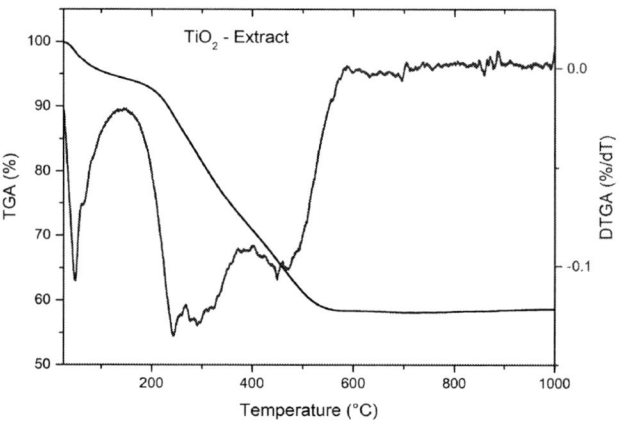

Fig. 2. TGA of TiO$_2$ with the extract.

## III. RESULTS AND DISCUSSION

The FTIR spectrum of Joannesia Princeps's fruit extract (Figure 1) displays OH bands (3221) cm$^{-1}$, C-O(H) at 1037 cm$^{-1}$, and functional groups capable of forming complexes with Ti$^{3+}$ ions, the precursor of TiO$_2$. The FTIR spectrum also reveals the presence of C=C double bonds in the extract. The residue obtained after thermogravimetric analysis indicates that the stable residue at 1000°C contains C=C and C≡C groups, suggesting the formation of stable cyclic species such as benzene rings, as indicated by the presence of the phenyl pattern (1558, 1568, 1438) cm$^{-1}$.

In Figure 2, the TGA of the extract together with TiO$_2$ is presented. It is possible to observe a mass loss below 100°C relative to humidity. Between 140 and 350°C, the loss of volatile compounds takes place, and from 300 to 600°C, the break of the main chain occurs. The two stages between 300 and 600°C can be attributed to a different secondary metabolic present in the vegetal extract. After the TGA, it was found that

979-8-3315-4064-7/24 $31.00 © 2024 IEEE

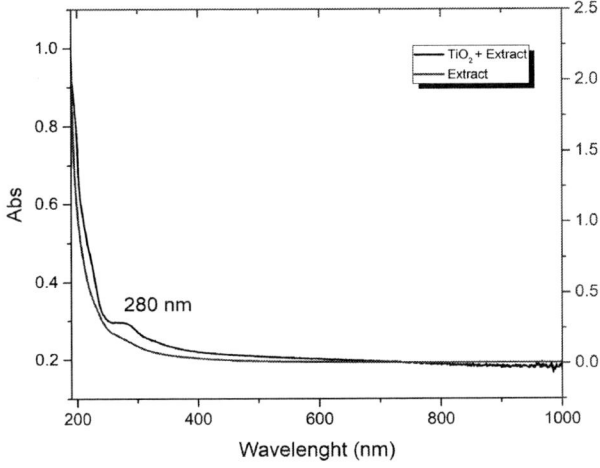

Fig. 3. UV-Vis of the pure extract and the extract with $TiO_2$.

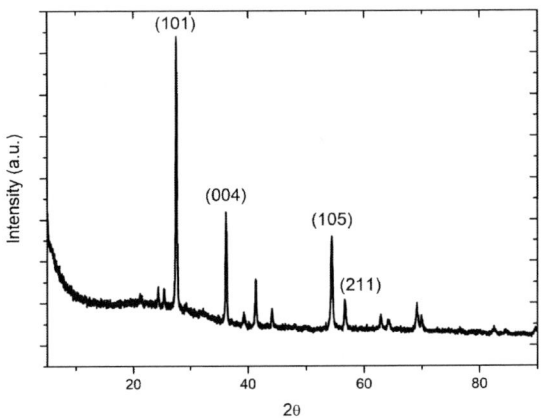

Fig. 4. XRD of the $TiO_2$ nanoparticles.

approximately 58% of the material remained.

Figure 3 presents the results of the UV-Vis measurements of the extract and the extract together with $TiO_2$. At 280 nm, absorption can be attributed to the high-energy electronic transition (n-sigma* antibonding). In the XRD analysis of the $TiO_2$ NPs (Figure 4), which resulted from a calcination process conducted on the extract together with $TiO_2$, peaks corresponding to anatase $TiO_2$ can be observed, along with some contamination related to a residue from the extract.

## IV. CONCLUSIONS

Titanium dioxide nanoparticles were synthesized through green synthesis using Joannesia Princeps fruit extract. The synthesis proved to be effective, producing titanium dioxide nanoparticles in the anatase phase. This methodology for preparing $TiO_2$ nanoparticles is important to avoid the use of expensive chemical reagents that can generate contamination when discarded. In future studies, it will be interesting to carry out calcination at higher temperatures, or to improve the washing process to reduce the contamination of nanoparticles by the extract, maintaining the anatase phase.

## ACKNOWLEDGMENT

The Brazilian agency CAPES and Fapemig for their financial support.

## REFERENCES

[1] F. H. Alharbi and S. Kais; Renewable and Sustainable Energy Reviews, 2015, 43, 1073–1089. https://doi.org/10.1016/j.rser.2014.11.101
[2] R. Parra, M. S. Góes, M. S. Castro, E. Longo, P. R. Bueno, and J. A. Varela; Chemistry of Materials, 2007, 20, 143–150. https://doi.org/10.1021/cm702286e
[3] Z. Zafar, R. Fatima, and J.-O. Kim; Environmental Research 2021, 197, 111120. https://doi.org/10.1016/j.envres.2021.111120
[4] M. Sedghi, R. Rahimi, and M. Rabbani; Inorganic Chemistry Communications, 2021, 126, 108486. https://doi.org/10.1016/j.inoche.2021.108486
[5] Brian O'Regan and Michael Grätzel; Nature, 1991, 353, 737-740. https://doi.org/10.1038/353737a0
[6] Dorian A. H. Hanaor and Charles C. Sorrell; Journal of Materials Science, 2010, 46, 855-874. https://doi.org/10.1007/s10853-010-5113-0
[7] M. Alfè, V. Gargiulo, M. Amati, V.-A. Maraloiu, P. Maddalena and S. Lettieri; Catalysts, 2021, 11, 795. https://doi.org/10.3390/catal11070795
[8] N. Laidani, G. Gottardi, R. Bartali, V. Micheli, R. Brusa, S. Mariazzi; Handbook of Modern Coating Technologies, 2021, 15, 509-554. https://doi.org/10.1016/b978-0-444-63237-1.00015-2
[9] X. Chen and S. S. Mao; Chemical Reviews, 2007, 107, 2891-2959. https://doi.org/10.1021/cr0500535
[10] U. Diebold; Surface Science Reports, 2003, 48, 53-229 https://doi.org/10.1016/s0167-5729(02)00100-0
[11] J. S. Lissau; Emerging Strategies to Reduce Transmission and Thermalization Losses in Solar Cells, M. Madsen, 2022
[12] E. F. A. Carvalho and M. J. F. Calvete; Revista Virtual de Química, 2010, 2, 192-203. https://doi.org/10.5935/1984-6835.20100018
[13] G. G. Sonai, M. A. M. Jr., J. H. B. Nunes, J. D. M. Jr., and A. F. Nogueira; Química Nova, 2015, 38, 1357-1365. https://doi.org/10.5935/0100-4042.20150148
[14] D. Cahen, G. Hodes, M. Grätzel, J. F. Guillemoles, and I. Riess; The Journal of Physical Chemistry B, 2000, 104, 2053-2059. https://doi.org/10.1021/jp993187t
[15] A.F. Oliveira, S.A.M. Silva, C.P. Rubinger, J. Ider, R.M. Rubinger, E.T.M. Oliveira, A.C. Doriguetto, H.B. de Carvalho, Materials Science and Engineering: B, 280, 2022, 115702. https://doi.org/10.1016/j.mseb.2022.115702
[16] SHAYANI, Rafael Amaral; OLIVEIRA, MAG de; CAMARGO, IM de T. Comparação do custo entre energia solar fotovoltaica e fontes convencionais. In: Congresso Brasileiro de Planejamento Energético (V CBPE). Brasília. 2006. p. 60.
[17] SHOU, Chunhui et al. Investigation of a broadband $TiO_2$/SiO2 optical thin-film filter for hybrid solar power systems. Applied energy, v. 92, p. 298-306, 2012. https://doi.org/10.1016/j.apenergy.2011.09.028
[18] ALVES, Helan Thaire de Albuquerque. Degradação de contaminantes petroquímicos via catalisadores ativados por luz visível. 2023. Trabalho de Conclusão de Curso.
[19] PARVEEN, Khadeeja; BANSE, Viktoria; LEDWANI, Lalita. Green synthesis of nanoparticles: Their advantages and disadvantages. In: AIP conference proceedings. AIP Publishing, 2016. https://doi.org/10.1063/1.4945168
[20] AKHTAR, Mohd Sayeed; PANWAR, Jitendra; YUN, Yeoung-Sang. Biogenic synthesis of metallic nanoparticles by plant extracts. ACS Sustainable Chemistry Engineering, v. 1, n. 6, p. 591-602, 2013. https://doi.org/10.1021/sc300118u

# SF$_6$/O$_2$ plasma for ICP/RIE SiC Etching

1st César, R. R.
*Center for Semiconductor Components and Nanotechnology - CCSNano*
rodrigo22cesar@gmail.com

2nd Mederos, M.
*Renato Archer Information Technology Center - CTI*
Campinas, Brazil
melissa.mederos@gmail.com

3rd Cioldin, F. H.
*Center for Semiconductor Components and Nanotechnology – CCSNano*
Campinas, Brazil
cioldin@unicamp.br

4rth Beraldo, R. M.
*Center for Semiconductor Components and Nanotechnology - CCSNano*
*Universidade Estadual de Campinas – UNICAMP*
Campinas, Brazil
renato.beraldo@hotmail.com

5th Teixeira, R. C.
*Renato Archer Information Technology Center - CTI*
Campinas, Brazil
rteixeir@cti.gov.br

6th Minamisawa, R. A.
*FHNW University of Applied Sciences and Arts Northwestern Switzerland School of Engineering*
Northwestern Switzerland
renato.minamisawa@fhnw.ch

7th Diniz, J. A.
*Center for Semiconductor Components and Nanotechnology - CCSNano*
*University of Campinas – UNICAMP*
Campinas, Brazil
jadiniz@unicamp.com.br

*Abstract*— In this study, we developed seven etching recipes using ICP/RIE plasma for etching the SiC substrate. We utilized scanning electron microscopy (SEM) to examine the angle and structure of the etched SiC wall, as well as to observe any particulates present after etching. Additionally, we employ the profilometry technique to determine the etching rate of each recipe. By conducting both analyses, we were able to study the seven recipes and determine which one has a rounded contact angle, higher etching rate, and lower residue/particulate formation. All recipes employed SF$_6$ gas with a flow of 20 sccm, O$_2$ with 5 sccm, ICP power of 1000 W, and a fixed time of 5 minutes at a temperature of 25°C. These parameters were consistent for all samples, with the variation occurring in the RIE power, working pressure, and DC Bias.

*Keywords—SF$_6$, SiC etching, SiC, ICP/RIE plasm*

## I. INTRODUCTION

Silicon carbide (SiC) has emerged as a highly promising material in the microelectronics industry due to its remarkable properties, including high thermal conductivity, a high breakdown voltage (3.8×106 V/cm), exceptional mechanical hardness, a wide band gap (greater than 2.4 eV), excellent chemical stability, high electron saturation speed (approximately 2×107 cm/s), and superior thermal conductivity [1]. Consequently, it is exceptionally well-suited for the development of devices intended for use in extreme power and temperature conditions. Examples include applications in satellites, nuclear reactors, and the aerospace industry at large [1].

With technological advancements and the increasing adoption of resilient materials like SiC, it becomes imperative to explore new techniques and manufacturing processes. Presently, we witness the integration of diverse components into chips, encompassing millions of MOS transistors, diodes, capacitors, and other passive components onto a single chip—a process known as very-large-scale integration (VLSI) [2]. Consequently, for SiC to be extensively utilized and integrated with other components, it's crucial to ensure the isolation of each component from one another [2-6]. The significance lies not only in the ability to etch SiC but also in understanding the quantity of by-products generated (which soil the etched area), the etching profile (whether it's anisotropic or not), and the profile of the corroded area concerning the non-corroded area (the wall). Among these challenges, the most critical aspect is the junction between the etched area (acting as insulation between devices) and the non-etched area (where the device is manufactured). A 90° angle at this junction leads to parasitic capacitance, signal reflections, and electrostatic interference, all of which impede the transmission of high-frequency signals in circuits [7]. Fig. 1 illustrates the side section of the SiC substrate, where Fig. 1A depicts the undesirable 90° angle that should be avoided, while Fig. 1B portrays the desired rounded profile. This rounded profile will be the focal point of our study.

Since SiC is chemically stable, wet etching proves inefficient and generates significant waste. Therefore, the alternative of plasma etching is favored for its cleanliness [8]. Although the inductively coupled plasma reactive ion etching (ICP/RIE) method has been established for some time, its calibration and refinement for SiC substrates are ongoing. Nonetheless, this method is employed for etching materials with high hardness values, such as GaN and diamond, alongside SiC. Literature already documents the utilization of ICP/RIE plasma for surface cleaning of SiC substrates (cleaning and "smoothing") in the manufacturing of Schottky diodes and certain stages of MEMS production [9-11]. Several studies address the etching of SiC, albeit often employing high power values (2500W) [12].

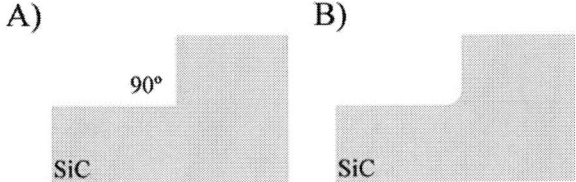

**Fig.1.** *Cross section of corroded SiC substrates, A) 90° angle and B) rounded angle.*

Hence, this study focuses on investigating the etching of SiC through ICP/RIE plasma, employing $SF_6$ gas and analyzing three key parameters: etching rate, waste generation, and the angle between the corroded and non-corroded regions. It is important to note that the utilized recipes feature low power values, with ICP set at 1000W and RIE at a maximum of 50W, conducted at room temperature (25°C). Scanning electron microscopy (SEM) was employed to observe the angle and structure of the etched SiC wall, as well as to detect any particulates post-etching, while profilometry was utilized to measure the etching depth.

## II. MATERIAL AND METHIDS

### A. MOS capacitor fabrication

N-Type <001> SiC samples were utilized in this study. Initially, RCA cleaning was performed on the samples. Subsequently, the lithography process was executed to fabricate grid structures for observing the etched SiC wall. After lithography, aluminum was deposited to act as a hard-mask.

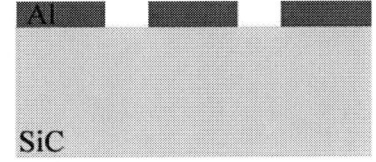

**Fig.2.** *Cross section of SiC samples for plams etch.*

Seven recipes were employed, and the samples were labeled with the identifier "A" followed by a number corresponding to the recipe used. Table I outlines the specific RIE and ICP power, working pressure, $SF_6$ and $O_2$ flow, temperature, time and DC Bias values employed for each recipe.

**Table I.** *Recipes used for SiC etching.*

| Parameters | A1 | A2 | A3 | A4 | A5 | A6 | A7 |
|---|---|---|---|---|---|---|---|
| ICP (W) | 1000 | | | | | | |
| RIE (W) | 50 | 25 | 25 | 5 | 50 | 50~30 | 50~22 |
| SF₆ (sccm) | 20 | | | | | | |
| O₂ (sccm) | 5 | | | | | | |
| Pressure (mTorr) | 10 | 10 | 20 | 20 | 20 | 20 | 20 |
| Time (min) | 5 | | | | | | |
| Temperature (°C) | 25 | | | | | | |
| DC Bias (V) | 115 | 115 | 115 | 25 | 150 | 150 | 150 |

The etching process was carried out using the Plasmalab System 100 from Oxford Instruments. SEM analysis was performed to assess the quality of the wall, quantify the products/dirt generated by etching, and analyze the angle formed between the wall and the floor. For SEM analysis, an SEM FEG Mira 3 XMU-TescanX instrument was utilized. Additionally, the height of the wall was determined using the Burke profilometer.

**Table II.** *Profilometry results for each ICP/RIE recipe.*

| Sample | Etch (um) |
|---|---|
| A1 | 1.49 |
| A2 | 0.97 |
| A3 | 1.00 |
| A4 | 0.70 |
| A5 | 1.20 |
| A6 | 1.05 |
| A7 | 1.00 |

Figure 2 displays SEM images of all recipes. It is evident that all recipes produce waste, with recipe A1 (Fig. 2A) generating the most. Figure 2B provides a magnified view of a residue. Energy-dispersive X-ray spectroscopy (EDS) analysis revealed that the residue was aluminum. In other words, the aluminum used as a hard-mask was corroding and depositing onto the region where the SiC was corroding.

Sample A2 exhibits bubbles along the corroded region (Fig. 2C). When subjected to the SEM electron beam, these bubbles burst, and some even began to bubble further. Figure 2D provides an amplified view of sample A2, enabling observation of the column's profile formed by etching. However, it is evident that certain sections feature a 90° angle, while others appear rounded.

Sample A3 displays a lesser quantity of bubbles, albeit they are still present along the corroded region (Fig. 2E). Similar to sample A2, when exposed to the SEM's electron beam, these bubbles burst and exhibit bubbling behavior. Figure 2F offers an amplified view of sample A3, revealing that the profile of the column formed by etching presents a steeper angle compared to sample A2.

Sample A4 exhibited no bubbles and had the least amount of residue (Fig. 2G). However, as depicted in Fig. 2H, there are still some prominent residues present. Unlike the other samples, this one displayed a more corroded region closer to the wall. Due to the lower etching rate, the wall was smaller, making observation in the SEM challenging. Nevertheless, upon magnification, it was possible to discern that the angle is 90°.

Sample A5 did not exhibit bubbles but displayed some granules (Fig. 2I). However, as depicted in Fig. 2J, the corroded area appears rough, with numerous ditches, a phenomenon not observed in the other samples. Additionally, it is noticeable that the profile of the wall changes upon contact with the corroded region: the upper part of the wall is linear, whereas it becomes granulated as it reaches the bottom. However, it is evident that the angle of this connection is not 90°.

Sample A6 did not display bubbles but exhibited numerous residues (Fig. 2K). These residues, as depicted in Fig. 2L, are observed to be in the form of pillars. From the EDS analysis, it was determined that these pillars are composed of Si and C, indicating that they originate from the SiC substrate itself. Therefore, the etching process is responsible for forming these pillars in certain areas, and it is also evident that the entire corroded wall has SiC pillars at its base. However, it can be observed that there are regions with both rounded and right angles.

Sample A7 did not exhibit bubbles but showed numerous residues (Fig. 2M). As depicted in Fig. 2N, these residues are observed to be pillars composed of Si and C, similar to sample A6. However, unlike the previous sample, these pillars are larger in both height and length. Upon observing the structure of the wall, it was noted that the base is composed of pillars, similar to the previous sample.

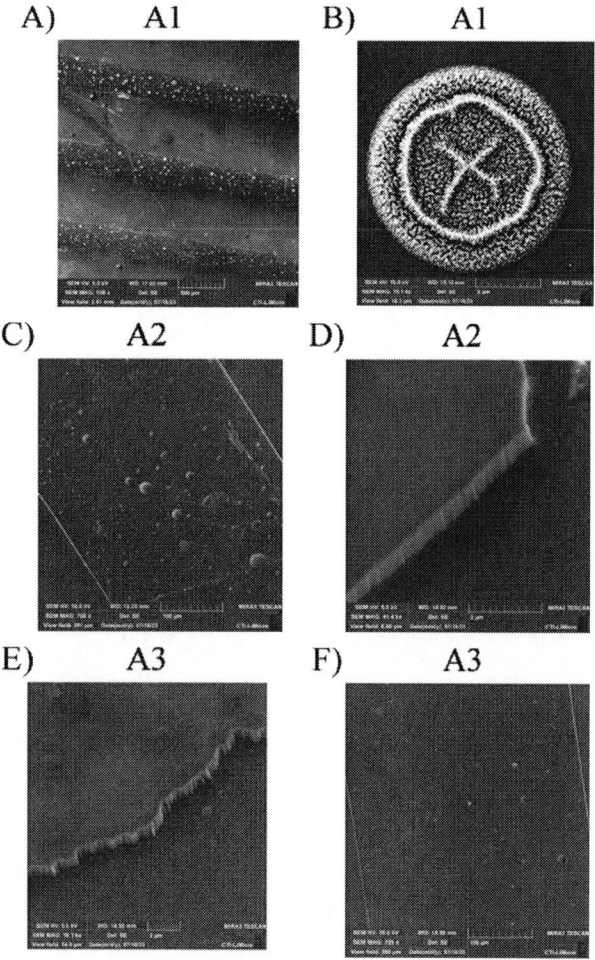

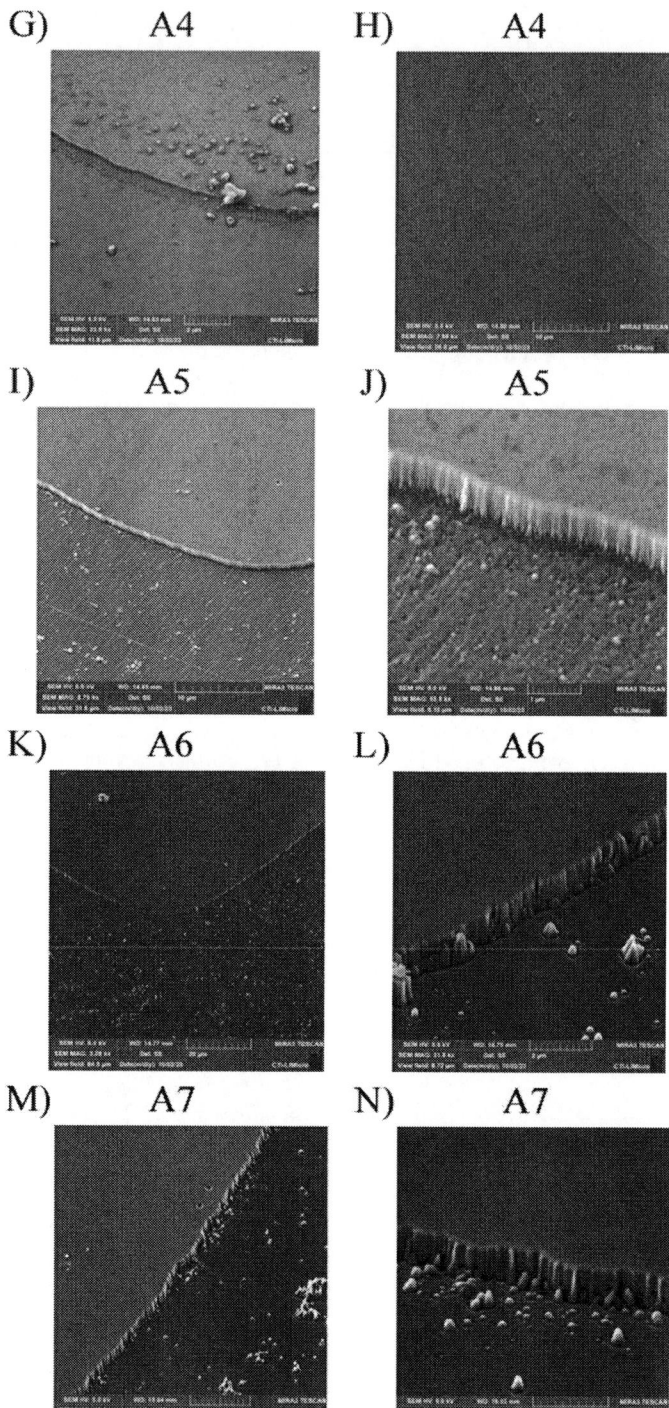

**Fig.2.** *SEM image of all samples, A1 A and B; A2 C and D; A3 E and F; A4 G and H; A5 I and J; A6 K and L; A7 M and N.*

Comparing these results with the literature [13-25], it can be noted that the results of this work are promising. It is observed that processes with high etching rates also have high values of temperature (300°C) [13] and ICP/RIE power of 4000W [14,15]. etching rates close to those presented in this work, 0.28 µm/min, use power and flow parameters similar to those used in this work [16-25]. Regarding gas flow and temperature, there was a significant difference; only one study used ambient temperature and similar flow, but with a power of 2000W [14]. In the vast majority of studies, straight or inclined walls were achieved, few achieved straight walls, and only the region between the wall and the substrate was rounded.

## CONCLUSION

In conclusion, all recipes exhibited certain drawbacks. Samples utilizing 50W power (A1 and A5) showed significant residue. Samples operated at 25W power (A2 and A3) displayed bubbles which, upon exposure to the electron beam, began to bubble and burst inside the SEM chamber. The sample with the lowest power, 5W (A4), demonstrated the lowest etching rate (0.7 μm) but had the largest residues. Recipes featuring power variation (A6 and A7) did not present bubbles but exhibited non-corroded SiC pillars.

Regarding the angle between the wall and the corroded area, samples A3, A3, and A5 exhibited an inclined and rounded angle, rendering them suitable candidates for device manufacturing. It is noteworthy that the next step involves filling this trench (corroded region) with $SiO_2$. Therefore, as long as the angle is rounded, the presence of pillars or bubbles may not interfere with the electrical measurements of the devices manufactured on the pillars.

## ACKNOWLEDGMENT

Authors would like to thank CCSNano staff for help in sample preparation and processing. This work was supported by ROTA 2030, grants number [28795-23].

This research used facilities of the Brazilian Nanotechnology National Laboratory (LNNano), part of the Brazilian Centre for Research in Energy and Materials (CNPEM), a private non-profit organization under the supervision of the Brazilian Ministry for Science, Technology, and Innovations (MCTI). The (names of the facilities) staff is acknowledged for the assistance during the experiments (20240001).

## REFERENCES

[1] Oliveira, A., R. Estudo da viabilidade de fabricação de dispositivos semicondutores baseados em filmes de carbeto de silício crescidos por PECVD. Tese. (Doutorado em Engenharia Elétrica) – Escola Politécnica da Universidade de São Paulo - USP, São Paulo, 2006.

[2] The History of the Integrated Circuit. Nobelprize.org. Archived from the original on 29 Jun, 2018. Retrieved 21 Apr 2012.

[3] Wang, J. J.; Lambers, E. S.; Pearton, S. J.; Ostling, M.; Zetteling, Z.; Grow, J. M.; Ren, F.; Shul, R. J. "ICP Etching of SiC." Solid-State Electronics Vol. 42, No. 12, pp. 2283±2288, 1998.

[4] Pan, W. S.; Steckl, A. J.; "Reactive Ion Etching of SiC Thin Films by Mixtures of Fluorinated Gases and Oxygen." J. Electrochem. Soc., Vol. 137, No. 1, January 1990, The Electrochemical Society, Inc

[5] Weitzel, C. E., Palmour, J. W., Carter, C. H.Jr., Moore, K., Nordquist, K. J., Allen, S., Thero, C. and Bhatanagar, M., "ICP etching on SiC." IEEE Trans. Electron. Dev., 1996, 43, 1732

[6] Ling, L.; Hua, X.; Li, X.; Oehrlein, G. S.; Celii, F. G.; Kirmse, K. H. R.; Jiang, P.; Wang, Y.; Anderson, H. M. "Study of $C_4F_8$/CO and $C_4F_8$/Ar/CO plasmas for highly selective etching of organosilicate glass over Si3N4 and SiC." J. Vac. Sci. Technol. A 22, 236–244 (2004).

[7] Manzillo, F. F.; Sauleau, R.; Capet, N.; Ettore, Mauro. "Mode Matching Analysis of an E-Plane 90∘ Bend With a Square Step in Parallel-Plate Waveguide." IEEE antennas and wireless propagation letters, vol. 16, 2017.

[8] Pirnaci, M. D.; Spitaleri, L.; Tenaglia, D.; Perricelli, F.; Fragala, M. F.; Bongiorno, C.; Gulino, A. "Systematic Characterization of Plasma-Etched Trenches on 4H-SiC Wafers." ACS Omega 2021, 6, 20667−20675.

[9] Guy, O.J.; Lodzinski, M.; Teng, K.S.; Maffeis, T.G.G.; Tan, M.; Blackwood, I.; Dunstan, P.R.; Al-Hartomy, O.; Wilks, S.P.; Wilby, T.; et al. "Investigation of the 4H-SiC Surface". Appl. Surf. Sci. 2008, 254, 8098–8105.

[10] Tsui, B.-Y.; Cheng, J.-C.; Yen, C.-T.; Lee, C.-Y. "Strong Fermi-level pinning induced by argon inductively coupled plasma treatment and post-metal deposition annealing on 4H-SiC." Solid State Electron. 2017, 133, 83–87.

[11] Judy, J.W. "Microelectromechanical systems (MEMS): Fabrication, design and applications." Smart Mater. Struct. 2001, 10, 1115–1134.

[12] Szmidt, K. R.; Stonio, B.; Zelazko, J.; Filipiak, M.; Sochacki, M. "A Review: Inductively Coupled Plasma Reactive Ion Etching of Silicon Carbide." Materials 2022, 15, 123. https://doi.org/10.3390/ma15010123

[13] Sung, H.-K. et al. Vertical and bevel-structured SiC etching techniques incorporating diferent gas mixture plasmas for various microelectronic applications. Sci. Rep. 7 (2017).

[14] Osipov, KYu. & Velikovskiy, L. E. Formation technology of through metallized holes to sources of high-power GaN/SiC high electron mobility transistors. Semiconductors 46, 1216–1220 (2012)

[15] Huang, Y., Tang, F., Guo, Z. & Wang, X. Accelerated ICP etching of 6H-SiC by femtosecond laser modifcation. Appl. Surf. Sci. 488, 853–864 (2019).

[16] Choi, J. H. et al. Fabrication of SiC nanopillars by inductively coupled $SF_6$/$O_2$ plasma etching. J. Phys. D Appl. Phys. 45, 235204 (2012).

[17] Okamoto, N. Elimination of pillar associated with micropipe of SiC in high-rate inductively coupled plasma etching. J. Vacuum Sci. Technol. A Vacuum Surf. Films 27, 295–300 (2009).

[18] Khan, F. A. & Adesida, I. High rate etching of SiC using inductively coupled plasma reactive ion etching in $SF_6$-based gas mixtures. Appl. Phys. Lett. 75, 2268–2270 (1999).

[19] Cho, H. et al. High density plasma via hole etching in SiC. J. Vac. Sci. Technol. A 19, 1878–1881 (2001).

[20] Jiang, L. Impact of Ar addition to inductively coupled plasma etching of SiC in $SF_6$/$O_2$. Microelectron. Eng. 73–74, 306–311 (2004).

[21] Camara, N. & Zekentes, K. Study of the reactive ion etching of 6H–SiC and 4H–SiC in $SF_6$/Ar plasmas by optical emission spectroscopy and laser interferometry. Solid-State Electron. 46, 1959–1963 (2002).

[22] Ahn, S. C., Han, S. Y., Lee, J. L., Moon, J. H. & Lee, B. T. A study on the reactive ion etching of SiC single crystals using inductively coupled plasma of $SF_6$-based gas mixtures. Met. Mater. Int. 10, 103–106 (2004).

[23] Beheim, G. Deep reactive ion etching for bulk micromachining of silicon carbide. in Te MEMS Handbook (ed. Gad-el-Hak, M.) Vol. 20013566 (CRC Press, 2001).

[24] Seok, O.; Kim, Y.; Bahng, W. "Micro-trench free 4H-SiC etching with improved SiC/SiO2 selectivity using inductively coupled $SF_6$/$O_2$/Ar plasma." Phys. Scr. 95 (2020) 045606 (5pp)

[25] Osipov, A.A.; Iankevich, G. A.; Speshilova, A. B.; Osipov, A. A.; Endiiarova, E. V. Berenzeko, V. I.; Tyurikova, I. A.; Tyurikov, K. S.; Alexandrov, S. E. "High-temperature etching of SiC in -SF6/O2 inductively coupled plasma." Scientifc Reports | (2020) 10:19977

# Experimentally Exploring the Performance of MOSFET Devices at Deep Cryogenic Temperatures

Lucas Stucchi-Zucchi
*DEEB/FEEC*
*Unicamp*
Campinas, Brazil
*Member, IEEE*
*l122975@dac.unicamp.br*

Marcelo Pavanello
*Electrical Eng. Dept*
*Centro Universitário FEI*
São Bernardo do Campo, Brazil
*Senior Member, IEEE*

Francisco Rouxinol
*DFMC/IFGW*
*Unicamp*
Campinas, Brazil

José Alexandre Diniz
*DEEB/FEEC*
*Unicamp*
Campinas, Brazil
*Member, IEEE*

Francisco Brito
*Electrical Eng. Dept*
*UFERSA*
Caraúbas, Brazil
*Member, IEEE*

*Abstract*—This study investigates commercially available 180 nm CMOS devices under deep cryogenic conditions, focusing on both short (180 nm) and long (600 nm) channel nMOS devices at temperatures down to 3.2 K. We meticulously measured key parameters including threshold voltage ($V_{th}$), carrier mobility ($\mu_n$), subthreshold slope (SS), on-current ($I_{on}$), and off-current ($I_{off}$). Notably, as temperatures decreased, $V_{th}$, $I_{off}$, and SS values declined, while $I_{on}$ and $\mu_n$ exhibited an increase. A notable observation was the dual $Z_t$ values for the short channel device, aligning with the significant rise in series resistance post carrier freeze-out at 77 K. Our results provide a foundational understanding for the extraction of compact model parameters, crucial for cryogenic CMOS simulation and circuit design. For short channel devices, $V_{th}$ evolved from 0.6 V at room temperature to 0.85 V at 3.8 K, carrier mobility increased from 200 to 400 cm²/V²s, and SS improved from 90 mV/dec to 20 mV/dec. Conversely, long channel devices, measured from 113 K to 3.2 K, showed $V_{th}$ alteration from 0.60 V to 0.64 V, mobility enhancement from 833 to 1220 cm²/V²s, and SS improvement from 30 mV/dec to 10 mV/dec.

*Index Terms*—CryoCMOS, cryogenic, CMOS, quantum computing, qubit

## I. INTRODUCTION

In the field of quantum computing, qubit coherence time is one of the most important limits on algorithm run time, and so is one of the challenges faced before large-scale quantum processing is feasible, especially in Circuit Quantum Electro Dynamics (cQED). This implementation of quantum circuitry, which is based on based on superconducting materials, boasts impressive characteristics in scalability, integration and installed fabrication capabilities, since much of its fabrication process is the same as traditional nanoelectronics. However,

Funding by CAPES, CNPq, INCT-NAMITEC, MCTI/Softex and FAPESP
FAPESP grant number 2017/08602-0
INCT-Namitec grant number CNPQ-406193/2022-3
MCTI/Softex grant number 01245.002636/2023-81

the benefits of cQED are not without drawbacks: having a device coupled more strongly with the measurement systems, which is what creates its ease of driving and measurements, takes a toll on the coherence times [1], [2]. One of the sources of environmental coupling that can lead to qubit decoherence is the introduction of electrical noise from the exterior environment to the cooled and shielded qubit through the electrical circuit that carries the control and readout signals [3]. This leads to an increased interest in operating as much of the control and readout circuitry inside the cryogenic apparatus in which qubits typically operate [4], [5]. To this end, many different technologies are being studied and characterized at cryogenic temperatures. Technologies such as Bipolar-Junction-Transistors (BJTs) [6], [7], Junctionless-Field-Effect-Transistors (JL-FET) [8], [9], but among those cryogenic implementations of CMOS (Cryo-CMOS) [4], [5], [10]–[14] are the most promising, especially for the ability to achieve large scale integration.

To transpose the many advantages of CMOS to the cryogenic realm, a thorough re-evaluation and re-characterization of the technology must be performed. In this work, various devices fabricated using the UMC 180 nm technologies were investigated during the cooling ramp of a Pulse Tube Refrigeration system that precools a Bluefors Dilution Refrigerator commonly used in quantum computing. The drain current ($I_D$) as a function of the gate voltage ($V_{GS}$) curves were measured at the operating temperature and during the dilution refrigerator warmup procedure to investigate the most important device parameters in multiple intermediary temperatures. The 180 nm technology was strategically chosen as an effective yet readily available architecture for cryo-CMOS ASICs, and results presented in this paper will form the basis for future circuitry design.

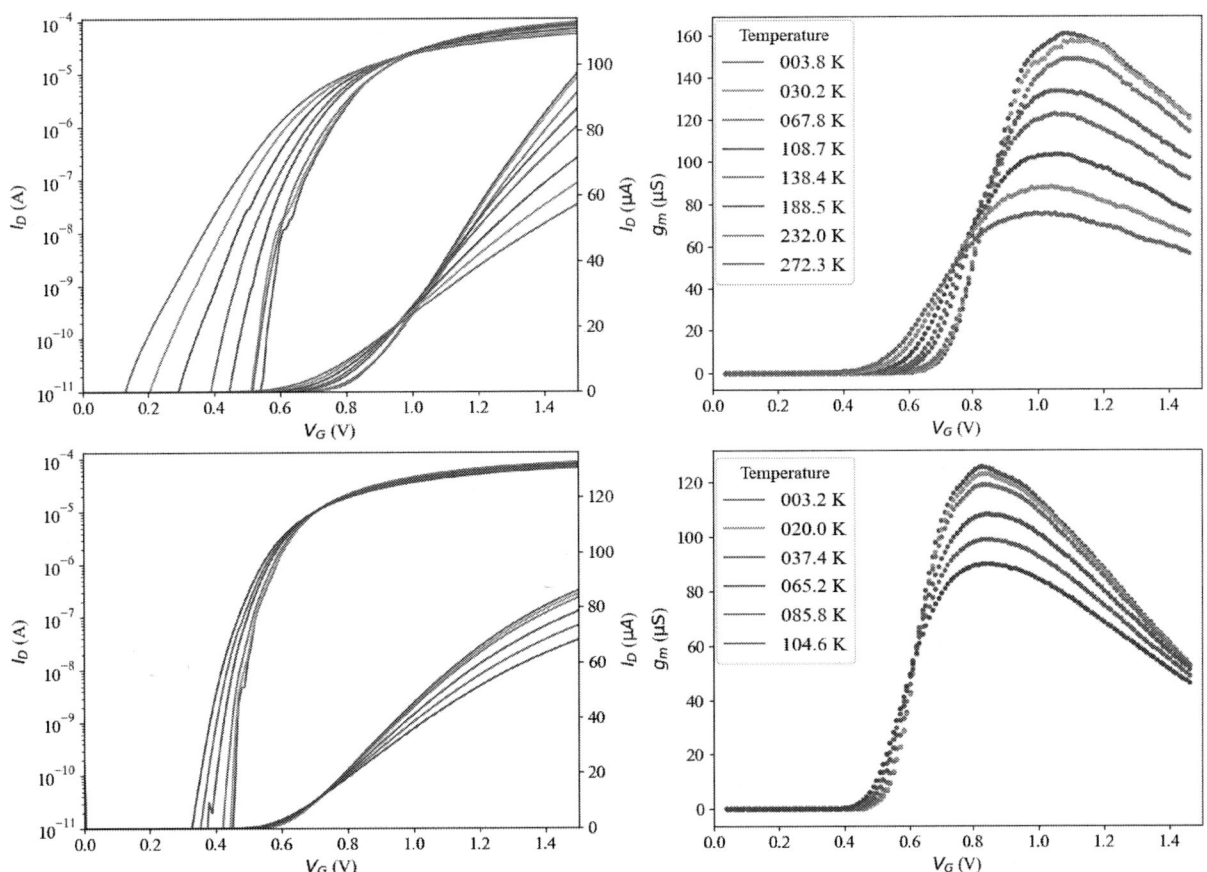

Fig. 1: Transistor behavior for multiple temperatures. Upper: Short device (L=180 nm and W=3 μm). Lower: Long device (L=600 nm and W=3 μm). Left: Linear and Log curves of Drain current ($I_D$) vs Gate voltage ($V_{GS}$). Right: transconductance ($g_m$) vs Gate voltage ($V_{GS}$) curves.

## II. SAMPLE PREPARATION

The designed chip was fixed on a copper sample carrier using a CMR-direct GE Varnish and wirebonded to a carrier printed circuit board. This sample as fully enclosed in an aluminum shielding box, which serves as an electromagnetic shield to guard both the devices from the operation of other circuitry inside the refrigerator and those circuitry from any electromagnetic radiation resulting of the transistor operation.

Two wide channel devices (W=3 μm) were measured to eliminate narrow channel effects. A long channel device with L=600 nm and a short channel one with L=180 nm were chosen to demonstrate channel length effect on parameters. This architecture presents pronounced effects on channel length not only through regular short channel effects, but also as a result of its halo implantations, which creates a channel doping concentration dependance on the channel length.

Measurements made at room temperature and below 4 K were taken after several hours of stabilization. Measurements taken during the warmup procedure had their temperature derived from the cold plate temperature at measurement start, this can introduce a small deviation from actual device op-

eration. To remediate this deviation, future experiments will implement an in-chip temperature measurement through metal interconnect resistance or diode I-V curve methods.

## III. DEVICE EVOLUTION DURING COOL DOWN PROCEDURE

Figure 1 show the drain current ($I_D$) and the transconductance ($g_m$) as a function of the gate voltage ($V_{GS}$), measured with a drain bias $V_{DS}$=25 mV for the long and the short channel transistors, respectively. The most notable parameters extracted from these curves are Vth, $I_{on}$, $I_{off}$ and SS. The threshold voltage was extracted by linear extrapolation method [15].

Figure 2a shows the extracted threshold voltage of long and short channel transistors as a function of temperature. In any temperature, the $V_{th}$ of the L=180 nm is larger than for the L=600 nm one, which is associated with the doping concentration increase in the channel with L reduction, known as the reverse short-channel effect [16], [17]. The temperature reduction increases inthe Vth regardless of the channel length. It is known that the threshold voltage of a MOSFET changes with the temperature as indicated in eqn. (1):

$$\frac{dV_{th}}{dT} = \frac{d\Phi_f}{dT} + \frac{1}{C_{OX}}\frac{dQ_{depl}}{dT} \qquad (1)$$

where $\Phi_F$ is the Fermi potential, $C_{OX}$ is the gate oxide capacitance per unit area and $Q_{depl}$ is the depletion change per unit area.

The increase in the Fermi potential as the temperature is lowered is associated with the reduction in the charge carrier concentration caused by the decreased number of ionized dopants, as the mean thermal energy of the atoms decreases, which demands a larger electrical potential in order for the semiconductor film to achieve inversion regime. The dashed line shows the linear regression for the $V_{th}$ on temperature. For the short-channel device the $V_{th}$ increases at a rate of $|dV_{th}/dT|$=0.71 mV/K whereas this rate reduces to $|dV_{th}/dT|$=0.44 mV/K for the long-channel transistor. This variation in the $|dV_{th}/dT|$ with the channel length is also associated with the higher channel doping concentration in the L=180 nm, which leads to larger $\Phi_F$, $d\Phi_F/dT$, and $dQ_{depl}/dT$ rates above 20 K. Below 20 K, the non-linearity can only be partly explained by the cryogenic characteristics of CMOS devices, further measurements are needed to fully explain this behavior, which could be connected to the series resistance introduced by the measurement apparatus, or simply measurement noise [14], [18].

Figure 2b shows the $\mu_n$ x T behavior, extracted by the maximum transconductance method and validated by the Y-function method [15]. The region for which $\mu_n$ decreases linearly with increases in temperature is the region for which electron-phonon interactions are the dominant scattering events. In this region, as temperature decreases there is also a decrease in the frequency and energy of phonon scattering events, increasing mobility. Below approximately 50 K, surface roughness scattering and Coulomb scattering become the dominant scattering events [10], [19]. Temperature has little to no influence on surface scattering events, but still some influence is exerted by temperature on Coulomb scattering, especially on number of ionized charges and thermal energy of electrons involved in scattering events.

In Figure 2c the SS dependence with temperature is presented. The decrease in SS is mostly explained by the decrease in the thermionic limit $ln(10)kT/q$, with a saturation behavior when near 20 mV/dec, with presents as an increase in the slope factor $n$ as it is commonly found in the literature. The slope factor $n$ is larger for the short-channel transistor in the whole temperature range, with $1.5 \leq n \leq 2$, whereas $n$ approaches 1 for the long channel transistor [5], [14].

Figures 2d and 2e shows the change in $I_{on}$ and $I_{off}$ with temperature, respectively. $I_{off}$ decreases as the charge carrier density plummets with the decrease in temperature, especially below the carrier freeze out temperature [4]. $I_{on}$ increases as the mobility increases, as a high transversal electric field introduced by the gate voltage is able to ionize enough charge carriers and achieve an inversion layer [10].

## IV. Conclusion and Future Works

In this work, we successfully demonstrated the capability of measuring cryo-CMOS devices under conditions pertinent to quantum computing applications. We plan on using the parameters extracted to design and test circuits for the control and readout of superconducting qubits, including voltage references, multiplexers, phase-locked loop (PLL) oscillators, arbitrary waveform generators and more.

Looking ahead, we plan to develop new electromagnetic (EM) shield and sample holders that will be fabricated fully in copper, which should improve the thermal transport characteristics of the systems and reduce noise.

## Acknoledgements

The authors would like to thank INCT-NAMITEC, CNPq, FAPESP and CAPES for the funding and the staff of CCSnano, LAMULT/IFGW, LPD/IFGW and LFDQ/IFGW for the support.

## References

[1] M. H. Devoret and R. J. Schoelkopf, "Superconducting circuits for quantum information: An outlook," *Science*, vol. 339, pp. 1169–1174, 3 2013.

[2] P. Krantz, M. Kjaergaard, F. Yan, T. P. Orlando, S. Gustavsson, and W. D. Oliver, "A quantum engineer's guide to superconducting qubits," *Applied Physics Reviews*, vol. 6, p. 021318, 6 2019.

[3] M. A. Nielsen and I. L. Chuang, *Quantum Computation and Quantum Information*. Cambridge University Press, 6 2012.

[4] H. Homulle, "Cryogenic electronics for the read-out of quantum processors," 2019.

[5] J. S. Park *et al.*, "A fully integrated cryo-cmos soc for qubit control in quantum computers capable of state manipulation, readout and high-speed gate pulsing of spin qubits in intel 22nm ffl finfet technology," vol. 64. Institute of Electrical and Electronics Engineers Inc., 2 2021, pp. 208–210.

[6] H. Homulle, L. Song, E. Charbon, and F. Sebastiano, "The cryogenic temperature behavior of bipolar, mos, and dtmos transistors in standard cmos," *IEEE Journal of the Electron Devices Society*, vol. 6, pp. 263–270, 1 2018.

[7] L. Song, H. Homulle, E. Charbon, and F. Sebastiano, "Characterization of bipolar transistors for cryogenic temperature sensors in standard cmos." Institute of Electrical and Electronics Engineers Inc., 1 2017.

[8] M. D. Souza, M. A. Pavanello, R. D. Trevisoli, R. T. Doria, and J. P. Colinge, "Cryogenic operation of junctionless nanowire transistors," *IEEE Electron Device Letters*, vol. 32, pp. 1322–1324, 10 2011.

[9] D. Madadi, "Investigation of junctionless fin-fet characterization in deep cryogenic temperature: Dc and rf analysis," *IEEE Access*, vol. 10, pp. 130 293–130 301, 2022.

[10] J. P. Colinge *et al.*, "Low-temperature electron mobility in trigate soi mosfets," *IEEE Electron Device Letters*, vol. 27, pp. 120–122, 2 2006.

[11] A. Akturk, M. Holloway, S. Potbhare, D. Gundlach, B. Li, N. Goldsman, M. Peckerar, and K. P. Cheung, "Compact and distributed modeling of cryogenic bulk mosfet operation," *IEEE Transactions on Electron Devices*, vol. 57, pp. 1334–1342, 6 2010.

[12] S. H. Hong, G. B. Choi, R. H. Baek, H. S. Kang, S. W. Jung, and Y. H. Jeong, "Low-temperature performance of nanoscale mosfet for deep-space rf applications," *IEEE Electron Device Letters*, vol. 29, pp. 775–777, 7 2008.

[13] A. H. Coskun and J. C. Bardin, "Cryogenic small-signal and noise performance of 32nm soi cmos." Institute of Electrical and Electronics Engineers Inc., 2014.

[14] A. Beckers, F. Jazaeri, A. Ruffino, C. Bruschini, A. Baschirotto, and C. Enz, "Cryogenic characterization of 28 nm bulk cmos technology for quantum computing." IEEE, 9 2017, pp. 62–65.

[15] A. Ortiz-Conde, F. J. García-Sánchez, J. Muci, A. Terán Barrios, J. J. Liou, and C.-S. Ho, "Revisiting mosfet threshold voltage extraction methods," *Microelectronics Reliability*, vol. 53, no. 1, pp. 90–104, 2013.

[16] B. Szelag, F. Balestra, and G. Ghibaudo, "Comprehensive analysis of reverse short-channel effect in silicon mosfets from low-temperature operation," *IEEE Electron Device Letters*, vol. 19, no. 12, pp. 511–513, 1998.

[17] R. Rios, W.-K. Shih, A. Shah, S. Mudanai, P. Packan, T. Sandford, and K. Mistry, "A three-transistor threshold voltage model for halo processes," in *Digest. International Electron Devices Meeting,*, 2002, pp. 113–116.

[18] Z. Chen, H. Wong, Y. Han, S. Dong, and B. L. Yang, "Temperature dependences of threshold voltage and drain-induced barrier lowering in 60 nm gate length mos transistors," *Microelectronics Reliability*, vol. 54, pp. 1109–1114, 2014.

[19] G. Pahwa, P. Kushwaha, A. Dasgupta, S. Salahuddin, and C. Hu, "Compact modeling of temperature effects in fdsoi and finfet devices down to cryogenic temperatures," *IEEE Transactions on Electron Devices*, vol. 68, pp. 4223–4230, 9 2021.

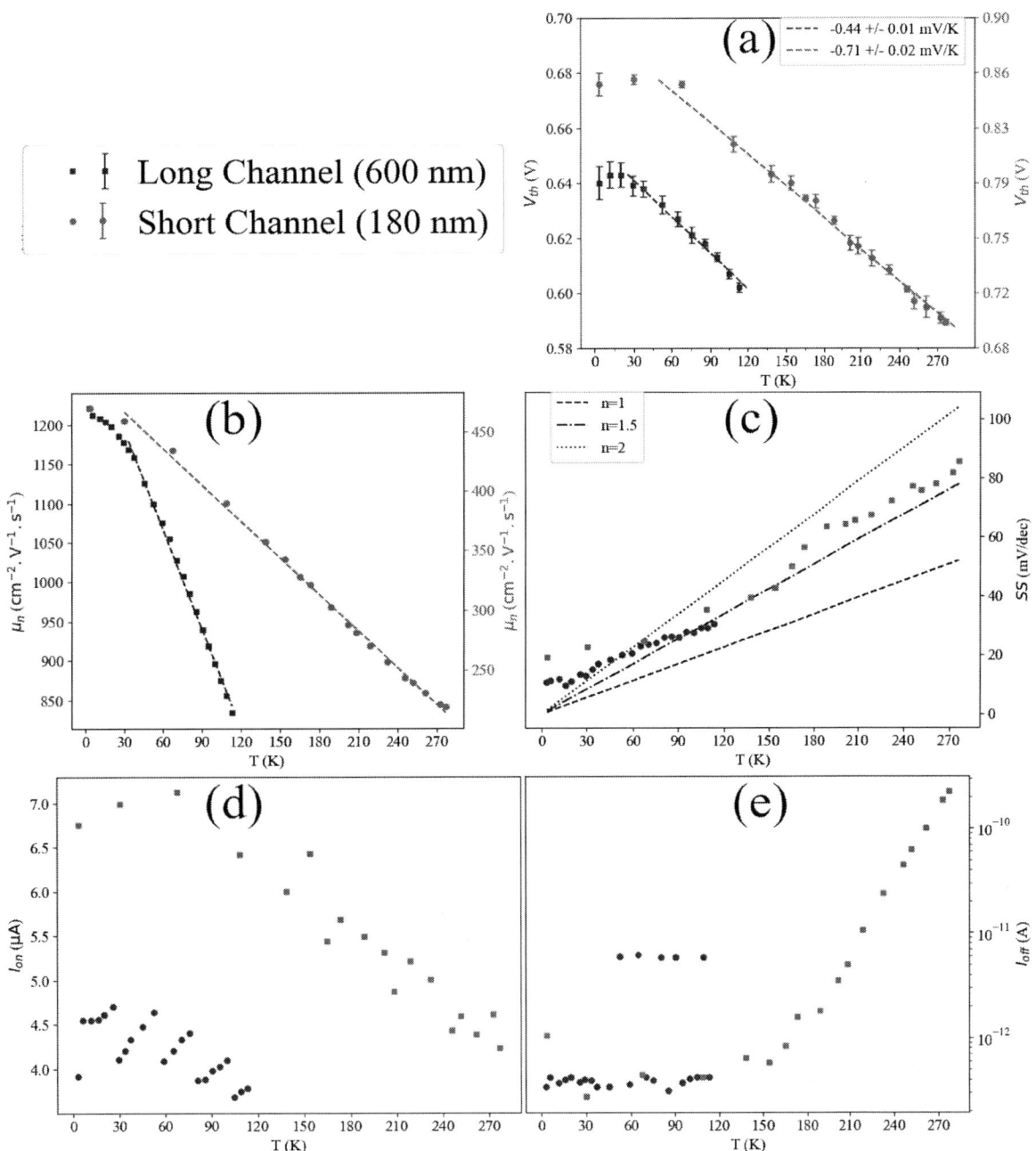

Fig. 2: Parameters extracted for the short (red) and long (blue) channel devices according to temperature: (a) $V_{th}$ X T curves. Dashed lines are a regression for the linear region (T > 20 K) (b) $\mu_n$ X T curves. Dashed lines are a linear regression for the phonon-electron dominated region. (c) $SS$ X T curves. Black lines are slope factor $n$=1, 1.5 and 2. (d) $I_{on}$ X T curves (e) $I_{off}$ X T curves

# Optical properties of extended inverted pyramids arrays for enhanced light absorption in silicon solar cells simulated with Comsol Multiphysics

Marcel Castilho Batista de Carvalho
LSI/PSI/EPUSP
University of São Paulo
São Paulo, Brazil.

Sebastião Gomes dos Santos Filho
LSI/PSI/EPUSP
University of São Paulo
São Paulo, Brazil.
https://orcid.org/0000-0002-0324-5703

*Abstract*— **Surface texturization plays a pivotal role in enhancing the efficiency of silicon solar cells by optimizing light absorption. Among various structures, extended inverted pyramids have emerged as a promising way due to their pyramidal pore structures. In this study, we employed 3D finite element simulations using the Comsol Multiphysics software with the Ray Optics module to investigate the optical characteristics of extended inverted pyramids texturization. Through systematic variations in pore depth, we identified a pore depth of 35 µm as optimal for enhanced infrared absorptance. Our simulations revealed consistent absorptance levels exceeding 96% across a broad spectrum in the range of 300 nm to 1200 nm, showcasing the potential of this texturization in improving the efficiency of silicon solar cells. Comparative analyses with alternative texturization techniques underscored the superior performance. Furthermore, accurate computation of lateral scattering effects proved to be essential in estimating absorptance with precision. While our findings highlight the promising attributes of this texturization, experimental validation is warranted to ascertain its practical feasibility. Overall, our study contributes valuable insights into advancing surface texturization techniques for enhancing the performance of silicon solar cells.**

*Keywords*— *Black silicon, porous silicon, solar cell texturization, extended inverted pyramids, silicon solar cells, ray optics simulation.*

## I. INTRODUCTION

The efficiency of silicon solar cells relies on light absorption, which can be notably enhanced through surface texturization techniques. Among these techniques, the creation of Random Pyramids (RP) stands out as one of the most prevalent methodologies. It involves the formation of vertically oriented pyramidal structures with square bases, characterized by random variations in height and placement across the surface.

This texturization is achieved by anisotropic etching of <100> wafers in alkaline solutions which confers ideally an angle between the pyramid base and face (base angle) of 54.74° due to the etching rate of (100) plane being higher than (111) plane. However, empirical observations have revealed that the actual base angle can vary from 40° to 55° [1]. Inverted pyramids (IP) are generated through analogous anisotropic etching processes on wafers featuring square-patterned masks on their surfaces, necessitating photolithographic techniques. This method typically yields experimental base angles ranging between 50° and 58° [1].

Using the ray tracing technique [2], a normal incidence ray that reaches a pyramid face of an RP or IP texturization produces two interactions with the surface as illustrated in Fig. 1 and Fig. 2, respectively. In the first interaction, the incident ray undergoes partial refraction into the bulk, where it is absorbed, and partially reflected to another face for a secondary interaction. In this subsequent interaction, the ray is again partially refracted into the bulk and partially reflected away from the solar cell.

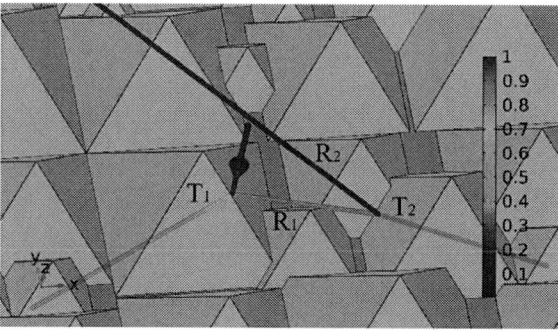

Figure 1 - Ray optics simulation for random pyramids (RP). Height ranging from 0.71 µm to 2.55 µm, base length ranging from 1 µm to 3.6 µm, base angle = 54.74°, closely packed random position and wavelength = 500 nm. Color scale showing the fraction of incident power. Depiction of reflected ray from first interaction ($R_1$), transmitted ray into the bulk from refraction of the first interaction ($T_1$), reflected ray from second interaction ($R_2$) and transmitted ray into the bulk from refraction of the second interaction ($T_2$). (Author's simulation).

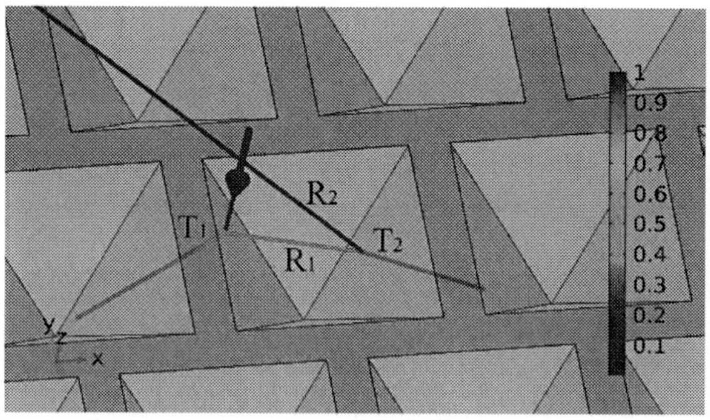

Figure 2 – Single ray tracing simulation for inverted pyramids (IP). Depth = 3.54 µm, base length = 5 µm, base angle = 54.74°, pitch = 6 µm and wavelength = 500 nm. Color scale showing the fraction of incident power. Depiction of reflected ray from first interaction ($R_1$), transmitted ray into the bulk from refraction of the first interaction ($T_1$), reflected ray from second interaction ($R_2$) and transmitted ray into the bulk from refraction of the second interaction ($T_2$). (Author's simulation).

Previous investigations employing ray optics simulations have emphasized that the angle at the top of the triangular cross-section is the most important factor in enhancing light absorption in closely packed rectangular pyramid structures [3]. A reduced top angle results in pyramids with higher aspect ratios, thereby augmenting the number of interactions and effectively trapping the light. However, conventional alkaline etching techniques are constrained with a typical top angle of 70.52° corresponding to a base angle of 54.74°.

A notable exception in the literature lies in the utilization of black silicon modulated macroporous structures, fabricated through anodic etching in hydrofluoric acid solutions of pre-patterned N-type wafers under back-side illumination. Previous studies have demonstrated the creation of closely packed pores with low top angles, similar to an extended inverted pyramid [4][5].

In this study, we employ simulations conducted using the Comsol Multiphysics ray optics module to investigate the optical properties of extended inverted pyramids (XP) and compare our findings with those derived from alternative texturization techniques.

## II. METHODOLOGY

3D finite element simulations were conducted employing the Comsol Multiphysics software with Ray Optics module [2] aiming to elucidate the optical characteristics of distinct surface texturizations. These include flat silicon; random pyramids (RP) with base lengths ranging from 1 μm to 3.6 μm, featuring a base angle of 54.74°; inverted pyramids (IP) with base length of 5 μm, a pitch of 6 μm and a base angle of 54.74°; and extended inverted pyramids (XP) exhibiting a conical pore tip, with a base length and pitch of 6 μm, as illustrated in Fig. 3. The geometric model under investigation consists of a quadrangular silicon prism, as shown in Fig. 3, possessing a base length of 200 μm, a thickness of 375 μm, and a planar rear surface.

Each incident ray is modeled as an unpolarized plane wave, with incidence normal to the surface originating 1 μm above the prism top face at a downward propagation direction. A total of 39601 rays are employed, equating to a density of approximately 1 ray per square micrometer. The maximum number of secondary rays (reflected rays) is set at 20 times the primary ray count, with the minimum computed reflected ray power being 1% of the initial individual ray power.

The refractive index of the surrounding medium was assumed to be equal to 1. The real and imaginary refractive indices, denoted as n and k, respectively, of silicon at 25 °C were sourced from the literature [6].

The simulations were time-dependent, with a maximum simulation duration of 200 ps. The simulated wavelength range varied from 250 nm to 1200 nm at 10 nm steps. A second prism encircling the first prism by a few microns distance was employed to compute the accumulated power of incident rays on each of its faces up to the final time step. This enables the obtention of the reflectance (R), transmittance (T), the fraction lost through the lateral faces, herein referred to as lateral loss (L) and, according to the equation 1, calculate the absorptance (A):

$$A = 1 - R - T - L \tag{1}$$

Accurate computation of the lateral loss (L) is paramount for simulations encompassing high aspect ratio geometries, as certain structures may significantly scatter rays laterally. Neglecting this component would engender an erroneous estimation of the absorptance.

An initial simulation was performed employing IP texturization, while systematically varying the prism's base dimension from 100 μm to 700 μm at steps of 25 μm, while maintaining a constant ray density, for wavelengths of 500 nm and 1100 nm. Comparative analysis between data obtained for a base dimension of 200 μm and an extrapolated value for 20000 μm reveals absorptance errors of 0.74%$_{abs}$ and 3.9%$_{abs}$ for wavelengths of 500 nm and 1100 nm, respectively.

## III. RESULTS

The XP texturization was initially simulated, with variations in pore depth, encompassing wavelengths of 300 nm and 1100 nm, as depicted in Fig. 4. Upon careful consideration of optimizing absorptance while minimizing pore depth, a range between 30 μm and 35 μm was deemed suitable. Notably, for enhanced infrared absorptance, a pore

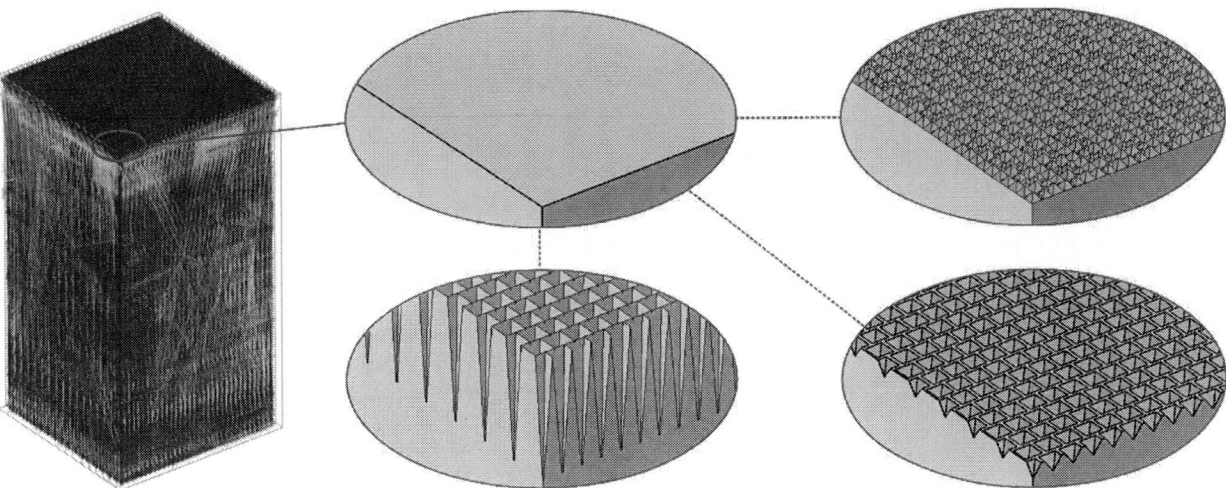

Figure 3 - Simulation results showing prism with ray trajectories. Zoom with geometric details of texturizations in this work. From left to right and top to bottom, flat surface (FLAT), random pyramids (RP), extended inverted pyramids (XP), and inverted pyramids (IP).

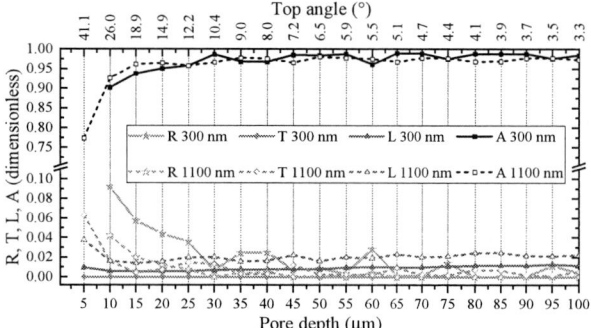

Figure 4 – Reflectance (R), Transmittance (T), Lateral loss (L), and Absorptance (A) spectra as a function of the pore depth and of the top angle for extended inverted pyramids texturization with a side length of 6 μm, pitch of 6 μm, and variable pore depth. Results for wavelengths of 300 nm and 1100 nm.

depth of 35 μm was selected for subsequent simulations, henceforth denoted as XP35. Fig. 5 illustrates a simulation wherein a single ray hits the structure, undergoing multiple interactions within the pore, thereby elucidating the fundamental principle of light trapping.

Fig. 6 provides a comparative spectral analysis among various texturizations and a planar frontal surface. It is noteworthy to observe that the performance of IP is inferior to that of RP considering the higher reflectance attributed to the presence of flat surfaces between inverted pyramids, which precludes close packing due to limitations imposed by photolithography. From a ray tracing perspective, the incident ray normal to this flat surface experiences only a single interaction upon striking it. In Fig. 6, attention is drawn to the observation that the XP35 texturization yielded the lowest reflectance, corresponding to the highest absorptance and the lowest transmittance within the simulated range.

Furthermore, a notable observation is the comparatively higher lateral scattering power of the XP35 structure, as evidenced by the elevated value of L. The absorptance remains consistently above 96% across the wavelength range of 300 nm to 1200 nm. However, it is imperative to acknowledge that higher errors are incurred in the infrared region, where the ray optics module does not account for diffraction considerations. Notwithstanding, the performance of the XP35 structure demonstrates promising attributes, whose experimental validation is a part of a doctorate work in progress [7].

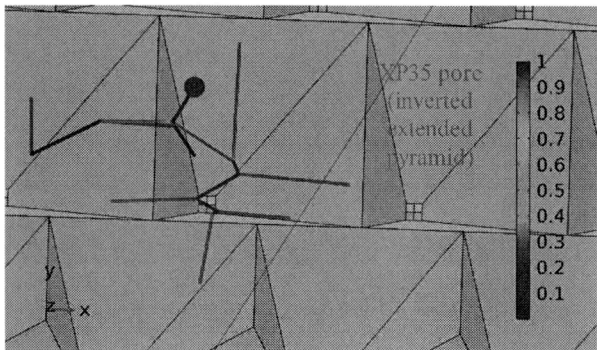

Figure 5 - Single ray tracing simulation for inverted extended pyramids (XP35). Depth = 35 μm, base length = 6 μm, pitch = 6 μm, conical tip and wavelength = 500 nm. Color scale showing the fraction of incident power.

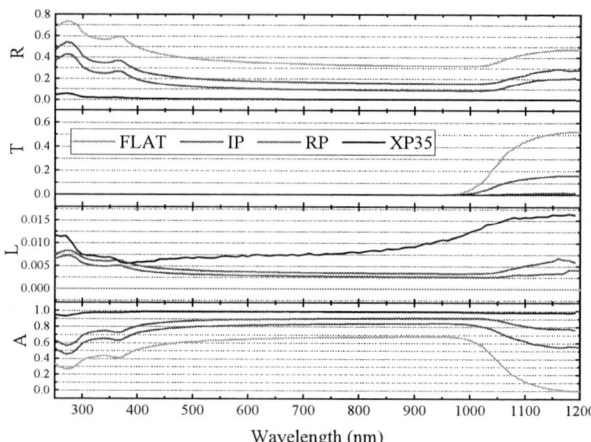

Figure 6 - Reflectance (R), Transmittance (T), Lateral loss (L), and Absorptance (A) spectra for flat surface (FLAT), inverted pyramids (IP), random pyramids (RP), and extended inverted pyramids with 35 μm depth (XP35).

Fig. 7 depicts a spectral comparison between the XP35 texturization and a structure extracted from the literature, as delineated in the inset of Fig. 7 [4][5]. Notably, XP35 exhibits superior performance across all parameters, reaffirming the auspicious characteristics of this texturization, coupled with an easier experimental manufacturing process that requires a linear photocurrent instead of a cyclic modulated photocurrent necessary to obtain rounded edges, as illustrated in inset of Fig.7 and in many other works in literature [4][5][8][9][10].

In Fig. 8, another pertinent graph is presented, wherein the XP35 texturization is simulated on a wafer with a thickness of 100 μm, designated as XP35-100. Both spectra exhibit comparable performance until the infrared region, where XP35-100 exhibits a decrease of approximately 1% in absorptance. This further underscores the favorable attributes of the XP35 texturization, even when applied to thinner wafers.

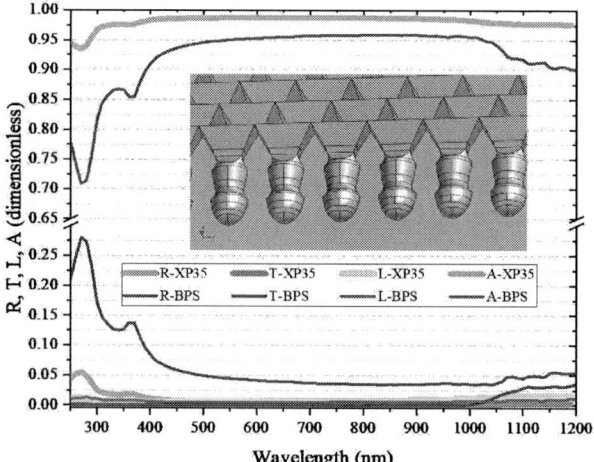

Figure 7 - Comparative simulated Reflectance (R), Transmittance (T), Lateral loss (L), and Absorptance (A) spectra for the XP35 structure and the BPS (Black Porous Silicon) structure with 500 μm thickness. Inset showing the simulated BPS structure, as per referenced papers [4][5].

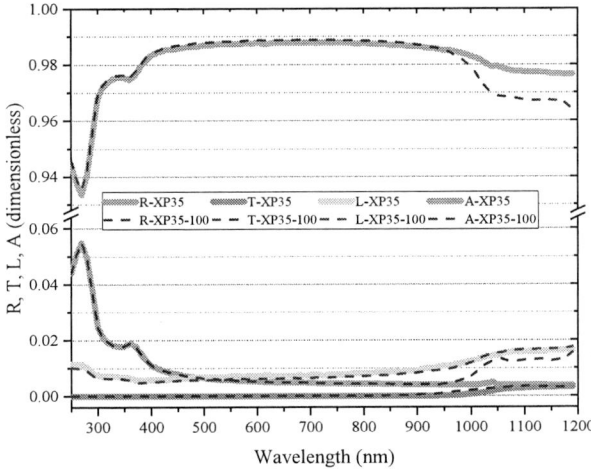

Figure 8 - Comparative spectra between extended inverted pyramid texturizations of 35 µm depth with silicon thickness of 375 µm (XP35) and 100 µm (XP35-100).

## IV. CONCLUSIONS

In conclusion, our study investigated the optical properties of various surface texturizations, with a particular focus on extended inverted pyramids, utilizing simulations conducted via the Comsol Multiphysics software with the Ray Optics module. By means of thorough examination, we observed that XP texturization, characterized by pyramidal pore structures, demonstrated promising attributes in enhancing light absorption across a broad spectrum of wavelengths, ranging from 300 nm to 1200 nm.

By optimizing pore depth, notably selecting a depth of 35 µm for enhanced infrared absorptance, our simulations illustrated consistent absorptance levels exceeding 96%. Comparative analyses with alternative texturization techniques revealed the superior performance of XP35, reaffirming its potential as a viable approach for optimizing the efficiency of silicon solar cells.

Furthermore, our findings emphasized the importance of accurately accounting for lateral scattering effects, particularly in simulations involving high aspect ratio geometries, to ensure precise estimation of absorptance.

Although our simulations provide valuable insights into the optical characteristics of XP texturization, experimental validation is imperative to corroborate these findings and ascertain their practical feasibility. The relatively straightforward experimental manufacturing process of XP35 further underscores its attractiveness as a potential candidate for enhancing the performance of silicon solar cells, particularly when applied to thinner wafers.

In summary, our study contributes to the ongoing research efforts aimed at advancing surface texturization techniques for improving the efficiency of silicon solar cells, with XP texturization emerging as a promising way warranting further investigation and validation.

## ACKNOWLEDGMENT

This study was financed in part by the Coordenação de Aperfeiçoamento de Pessoal de Nível Superior - Brasil (CAPES) - Finance Code 001

## REFERENCES

[1] Manzoor, S., Filipič, M., Onno, A., Topič, M. & Holman, Z. C. Visualizing light trapping within textured silicon solar cells. **J. Appl. Phys.** 127, (2020).

[2] The Ray Optics Module User's Guide, Version: Comsol 6.1, 1998-2022 (Part Number CM24201).

[3] Hua, X. S., Zhang, Y. J. & Wang, H. W. The effect of texture unit shape on silicon surface on the absorption properties. **Sol. Energy Mater. Sol. Cells** 94, 258–262 (2010).

[4] Ao, X. *et al.* Black silicon with controllable macropore array for enhanced photoelectrochemical performance. **Appl. Phys. Lett.** 101, (2012).

[5] Huang, W., Xue, Y., Wang, X. & Ao, X. Black silicon film with modulated macropores for thin-silicon photovoltaics. **Opt. Mater. Express** 5, 1482 (2015).

[6] Green, M. A. Improved silicon optical parameters at 25°C, 295 K and 300 K including temperature coefficients. **Prog. Photovoltaics Res. Appl.** 30, 164–179 (2022).

[7] Carvalho, M. C. B. Doctorate in progress. Escola Politécnica da USP, São Paulo.

[8] Matthias, S., Müller, F., Schilling, J. et al. Pushing the limits of macroporous silicon etching. **Appl. Phys. A** 80, 1391–1396 (2005).

[9] Matthias, S. *et al.* Large-Area Three-Dimensionsional Structuring by Electrochemical Etching and Lithography. **Adv. Mater.** 16, 2166–2170 (2004).

[10] Müller, F. *et al.* Membranes for Micropumps from Macroporous Silicon. **Phys. Stat. Sol. (a)** 182, 585-590 (2000).

# Transfer of InGaP/GaAs thin-film to unprecedented flexible polymeric bases of PVC:PMMA:DOP modified with EG for solar cell applications

Graciana S. Sousa
*Instituto de Física*
*Universidade Federal do*
*Rio de Janeiro*
Rio de Janeiro, Brazil
grfisica2010@gmail.com

Luciana D. Pinto
*Laboratório de*
*Semicondutores (LabSem)*
*Pontifícia Universidade*
*Católica do Rio de Janeiro*
Rio de Janeiro, Brazil
dornnelas@yahoo.com.br

Fabiele C. Tavares
*Campus Duque de Caxias*
*Universidade Federal do*
*Rio de Janeiro*
Duque de Caxias, Brazil
fabieletavares@hotmail.com

Guillermo J. N. Soares
*Campus Duque de Caxias*
*Universidade Federal do*
*Rio de Janeiro*
Duque de Caxias, Brazil
guillermo.nog9@gmail.com

Rudy M. S. Kawabata
*Laboratório de Semicondutores*
*(LabSem)*
*Pontifícia Universidade*
*Católica do Rio de Janeiro*
Rio de Janeiro, Brazil
rudykawarudykawa@gmail.com

Rogério Valaski
*Laboratório de*
*Fenômenos de*
*Superfície/DIMAT*
*Instituto Nacional de*
*Metrologia*
Rio de Janeiro, Brazil
rvalaski@inmetro.gov.br

Alexander. M. Silva
*Laboratório de Fenômenos*
*de Superfície/DIMAT*
*Instituto Nacional de*
*Metrologia*
Rio de Janeiro, Brazil
amartins@inmetro.gov.br

Maurício P. Pires
*Instituto de Física*
*Universidade Federal do*
*Rio de Janeiro*
Rio de Janeiro, Brazil
pires@if.ufrj.br

Roberto Jakomin
*Campus Duque de Caxias*
*Universidade Federal do*
*Rio de Janeiro*
Rio de Janeiro, Brazil
roberto.jakomin@gmail.com

Patrícia L. Souza
*Laboratório de Semicondutores*
*(LabSem)*
*Pontifícia Universidade*
*Católica do Rio de Janeiro*
Rio de Janeiro, Brazil
plustoza@puc-rio.br

*Abstract* — Flexible bases for solar cells, expanding the range of applicability while also lowering manufacturing, freight, and installation expenses.. We are developing and characterizing polymeric blends of PVC:PMMA:DOP modified with exfoliated graphene, intended for use as flexible substrates in III-V photovoltaic applications. The exfoliated graphene was employed to enhance properties such as thermal and mechanical resistance of the polymeric bases. InGaAs/GaAs heterostructures were transferred from the GaAs substrate onto two compositions of the blends. Photoluminescence measurements were conducted before and after the transfer. The results demonstrated that the transfer methodology was successfully implemented. Subsequently, complete solar cell structures will be transferred.

*Keywords* — *photovoltaic, polymeric blends, exfoliated graphene*

## I. INTRODUCTION

Photovoltaic solar energy is the clean and renewable energy source that has been experiencing the most significant global growth in recent years. III-V thin-film solar cells are identified as the most promising devices for the future photovoltaic industry, as they currently hold the word records for solar energy conversion and are still far from reaching the theoretical limits [1]. Gallium Arsenide (GaAs) and other rigid monocrystalline substrates are used for producing III-V films, primarily through Metalorganic Vapor Phase Epitaxy (MOVPE). The expense of manufacturing these materials is significantly elevated, as a substantial portion of the outlay is attributed to the monocrystalline substrate [2]. However, following epitaxy, the substrate operates exclusively as structural backing for the thin-film [3]. The manufacturing of solar cells on lightweight and inexpensive flexible substrates, such as polymers, can lower production and installation costs, as well as enable the creation of foldable and portable solar panels, further expanding the possibilities of applications. The MOVPE growth technology does not allow III-V materials to be directly deposited onto flexible polymeric substrates [4].The solution is to employ a methodology that enables the transfer of III-V materials from the GaAs substrate to new bases after growth.. Significant research has been dedicated to the transfer of III-V photovoltaic structures from GaAs to flexible bases, as evidenced by a growing body of literature [5-8].

Polymeric blends are a cheap alternative to be used as flexible bases for solar cells. In previous studies [9], we developed polymeric bases using poly(vinyl chloride) (PVC) [10-13] and poly(methyl methacrylate) (PMMA) [10-12], with the plasticizers dioctyl adipate (DOA) [13] and dioctyl phthalate (DOP) [13]. The blends exhibited flexibility but did not withstand temperatures higher than 60°C. Studies show high mechanical strength and thermo-mechanical stability in flexible thin films of PVC [14, 15] and other polymers with the incorporation of graphene [16, 17]. In this work, exfoliated graphene (EG) was introduced into the blend, with the objective of increasing the mechanical and thermal resistance. Blends containing exfoliated graphene are optimal candidates as substrates for solar cells, indicating a new opportunity for reducing the price of high-efficiency devices and broadening the range of photovoltaic applications.

## II. EXPERIMENTAL DEVELOPMENT

Polymeric blends of polyvinyl chloride (PVC) and polymethyl methacrylate (PMMA) modified with exfoliated graphene (EG) were prepared by the solution-casting method, which involves controlled evaporation of a solvent. Dimethylformamide (DMF) was used as the solvent, along with the plasticizer dioctyl phthalate (DOP) [9]. The blends were prepared with 20 mg of EG in 40 mL of DMF after 2 hours in an ultrasonic bath. The polymeric membranes are formed after drying in a low-temperature vacuum oven (60 °C) for slow solvent evaporation [9]. Various proportions of PVC, PMMA, and the plasticizer DOP were attempted to achieve flexibility with high mechanical and thermal resistance.

The blends were analyzed by Thermogravimetric Analysis (TGA) using a PerkinElmer TGA 7, with a temperature range of 25 to 600 °C and a heating rate of 10.0 °C/min under a nitrogen atmosphere. FT-IR analysis of the blends was conducted using a Perkin-Elmer FT-IR 2000, covering 600–4000 cm⁻¹, with spectra acquired using the Universal ATR accessory. Raman spectroscopy was performed on a Witec Alpha 300 using a 514.5 nm laser, 0.5 mW power, and a 100x objective, with spectra collected over 250–3600 cm⁻¹. Dynamic mechanical analysis was conducted with a Mettler-Toledo DMA/SDTA861e at 1 Hz in tension mode, using rectangular specimens of 40 × 6 × 0.5 mm..

979-8-3315-4064-7/24 $31.00 © 2024 IEEE

After the characterizations of the blends, two InGaP/GaAs heterostructures were transferred from their original GaAs substrate to the optimized PMMA:PVC:DOP + EG blends. The photoluminescence (PL) measurements were performed on the rigid samples and after the transfer process to novel flexible polymeric bases. The PL experimental setup consisted of a Nd:YAG laser (MatchBox Series model), emitting at 532 nm with 50nW and some optical components for the visible range spectra (lenses and mirrors) to guide the laser and the PL emission to a monochromator and a wide range germanium photodetector. A 570 nm longpass optical filter was used to block the laser from the fotodetector and to avoid second-harmonic diffraction effects from the diffraction grating.

## III. RESULTS AND DISCUSSION

### A. Preparation and characterization of polymeric blends modified with EG

TABLE I shows the composition, physical characteristics and results of DMA of two blends previously prepared without graphene [9] and four different compositions of blends modified with EG.

TABLE I. Compositions, characteristics and DMA results of the blends.

|  | Composition of blends (mass grams) | Physical characteristics (softening point* ) | DMA (Mpa) |
|---|---|---|---|
| B1 | PVC:PMMA (1.0:1.0) | Flexible (60 °C) | 4.9 |
| B2 | [PVC:PMMA]:DOP (1.5:0.5:1.0) | Flexible, elastic (60 °C) | 8.3 |
| B3 | [PVC:PMMA]:DOP:EG (1.5:0.5:0.5) | Flexible, elastic (140 °C) | 791 |
| B4 | [PVC:PMMA]:DOP + EG (1.5:0.5:1.0) | Flexible, elastic (140 °C) | 226 |
| B5 | [PVC:PMMA]:DOP:EG (1.0:1.0:0.5) | Flexible, elastic (140 °C) | 831 |
| B6 | [PVC:PMMA]:DOP:EG (1.0:1.0:1.0) | Flexible, elastic (140 °C) | 121 |

\* The softening point is the temperature at which the polymer starts to soften and lose its mechanical strength.

As blends B1 and B2 demonstrate the desired flexibility, they begin to soften and lose their mechanical strength when exposed to temperatures exceeding 60 °C. To increase both mechanical and thermal resistance, EG was incorporated into the blend. All four EG-containing blends (B3, B4, B5, B6) exhibited flexibility and good adherence to the Petri dishes on which they were prepared and withstood heating up to 140 °C. In Fig. 1, we can see the flexibility of the blends without graphene (B1, B2) and with graphene (B3, B5).

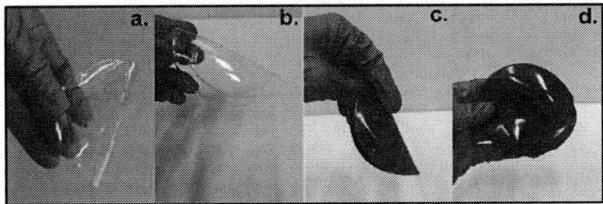

Fig. 1. Flexibility of the blends without EG: (a) B1; (b) B2; and with EG: (c) B3 and (d) B5.

The blends were investigated by Dynamic Mechanical Analysis (TABLE I and Fig. 2a). A significant increase in storage modulus with the addition of EG to the PVC:PMMA:DOP blends can be observed in Fig. 2a. Comparing blends with the same concentration of

PVC:PMMA, namely B3 with B4 and B5 with B6, we observe that in both cases, the higher concentration of plasticizer DOP led to a reduction in mechanical strength. Further analysis will be performed to investigate whether the excess plasticizer might be affecting the interactions between graphene and the blends. Atomic Force Microscopy and Scanning Electron Microscopy measurements will be conducted to investigate the distribution of graphene within the polymer matrix of the blends.

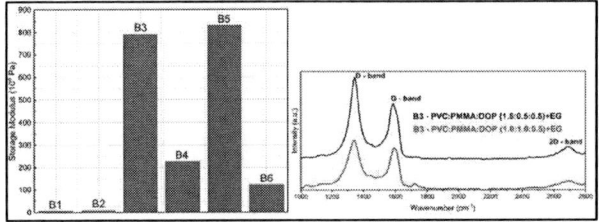

Fig. 2. a. DMA of the blends without graphene (B1, B2) and with graphene (B3 and B4; B5 and B6); b. Raman spectrum of the blends B3 and B5.

The presence of graphene in blends B3 and B5, as shown in Fig. 2b, and in all blends with EG, was confirmed by Raman spectroscopy. The spectra of these blends display the characteristic graphene bands: the D-band (1349 cm⁻¹), G-band (1582 cm⁻¹), and 2D-band (2700 cm⁻¹). The D-band indicates structural defects or edges of graphene nanosheets, the G-band represents the sp² bonded carbon in graphene, and the 2D-band is an overtone of the D-band [18,19].

The thermogravimetric Analysis (TGA) of blends B3, B4, B6 blends B3, B4, B5, B6 are shown in Fig. 3.

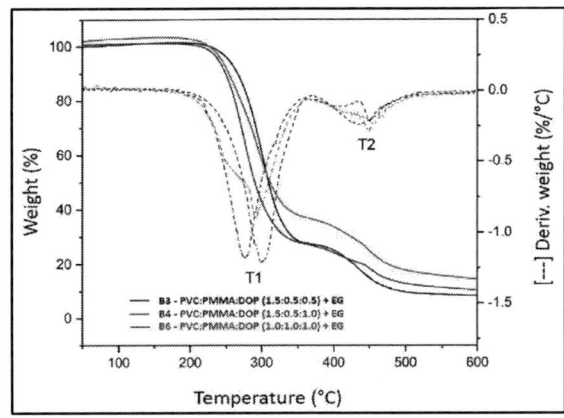

Fig. 3. Thermogram of blends B3, B4, B6.

The TGA of B5 will be done later. The thermogram of the blends with EG demonstrates thermal stability up to 200 °C, indicating that the blend can withstand the temperatures required for the transfer process (120 °C). T1 represents the breakage of $C - H$ and $C - Cl$ bonds, resulting in the formation of HCl. T2 represents the degradation of the polyene structure and the formation of volatile aromatic compounds. The lack of mass loss below 120 °C (Fig. 3) further suggests that there is no residual solvent present in the blend's structure, demonstrating a strong interaction between the polymers and the plasticizer molecules. These results correspond to the observations from the FTIR spectra (Fig. 4), which indicate that the strong intermolecular interactions between polymers and plasticizers enhance the evaporation of the solvent during the drying process.

Financiadora de Estudos e Projetos (FINEP) - Brazil, grant # 01.21.0110.00; Fundação de Amparo à Pesquisa do Estado do Rio de Janeiro (FAPERJ) - Brazil, grant # E-26/210.787/2021; Coordenação de Aperfeiçoamento de Pessoal de Nível Superior (CAPES) - Brazil, grant # 23038.004236/2019-10; Conselho Nacional de Desenvolvimento Científico e Tecnológico (CNPq) - Brazil, grant # 301838/2016-0.

The FT-IR spectra (Fig. 4), presented in transmittance mode over the 600-2000 cm⁻¹ frequency range, reveal that while the stretching vibration of PMMA (1193 cm⁻¹) remains unchanged, there is a notable shift in the carbonyl band of PMMA (1734 cm⁻¹) to lower wavenumbers in all blends. Specifically, the shifts are to 1723 cm⁻¹ for B3 and B6, 1722 cm⁻¹ for B4, and 1725 cm⁻¹ for B5. These shifts suggest the formation of hydrogen bonds between the carbonyl group of PMMA and the H⁺ ions from PVC. Additionally, the FTIR spectra of the four blends reveal aromatic ring vibrations: a double peak for stretching at 1600 cm⁻¹ and 1580 cm⁻¹, and angular deformation at 743-747 cm⁻¹. The presence of these aromatic bands, characteristic of the DOP plasticizer, confirms its successful incorporation into the blends and indicates the miscibility of PVC and PMMA.

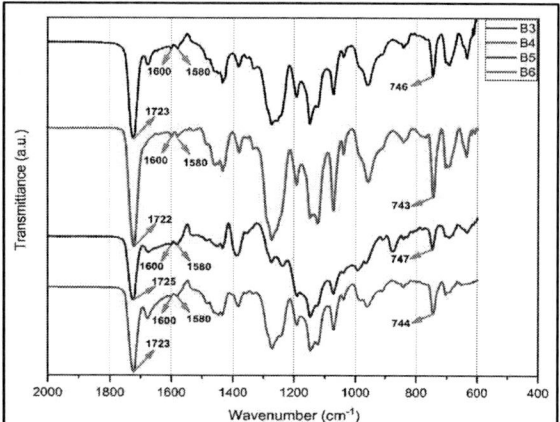

Fig. 4. FTIR spectra of B3, B4, B5 and B6.

### B. Transfer of InGaP/GaAs from the GaAs substrates to polymer blends with EG

The two blends with higher mechanical strength, B3 and B5 (Fig. 2b), were utilized to transfer the InGaP/GaAs heterostructures, serving as the flexible bases. Fig. 5 illustrates the general scheme for transferring III-V materials from rigid GaAs substrates to flexible blends. [7,9].

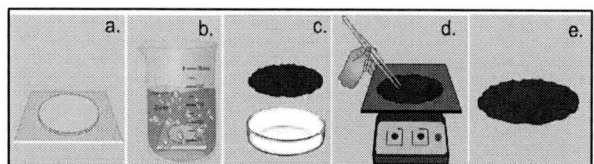

Fig. 5. Transfer Procedure. a) Bonding a temporary substrate (glass); b) etching the GaAs substrate; c) depositing the flexible base; d) removing the temporary substrate; e) sample flexible [7,9].

The transfer was carried out following the methodology illustrated in Fig. 5, after complete etching of the GaAs substrate [17]. A temporary glue was used to bond the III-V sample on a glass slide [22]. After the etching process, the III-V thin-film is stable on the temporary glass substrate, and different flexible bases can be adhered to it. Two different compositions of blends with EG were prepared by solution-casting in accordance with the experimental procedures outlined in the Experimental Development section. The adhesion of the blends to the III-V thin films occurs during the drying of the blend in a vacuum oven at 60 °C for 48 hours. The removal of the temporary glass substrate is performed by heating at 120 °C [22]. Blends previously prepared without the addition of graphene did not withstand temperatures above 60 °C; they softened, deformed, and did not provide the necessary mechanical support to the III-V materials during the removal of the temporary glass substrate. Fig. 6a shows the III-V material completely destroyed after the heating process for the removal of the glass substrate. In Fig. 6b, we observe two intact InGaP/GaAs samples transferred to blends modified with EG. The blends with EG perfectly withstood the heating process without compromising the III-V structures and the III-V thin films remained perfectly adhered to the flexible bases. The blends with EG was show to be good candidates as a mechanical support for the semiconductor thin-film. These results demonstrate that the presence of EG in the blend composition brings additional thermal resistance to these polymers, allowing III-V solar cells to be transferred to these novel polymeric bases.

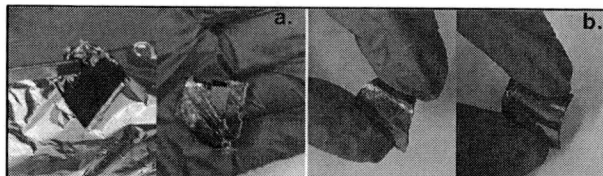

Fig. 6. a. GaAs transferred to blends without graphene; b. InGaP transferred to blends with graphene.

### C. Photoluminescence analysis

The photoluminescence (PL) measurements were performed on the thin film samples in two distinct moments: while still on the rigid GaAs substrate and after the transfer process onto the novel flexible polymeric bases. The samples (A1 and A2) consisting of InGaP/GaAs thin films, as detailed in Fig. 7, were grown on GaAs substrates by MOVPE located at the Semiconductor Laboratory at PUC-Rio. A1 and A2 samples have InGaP thicknesses of 380 and 490 nm respectively. Thicknesses above 100 nm are considered bulk, therefore the energy value found in PL depends solely on the material composition, which is the same in both samples.

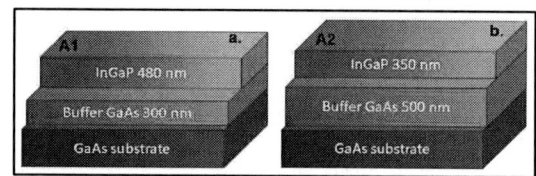

Fig. 7. Schematic illustration of the InGaP/GaAs heterostructures: a. Sample A1; b Sample A2.

In Fig. 8, we present the PL spectra of the samples A1 and A2 measured at room temperature (300 K), both on the GaAs substrate (in black) and after the transfer to the polymeric blends (in red). The intensity differences between A1 and A2 and between the rigid samples and their respective flexible samples cannot be considered. The units are arbitrary; peak energy and full width at half maximum (FWHM) should be compared. It can be observed that the emission peak energy of InGaP remains around 1.85 eV, corresponding to the bandgap energy of bulk InGaP. This suggests that the etching of the GaAs substrate did not damage the thin film, which would have resulted in crystalline defects exhibiting PL peaks at energies lower than the bandgap energy or a widening of the emission's FWHM.

979-8-3315-4064-7/24 $31.00 © 2024 IEEE

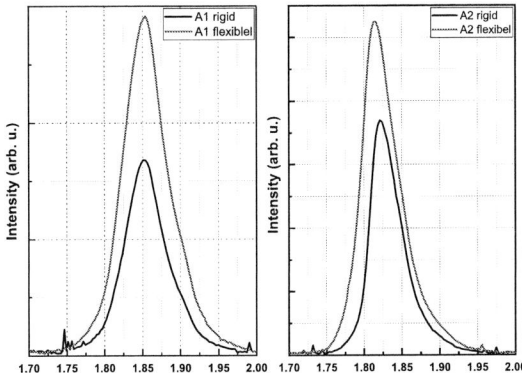

Fig. 8. Photoluminescence spectra of samples A1 and A2 on rigid and flexible substrates.

These results demonstrate that PMMA:PVC:DOP blends prepared with EG have great potential to be used as flexible substrates for III-V solar cells with no damage to their optical properties.

## IV. CONCLUSIONS

Polymeric blends of PVC:PMMA:DOP + EG were prepared and characterized to be used as innovative flexible bases for III-V solar cells. Different proportions of polymers and plasticizer were tested. Results from FTIR and TGA demonstrated the formation of the blends and the presence of the plasticizer in the polymeric structures. Raman spectroscopy measurements showed the presence of EG in the blend structure. Blends without EG do not withstand temperatures above 60 °C, compromising the final step of the transfer process, which needs to be carried out at 120 °C for the removal of the temporary glass substrate. The thermogram of the blends with EG demonstrates thermal stability up to 200 °C, and DMA measurements showed a significant increase in mechanical resistance. Two III-V structures were successfully transferred to two blends with higher mechanical strength. The blends adhered perfectly to the III-V semiconductor films, providing the necessary mechanical support. Photoluminescence measurements confirmed that the III-V structures were not damaged by the transfer process. These results demonstrate that the addition of EG increased the thermal and mechanical resistance of the blends. Therefore, we can conclude that PMMA:PVC:DOP blends modified with EG have great potential to be used as a flexible base for III-V solar cells. Further characterizations will be conducted, and other blend compositions will be tested. Solar cells grown on GaAs will be processed and transferred to the blend compositions with higher thermal and mechanical resistance.

### ACKNOWLEDGMENT

We would like to thank the financial support from: Financiadora de Estudos e Projetos (FINEP) - Brazil, Fundação de Amparo à Pesquisa do Estado do Rio de Janeiro (FAPERJ) - Brazil; Coordenação de Aperfeiçoamento de Pessoal de Nível Superior (CAPES) - Brazil; Conselho Nacional de Desenvolvimento Científico e Tecnológico (CNPq) - Brazil.

### REFERENCES

[1] NREL.gov, "Best Research-Cell Efficiency Chart," National Renewable Energy Laboratory, 2023, <https://www.nrel.gov/pv/cell-efficiency.html>. Accessed 10 April 2023.

[2] J. S. Ward, et al., "Techno-economic analysis of three different substrate removal and reuse strategies for III-V solar cells," Progress in Photovoltaics: Research and Applications, vol. 24, n. 9, pp. 1284-1292, September 2016.

[3] EL-ATAB, Nazek; HUSSAIN, Muhammad M. Flexible and stretchable inorganic solar cells: Progress, challenges, and opportunities. MRS Energy & Sustainability, vol. 7, 2020.

[4] C. H. Lee, D. R. Kim, and X. Zheng, "Transfer printing methods for flexible thin film solar cells: Basic concepts and working principles," ACS Nano, Vol. 8, n. 9, pp. 8746 – 8756, September 2014.

[5] M. B. Schubert, and J. H. Werner, "Flexible solar cells for clothing," Materialstoday, vol. 9, n. 6, pp. 42-50, June 2006.

[6] T. Masuda et al., "Highly Decorative, Lightweight Flexible Solar Cells for Automotive Applications," SAE Technical Paper, n. 2019-01-0863 April 2019.

[7] M. O. Silva (2021), "Células de multijunções de alta eficiência: Metodologias para transferência de materiais semicondutores III-V de forma reprodutível para substrato flexível," (master's dissertation). Departamento de Engenharia Elétrica, PUC-Rio.

[8] V. V. Soman, and D. S. Kelkar, "FTIR Studies of Doped PMMA - PVC Blend System". Macromolecular Symposia: POLYCHAR – 16 World Forum on Advanced Materials, vol. 277, pp. 152-161, March 2009.

[9] G. S. Sousa et al., "Preparation and characterization of PVC-PMMA polymer blends as flexible bases for III–V photovoltaics," 2023 37th Symposium on Microelectronics Technology and Devices (SBMicro), Rio de Janeiro, Brazil, 2023, pp. 1-4, doi: 10.1109/SBMicro60499.2023.10302464.9. Sbmicro 2024

[10] M. S. Khan, R. A. Qazi, and M. S. Wahid, "Miscibility studies of PVC/PMMA and PS/PMMA blends by dilute solution viscometry and FTIR," African Journal of Pure and Applied Chemistry, vol. 2, pp. 41 – 45, April 2008.

[11] S. Ramesh, and CW. Liew, "Development and investigation on PMMA– PVC blend-based solid polymer electrolytes with LiTFSI as dopant salt," Polymer Bulletin, vol.70, pp. 1277–1288, April 2013.

[12] D. S. Rosa, et al., "Estudo do efeito da incorporação de plastificante de fonte renovável em compostos de PVC," Polímeros: Ciência e Tecnologia, vol. 23, n. 4, pp. 570-577, 2013.

[13] G. Duan, et al., "Preparation and Characterization of Mesoporous Zirconia Made by Using a Poly (methyl methacrylate) Template," Nanoscale Research Letters; vol. 3, pp. 118 – 122, February 2008.

[14] Sajini Vadukumpully, Jinu Paul, Narahari Mahanta, Suresh Valiyaveettil, Flexible conductive graphene/poly(vinyl chloride) composite thin films with high mechanical strength and thermal stability, Carbon, Volume 49, Issue 1, 2011,

[15] Mudassir Hasan, Moonyong Lee, Enhancement of the thermo-mechanical properties and efficacy of mixing technique in the preparation of graphene/PVC nanocomposites compared to carbon nanotubes/PVC, Progress in Natural Science: Materials International, Volume 24, Issue 6, Pages 579-587, 2014.

[16] Pingan Song, Zhenhu Cao, Yuanzheng Cai, Liping Zhao, Zhengping Fang, Shenyuan Fu, Fabrication of exfoliated graphene-based polypropylene nanocomposites with enhanced mechanical and thermal properties, Polymer, Volume 52, Issue 18, 2011, Pages 4001-4010,

[17] Dhaiwat N. Trivedi, Nikunj V. Rachchh, Graphene and its application in thermoplastic polymers as nano-filler- A review, Polymer, Volume 240, 2022,

[18] Thermo Fisher Scientific Inc. Characterizing graphene with Raman spectroscopy. < https://encurtador.com.br/ehmpu >. Accessed 06 October 2023.

[19] T. S. Tran, S. J. Park, S. S. Yoo, T.-R. Lee and T. Kim, "High shear-induced exfoliation of graphite into high quality graphene by Taylor–Couette flow", RSC Adv., 6, 12003–12008, 2016.

[20] Guorong Duan, et al. Preparation and Characterization of Mesoporous Zirconia Made by Using a Poly (methyl methacrylate). Nanoscale Res Lett.; 3(3): 118–122, 2008.

[21] C. Bryce, and D. Berk, "Kinetics of GaAs Dissolution in H2O2– NH4OH– H2O Solutions," Industrial & Engineering Chemistry Research, vol. 35, n. 12, pp. 4464 – 4470, December 1996.

[22] BREWER SCIENCE, "WaferBOND HT-10.10," Brewer Science Company, 2016, <https://encurtador.com.br/jkuGU>. Accessed 13 Setember 2021.

# Development of Epoxy Bonding Techniques for III-V on Silicon Tandem Solar Cells

1st Willian M. M. Bazilio
*Departamento de Engenharia Elétrica*
*Pontifícia Universidade Católica do Rio de janeiro*
Rio de Janeiro, Brazil
willian.m.bazilio@aluno.puc-rio.br

2nd Rudy M. S. Kawabata
*Departamento de Engenharia Elétrica*
*Pontifícia Universidade Católica do Rio de janeiro*
Rio de Janeiro, Brazil
rudykawarudykawa@gmail.com

3rd R. T. Mourão
*Instituto Federal de Educação, Ciência e Tecnologia Fluminense*
*Campus Quissamã*
Rio de Janeiro, Brazil
renato.mourao@iff.edu.br

4th Guilherme M. Torelly
*Departamento de Engenharia Elétrica*
*Pontifícia Universidade Católica do Rio de janeiro*
Rio de Janeiro, Brazil
torelly@puc-rio.br

5th Patricia L. de Souza
*Departamento de Engenharia Elétrica*
*Pontifícia Universidade Católica do Rio de janeiro*
Rio de Janeiro, Brazil
plustoza@puc-rio.br

*Abstract*—**Tandem solar cells, integrating III-V and silicon materials, represent a promising approach for enhancing the efficiency of silicon solar cells.Several challenges, such as optimizing the thickness of the active region, removing the substrate and processing the GaAs top solar cell, ensuring long-term stability (mechanical and electrical), and reducing costs, must be overcome to achieve this goal, including bonding both materials. Bonding III-V and Si with epoxy is a less complex to heteroepitaxy. In this work we study the epoxy application technique and characterize the optical properties of the a Si SC with epoxy, glass and GaAs materials, to assess their impact on multi-junction solar cell performance. The epoxy (EPO TEK 353 ND) and glass slides, were selected for their high transmittance (above 95%) and cost-effectiveness. This process resulted in a reduction of short circuit current density below 4.7%.**

**Keywords — solar cell; III-V on Si; Epoxy; Tandem solar cell**

## I. INTRODUCTION

The integration of III-V semiconductor materials and silicon (Si) is a research field with significant interest for optoelectronics due to what each of these materials offers in terms of their optical properties or technological development. III-V semiconductors present high photon absorption efficiency by virtue of their high crystalline quality and energy bandgap configuration, although their fabrication is costly and complex. Silicon has less ideal optical properties, but has low production cost, due to its technological maturity stage and large scale

production. For these reasons, the idea of integrating III-V semiconductors and Si for photovoltaic devices emerged in the 1980's [1, 2], coinciding with the advancement of epitaxial growth techniques that produced high crystal quality III-V based optoelectronic devices, such as solar cells, light emitting diodes and photodetectors. Despite presenting high incident light to electrical current conversion efficiency, their high-cost production limited III-V devices to niche applications, such as solar cells for satellites and aerospace industries.

Even though the commercially available solar cell (SC) market has been dominated by Si SC's, their efficiency has presented only marginal growth in the past decades, due to being near its theoretical limit (below 30%). In this same period, III-V SC's presented a conversion efficiency increase of approximately 1% per year, while the production cost diminished significantly, even though it remains high when compared with other technologies [3]. In order to surpass the low conversion efficiency of single junction Si SC's, a multi-junction SC design (or tandem SC) was proposed, integrating Si with other SC technologies. One of the most promising candidates are monocrystalline III-V materials, which currently hold the highest demonstrated solar cell efficiencies [4, 5]. Epitaxial growth of III-V semiconductor directly on top of Si has been tested but is yet to deliver satisfactory results[6]. One viable approach is to grow III-V SC's on III-V substrates and transfer the active layers to a Si SC. The simplest method for this approach is to attach or bond the III-V semiconductor to the Si through a glass slide and epoxy adhesive [7]. In this method, the III-V on Si tandem SC is

979-8-3315-4064-7/24 $31.00 © 2024 IEEE

in a four terminal configuration, being optically coupled but electrically isolated. In 2015, Essig at al.[8], used a transparent and non-conductive epoxy adhesive (TRA-BOND-931-1) to bond an indium gallium phosphide (InGaP) SC to a Si SC and created a 27% efficient InGaP / Si tandem SC. In 2017, with the addition of a glass slide and improved methods, a double junction (GaAs / Si) SC with 32.8% efficiency and a triple junction (InGaP / GaAs / Si) SC with 35.9% were built [9, 10].

In this work, we developed a method to bond a glass slide to a Si SC using epoxy adhesive. After bonding, we measured the optical and electrical properties of the device. We compared the silicon solar cell before and after the application of the glass slide and epoxy adhesive. Additionally, we tested a polished GaAs substrate as a filter over the Si cell, mimicking the optical properties of a GaAs solar cell. We present conventional figures of merit for solar cells, such as short-circuit current density ($J_{SC}$), open-circuit voltage ($V_{OC}$), fill factor (FF), maximum power ($P_{MAX}$), conversion efficiency ($\eta$), and quantum efficiency spectrum. The study of epoxy application is crucial for the future coupling of GaAs and silicon solar cells. The collected data will allow us to optimize the bonding process and identify the cause of defects such as surface bubbles and non-uniformity in epoxy layer thickness, factors that directly impact the solar cell efficiency.

## II. METHODOLOGY

To bond the Si SC to the glass slide, we selected EPO - TEK 353 ND epoxy (Epoxy Technology) due to its low cost and high transmittance in the visible and near-infrared spectrum. The Si SC's used in this study are commercially available, low-cost polycrystalline cells measuring 20 x 20 x 0.2 mm³. Similarly, the glass slide chosen is a commercially available product, with dimensions of 18 x 18 x 0.1 mm³, precisely to cover the entire area of the Si SC except for the busbar, where we will make the electrical contact for measurements.

Two methods for applying the epoxy to the Si SC's were studied. One method consists in applying a drop of epoxy on top of the solar cell, then carefully placing the glass slide, on top of which a calibrated weight is used to apply pressure (weight method). The other method employs spin coating to obtain an homogeneous and thin film of epoxy on top of the Si SC, then place the glass slide on the epoxy (spinner method). Fig. 1 shows an schematic of the methods. For the weight method, four calibrated weights ranging from 10 to 75 g were used. For the spinner method, the sample was spun at 5000 rpm for 10 to 60 s. In both methods, after the epoxy application, the samples are cured in a hot plate at 120 °C for 5 minutes. The thicknesses of the resulting epoxy film in both methods were measured using a micrometer screw gauge. The transmittance and quantum efficiency of the samples were measured using the setup presented in Fig. 2. An SF300A solar simulator (Sciencetech, On, Canada) was used as a bias light source. The electrical parameters and figures of merit were measured with an HP 4145B Semiconductor Parameter

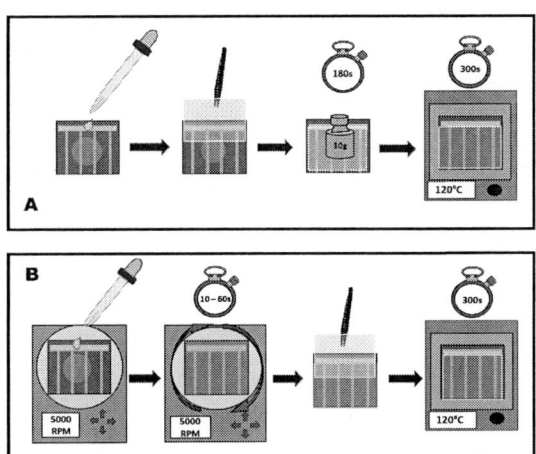

Fig. 1. Schematic for the Si SC and glass slide bonding procedure using the epoxy adhesive with a weight method (A) and a spin coating method (B).

Analyzer. Transmittance measurements were performed using monochromatic light obtained from an incandescent lamp filtered by a monochromator coupled to a chopper. During tests, the incident light power was measured with a calibrated Si photodiode (Thorlabs, model SM05PD3A). In order to mimic a III-V solar cell on top of the Si one, a 300 $\mu$m - thick, polished, GaAs substrate was used as a filter.

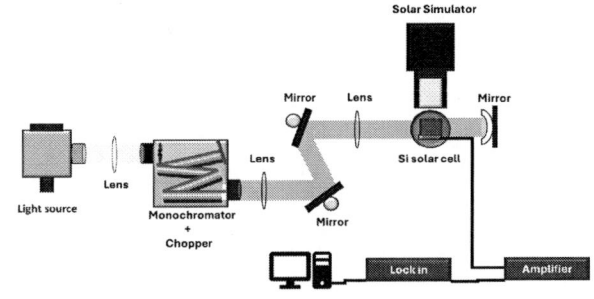

Fig. 2. Experimental setup to measure solar cell quantum efficiency, conversion efficiency and epoxy transmittance.

## III. RESULTS AND DISCUSSIONS

Fig. 3 shows the Si SC both before and after bonding to the glass slide with the epoxy adhesive. In the initial stage (left), the multicrystalline Si solar cell presents its characteristic blue hue. Following bonding, the sample's color darkens, displaying a brownish tone. Notably, air bubbles may become entrapped between the glass and the Si, as depicted in the middle image. Conversely, when the epoxy is applied correctly, the occurrence of air bubbles in the final sample is minimized, as evidenced in the image on the right. The transmittance of a glass slide bonded with epoxy adhesive was measured by comparing the photocurrent of the Si solar cells before and after the bonding process, allowing for the calculation of the change in transmittance. Fig. 4 presents the transmittance spectra of eight solar cell samples made using both methods.

979-8-3315-4064-7/24 $31.00 © 2024 IEEE

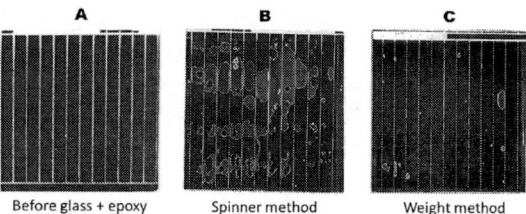

**Fig. 3.** Photos of the Si SC before the epoxy (A), after a non-optimized epoxy (B) and after an optimized epoxy (C). Bubbles highlighted in yellow.

The inset shows the transmittance of individual components: glass, epoxy, and polished GaAs subtrate.

It is possible to assert that there is a significant drop in transmittance in the 400 to 600 nm range, which is in accordance with the epoxy's specifications. Outside this range, transmittance remains above 80%, increasing in the near-infrared to almost 100%. This transmittance is high enough to ensure that most photons that pass through the III-V SC reach the Si SC. Considering a GaAs top cell, only light with wavelength longer than 850 nm will reach the Si SC, which corresponds to the spectrum range where transmittance is closer to 100%. If we envision an InGaP/Si SC, the Si SC would work in the range above 635 nm corresponding to a transmittance above 80%. Table I and II present the results

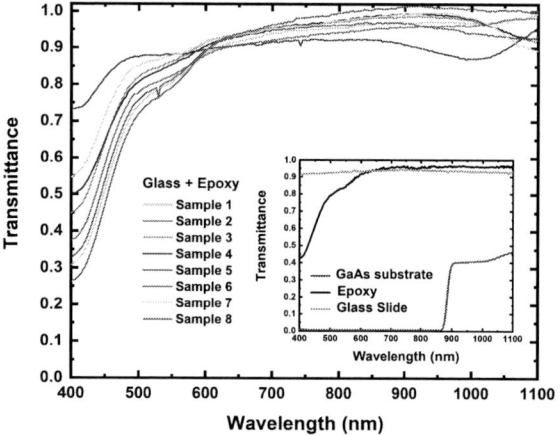

**Fig. 4.** Transmittance spectrum measurement through the glass slide and epoxy adhesive for different bonding methods. In the inset it is the individual transmittance for the glass slide, the epoxy and a GaAs substrate.

for eight Si SC, tested using both epoxy application methods, with their respective $J_{SC}$ and $V_{OC}$, before and after the epoxy application, as well as the epoxy thickness. An analysis of the thickness change as a function of the weight or spinning duration is not conclusive, requiring more samples to properly define the influence of each parameter on the epoxy layer thickness. Batch processing of many samples fabricated in similar conditions might help reduce the thickness variation. Regarding the SC's figures of merit before and after the epoxy application, the $V_{OC}$ is reduced by approximately 10 mV or less for the samples obtained using the weight method,

while the spinner method reduced the $V_{OC}$ by up to 30 mV, demonstrating it's low sensitivity to the epoxy application. The $J_{SC}$ is more sensitive to the application of epoxy because it is directly proportional to the amount of light that reaches the surface of the SC. Therefore, if the intensity of the incident light decreases, meaning the number of photons per unit area is reduced, the $J_{SC}$ will be negatively affected. Using the weight method a reduction ranging from 0.3% to 4.7% was observed. The spinner method lead to worse results, with a reduction in $J_{SC}$ between 1.6% and 13.5%. This result may be explained by the appearance of many bubbles on the spin-coated samples, which indicate that the spinner method is not forming an homogeneous thin film. Immediately after being spun onto the solar cell, the epoxy coagulates due to surface tension and it's low viscosity. The weight method more uniformly distributes the epoxy on top of the solar cell, avoiding the entrapment of bubbles.

TABLE I
THICKNESS, SHORT CIRCUIT CURRENT AND OPEN CIRCUIT VOLTAGE FOR SAMPLES FABRICATED USING THE WEIGHT METHOD

| Weight Method | | Before epoxy | | After epoxy | | Reduction | |
|---|---|---|---|---|---|---|---|
| Sample | Mass (g) | Epoxy Thickness ($\mu$m) | $J_{sc}$ (mA/cm²) | $V_{oc}$ (V) | $J_{sc}$ (mA/cm²) | $V_{oc}$ (V) | $\Delta J_{sc}$ (%) | $\Delta V_{oc}$ (%) |
| 1 | 10 | 9 | 28.8 ± 0.6 | 0.56 ± 0.01 | 28.7 ± 0.2 | 0.55 ± 0.01 | 0.3% | 1.8% |
| 2 | 25 | 27 | 29.1 ± 0.6 | 0.56 ± 0.01 | 28.4 ± 0.2 | 0.56 ± 0.01 | 2.2% | 0% |
| 3 | 50 | 20 | 29.4 ± 0.6 | 0.55 ± 0.01 | 28.1 ± 0.2 | 0.55 ± 0.01 | 4.7% | 0% |
| 4 | 75 | 10 | 29.8 ± 0.6 | 0.56 ± 0.01 | 28.8 ± 0.2 | 0.55 ± 0.01 | 3.5% | 1.8% |

TABLE II
THICKNESS, SHORT CIRCUIT CURRENT AND OPEN CIRCUIT VOLTAGE FOR SAMPLES FABRICATED USING THE SPINNER METHOD

| Spinner Method | | Before epoxy | | After epoxy | | Reduction | |
|---|---|---|---|---|---|---|---|
| Sample | Time(s) | Epoxy Thickness ($\mu$m) | $J_{sc}$ (mA/cm²) | $V_{oc}$ (V) | $J_{sc}$ (mA/cm²) | $V_{oc}$ (V) | $\Delta J_{sc}$ (%) | $\Delta V_{oc}$ (%) |
| 5 | 10 | 23 | 29.5 ± 0.6 | 0.55 ± 0.01 | 25.5 ± 1.0 | 0.53 ± 0.01 | 13.5% | 3.6% |
| 6 | 20 | 38 | 29.7 ± 0.6 | 0.56 ± 0.01 | 27.8 ± 1.0 | 0.54 ± 0.01 | 6.4% | 3.6% |
| 7 | 30 | 13 | 32.2 ± 0.6 | 0.57 ± 0.01 | 31.1 ± 1.0 | 0.54 ± 0.01 | 3.5% | 5.3% |
| 8 | 60 | 16 | 29.2 ± 0.6 | 0.54 ± 0.01 | 28.8 ± 1.0 | 0.54 ± 0.01 | 1.6% | 0% |

Aiming to further explore the cause of the reduction in $J_{sc}$ after the epoxy application, the quantum efficiency spectrum of a sample was measured, as shown in in Fig. 5, presenting the Si SC with and without the epoxy and glass slide. One can clearly see that between 400 and 700 nm there is a reduction of efficiency due to the reduced epoxy transmittance. Ranging from 600 nm to longer wavelengths, this reduction is less than 10%, compared to the curve without the epoxy, and beyond 850 nm the absorption is negligible. Considering a GaAs/Si tandem SC, represented by a GaAs substrate put on top of the Si SC, the quantum efficiency is significantly reduced. Since the epoxy and glass are transparent for wavelength beyond 850 nm, introduction of the GaAs substrate is responsible for a 70% lower quantum efficiency. Due to the high refractive index of GaAs in the interest range ($n \approx 3.6$), losses due to reflection are significant. The 300 $\mu$m thickness of the GaAs substrate also contributes to absorption. In future studies, a thin III-V SC ($d \approx 5\,\mu$m) will be used, together with antirreflective coatings to reduce Fresnel losses on the interfaces, which should improve the overall quantum efficiency of the Si SC.

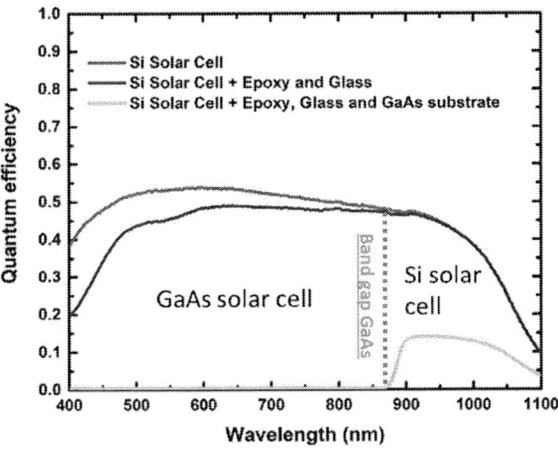

Fig. 5. Quantum efficiency spectrum for the Si solar cell with and without the glass slide and epoxy adhesive, as well as with a GaAs substrate on top, along with the spectral range that will be utilized by each cell after tandem integration.

Fig. 6 shows the current density versus applied voltage for the Si SC with the GaAs substrate on top of the glass and epoxy, among the SC's figures of merit (short circuit current density, $J_{SC}$, $V_{OC}$, FF and $\eta$) for comparison to the results without GaAs (see Table I).In comparison to the results obtained without the inclusion of GaAs, there was a notable reduction of 85% in the short-circuit current density $J_{SC}$ and a 10% decrease in the open-circuit voltage $V_{OC}$, consequently resulting in a significant decline in conversion efficiency from 8.0% to 1.1%. Notably, the reduction in conversion efficiency, approximately 86%, closely aligns with the reduction observed in $J_{SC}$. This suggests that the presence of GaAs predominantly influences $J_{SC}$ compared to other critical performance indicators essential for the design of tandem III-V/Si solar cells.

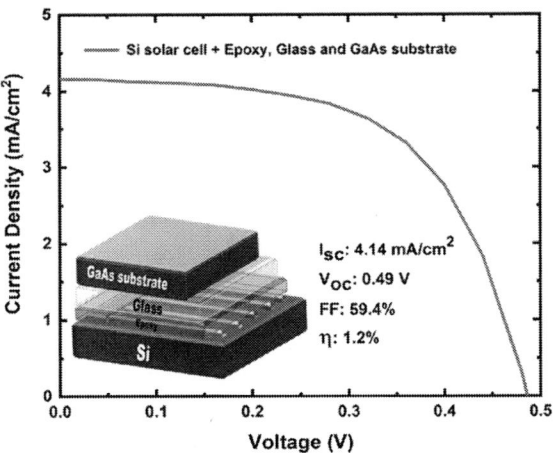

Fig. 6. Current density versus applied voltage curve for the Si solar cell with epoxy, glass and GaAs substrate on top with its respective figures of merit.

## IV. CONCLUSION

This work demonstrates a viable, cost-effective method for bonding III-V semiconductors to silicon solar cells using epoxy and a glass slide as support. In the final device, the GaAs substrate will be removed, leaving only the active region, paving the way for high-efficiency III-V/Si tandem solar cells and potentially overcoming single-junction silicon limitations.

Two bonding methods were tested. The Weight Method resulted in a 4.7% reduction in $J_{sc}$, and the Spinner Method in 13.5%. Due to minimal variation in $J_{sc}$ and the simplicity and lower cost, the Weight Method was selected for further refinement.

Transmission curves of glass and epoxy showed a 10% variation in the visible range, also seen in the quantum efficiency (QE). In the Si SC's absorption range (850-1200,nm), QE decreased by 2.5%. Although epoxy and glass alter transmittance and QE in the 400-700,nm range, this won't significantly affect Si SC performance when bonded to a III-V SC, as this range will be absorbed by the III-V layer. The main impact on Si SC performance will be the shading caused by the III-V SC. Testing with the GaAs substrate, about 300,$\mu$m thick, showed an 85% reduction in $J_{sc}$ in the Si SC.

These results enable our group to focus on optimizing the III-V SC for a tandem design. Future research will refine epoxy application to minimize thickness variations and eliminate bubbles. Additionally, exploring anti-reflective coatings and thin-film III-V SC integration will be crucial for maximizing tandem efficiency.

### ACKNOWLEDGMENT

This work was partially supported by the Brazilian funding agencies FINEP, FAPERJ, CAPES and CNPq.

### REFERENCES

[1] Gee, J. M. et al. A 31%-efficient GaAs/silicon mechanically stacked, multijunction concentrator solar cell. Conference Record of the Twentieth IEEE Photovoltaic Specialists Conference, vol. 1, p. 754-758, 1988.

[2] Jain, Nikhil et al. III-V Multijunction Solar Cell Integration with Silicon: Present Status, Challenges and Future Outlook. Energy Harvesting and Systems, vol. 1, no. 3-4, p. 121-145, 2014.Rado and H. Suhl, Eds. New York: Academic, 1963, pp. 271–350.

[3] NREL chart at https://www.nrel.gov/pv/cell-efficiency.html

[4] M. A. Green, E. D. Dunlop, J. Hohl-Ebinger, M. Yoshita, N. Kopidakis, X. Hao, Prog. Photovoltaics 2021, 29, 657

[5] P. Schygulla, R. Müller, D. Lackner, O. Höhn, H. Hauser, B. Bläsi, F.Predan, J. Benick, M. Hermle, S. W. Glunz, F. Dimroth, Prog. Photovol.:Res. Appl. 2021, 30, 869.

[6] Green MA, Dunlop ED, Yoshita M, et al. Solar cell efficiency tables (version 62). Prog Photovolt Res Appl. 2023; 31(7): 651-663. doi:10.1002/pip.3726

[7] P. Zhang, C. Li, M. He, Z. Liu, X. Hao, The Intermediate Connection of Subcells in Si-based Tandem Solar Cells. Small Methods 2024, 8, 2300432. https://doi.org/10.1002/smtd.202300432

[8] S. Essig, S. Ward, M. A. Steiner, D. J. Friedman, J. F. Geisz, P. Stradins, D. L. Young, Energy Procedia 2015, 77, 464.

[9] S. Essig, M. A. Steiner, C. Allebé, J. F. Geisz, B. Paviet-Salomon, S.Ward, A. Descoeudres, V. LaSalvia, L.Barraud, N. Badel, A. Faes, J.Levrat, M. Despeisse, C. Ballif, P. Stradins, D. L. Young, IEEE J Photo volt 2016, 6, 1012

[10] S. Essig, C. Allebé, T. Remo, J. F. Geisz, M. A. Steiner, K. Horowitz,L. Barraud, J. S. Ward, M. Schnabel, A. Descoeudres, D. L. Young, M.Woodhouse, M. Despeisse, C. Ballif, A. Tamboli, Nat. Energy 2017, 2,17144

# Impact of Ionizing Radiation on the Behavior of Pseudo-Resistors with Temperature-Dependent Analysis

Antonio Aurélio de Sousa Gomes
*Centro Universitário FEI*
São Bernardo do Campo, Brazil
antonio.aurelio06@gmail.com

Cleiton Felix Pereira
*Centro Universitário FEI*
São Bernardo do Campo, Brazil
cleitonfp@fei.edu.br

Marcilei A. Guazzelli
*Centro Universitário FEI*
São Bernardo do Campo, Brazil
marcilei@fei.edu.br

Alexis C. Vilas Boas
*Centro Universitário FEI*
São Bernardo do Campo, Brazil
alexiscvb@gmail.com

Ricardo Germano Stolf
*Centro Universitário FEI*
São Bernardo do Campo, Brazil
rstolf@fei.edu.br

Renato Camargo Giacomini
*Centro Universitário FEI*
São Bernardo do Campo, Brazil
renato@fei.edu.br

*Abstract*—Microelectronics has made significant progress in recent decades. As a result of these advances, various devices derived from MOSFETs have emerged, including the pseudo-resistor. This device is an MOS transistor with a specific connection between its terminals. As such, it behaves like a MOS transistor for a specific voltage range, like a bipolar transistor in another range, and like a resistor in a narrow range of tens of millivolts. When another MOS transistor is placed in a "back-to-back" configuration with the first, symmetrical behavior is achieved across the full voltage range. In this work, the behavior of the pseudo-resistor was analyzed concerning its operating temperature variation and exposure to ionizing radiation, evaluating the phenomenon of total ionizing dose (TID). It was found that the device exhibits different resistance values at different temperatures. For exposure to ionizing radiation such as X-rays, its equivalent resistance did not vary significantly, regardless of the total accumulated dose. Even after the irradiation ended, the resistance dependence remained constant.

Keywords — Pseudo-resistor, resistance, temperature, ionizing radiation.

## I. INTRODUCTION

The brief technological progress in microelectronics primarily occurred due to a significant milestone in the 1970s when the study of silicon oxidation processes led to the creation of field-effect transistors (FETs). Since then, the need to understand oxides' mechanisms, silicon and oxides' interfaces, and other semiconductor compounds has emerged. Concurrently, research on the reliability and the effects of temperature and radiation on electronic devices has also progressed. These studies require a basic understanding of semiconductor physics and electronic structures and a solid knowledge of material properties. The effects of temperature tend to alter the electrical parameters related to free carriers in the semiconductor, causing momentary variations due to a specific temperature. On the other hand, the effects of radiation on these devices can modify the trapped charges at the silicon-oxide interfaces of electronic components, resulting in degradation and temporary or permanent functional failures. This has been a significant concern for engineers designing and manufacturing integrated circuits (ICs) [1], [2], [3].

In this study, the pseudo-resistor was chosen to investigate its behavior under the influence of temperature variations and exposure to ionizing radiation. This electronic device, known

as a pseudo-resistor, was introduced in 1994 by Delbruck and Mead as an "adaptive element" [4]. Its initial application occurred in a photoreceptor circuit with a wide dynamic reception range. Fig.1 shows the connection diagram of this adaptive element [4], [5], [6], [7], [8].

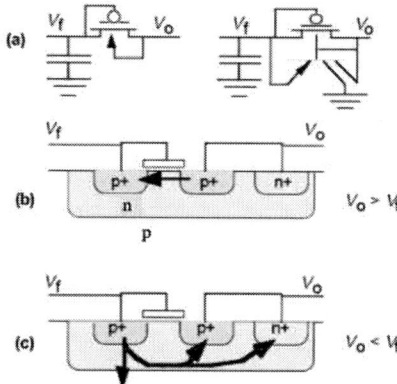

Fig.1. Adaptive element (pseudo-resistor) [5].

From the representation in Fig. 1, we can see that the principle of a pseudo-resistor is a MOS transistor, where the body is connected to the source and the gate to the drain. It can be composed of a PMOS or NMOS, but a PMOS was used for this study [5], [6], [7], [8].

In the connection diagram illustrated above, in part (a), a capacitor operates together to maintain the adaptation state. $V_0$ represents the output voltage and $V_f$ denotes the voltage across the capacitor. In part (b), when forward bias occurs ($V_0 > V_f$), the MOS diode operates as if it were on, that is, there is a current flowing between drain and source, while the bipolar diode remains off. In the case of reverse bias, in part (c), the MOS diode turns off and the bipolar diode starts to operate, with these currents appear between the P and N doping, mainly between the body and the drain. The term "adaptive" is justified by the variable resistance of this element, which changes according to the applied voltage level. This characteristic can generate variations of up to two decades in the signals detected by the Delbruck and Mead photoreceptor circuit [5], [6], [7], [8].

The adaptive element behaves similarly to a resistor when the voltage applied to it is very close to 0V, where a reverse current appears between the drain and the body of the transistor. At this point, it has a monotonic I-V (current-voltage) relationship. This results in extremely low voltage and current, and very high resistance [5], [8].

As mentioned, the behavior of the pseudo-resistor differs for positive and negative voltages. Often, it is necessary to achieve current symmetry between the two directions of voltage polarization. To solve this problem, they are connected in series, but in opposite directions or "mirrored." Fig.2 shows this configuration, known as "back-to-back" [5], [6], [7], [8].

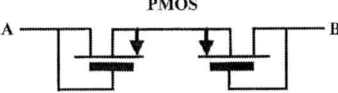

Fig.2 Pseudo-resistors in back-to-back configuration [5].

## II. METHODOLOGY

### A. Pseudo-resistor ($R_P$) characterization

The experimental characterization circuit was implemented using TSMC 180 technology. Given the number of terminals in the circuits and the need to reduce noise levels, the integrated circuit (IC) was encapsulated, and tests were conducted on a dedicated board for powering and biasing the device. Due to the high equivalent resistance values of the pseudo-resistor, which can reach the order of tera-ohms, direct measurements of resistance or current were not feasible, and even the air resistance between contacts could interfere with the measurements [5], [6], [7], [8].

To address such an issue, a circuit was developed to characterize the pseudo-resistor indirectly. Fig.3 shows the circuit characterizing the pseudo-resistor and its implementation layout. Characterization relies on the time constant $R_PC$ (the product of the resistance of the pseudo-resistor $R_P$ and the capacitance C). Knowing the value of capacitance C, one can calculate the equivalent resistance $R_P$ through the $R_PC$ time constant by applying a voltage step at the circuit's input in a time interval ($\Delta t$), which in this case was 1ms. Transistors M1 and M2 were designed as source follower amplifiers (buffers), because it presents a high input impedance and low impedance allowing isolation of the characterizing $R_PC$ circuit [8], [9], [10].

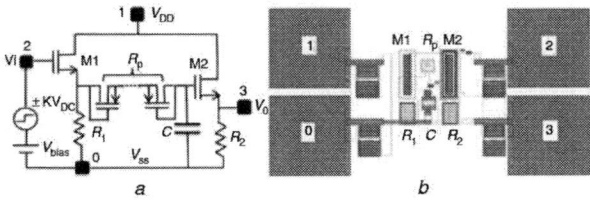

Fig.3 Pseudo-resistor ($R_P$) characterization circuit [9].

a. Pseudo-resistor evaluation circuit

b. Implemented layout

Table I shows the technology parameters presented in Fig3.

TABLE I.    CHARACTERISTIC PARAMETERS

| Transistor | $\lambda$ | $\gamma$ | k' | $V_A'$ | $V_{TO}$ |
|---|---|---|---|---|---|
| NMOS | 0.09μm | 0.580V$^{0.5}$ | 223μA/V$^2$ | 15V/μm | 0.373V |
| PMOS | 0.09μm | -0.576V$^{0.5}$ | 94.3μA/V$^2$ | 10V/μm | -0.395V |

To verify the $R_P$ of this device, were considered, including the bias voltage ($V_{bias}$) with 1.25V, supply voltage ($V_{DD}$) with 2.5V, input voltage over time, being a square waveform ($V_{STEP}$) with 200mV$_{pp}$, and input signal frequency (f) being of 500mHz [9].

The analysis parameters for calculating the equivalent resistance value of the pseudo-resistor are based on the method of transient response of the charge and discharge of a capacitor, RC circuit (1) [10].

$$V_{C(t)} = V_\infty - (V_\infty - V_0^+). e^{-\frac{t}{R.C}} \quad (1)$$

In this context, we must consider a time variation ($\Delta t$) during this transient period and measure the voltage at the initial moment ($V_{Ci}$) of this interval and at the final moment ($V_{Ci+1}$), which is the primary region of interest for analyzing resistance behavior, where it is understood that the resistance has a constant value. Another important data is the value of maximum voltage level ($V_{STEP}$). Making the substitution of the parameters mentioned in (1) will result in (2) [9], [10], [11].

$$V_{C(i+1)}(t) = V_{STEP} - (V_{STEP} - V_{Ci}(t)). e^{-\frac{\Delta t}{R_p.C}} \quad (2)$$

Knowing the other parameters and without knowing the value of $R_P$, given that C has already been analyzed and found to have a value of 10 pF. So, we can determine the equivalent resistance of the device using (2) rearranged into (3) [9], [10], [11].

$$R_{pi} = \frac{\Delta t}{C.ln\left(\frac{V_{STEP}-V_{Ci}(t)}{V_{STEP}-V_{C(i+1)}(t)}\right)} \quad for\ i = (0, 1, 2, 3, ... n-1) \quad (3)$$

### B. Behavioral analysis of the pseudo-resistor considering temperature variation

To collect data on the equivalent resistance of the device considering temperature variation, the device was placed in a thermal chamber with temperature control in tenths of a degree Celsius. Once the desired temperature was reached, a waiting period of 30 minutes was observed to ensure that the internal temperature was uniform at all points in the chamber. Data on the resistance of the pseudo-resistor were collected at the following temperatures: 30, 40, 50 and 60°C [12].

### C. Behavioral analysis of the pseudo-resistor considering exposure to ionizing radiation

The A Shimadzu XRD-6100 diffractometer was used as a 10keV effective energy X-ray source, incident frontally on the device. The interaction of this ionizing radiation, primarily with the oxide layers and oxide/Si interface of the basic devices comprising the pseudo-resistor, induces alterations in its functionality. The effects of total ionizing dose (TID) due to X-rays are cumulative and thus depend on the radiation exposure time [13].

The X-ray emission equipment allows only the analysis of Total Ionizing Dose (TID). The TID effect occurs when radiation passes through an oxide layer, generating electron-hole pairs that may or may not recombine. If they do not recombine, the device terminals will collect the electron due to its high mobility. In contrast, due to its low mobility, the hole will likely become trapped in a material's defect. Besides being trapped in oxide defect traps, these holes can also induce states at the silicon-oxide interface. Given that the system is primarily composed of PMOS-type transistors, current carriers in the channel will always be repelled by the resulting electric field at the gate, leading to an increase in the threshold voltage. Consequently, the carrier mobility controlling the system's current may change [13].

For the TID experiment, the device was kept at a constant temperature of 21°C and exposed to a dose rate of 100krad/h until reaching an accumulated dose of 22krad. To verify that with this small dose there is already a modification in the device's behavior. The characteristics of the pseudo-resistor were analyzed before, during, and after irradiation.

The Rohde & Schwarz RTO1012 oscilloscope was used to collect the waveforms of the $V_I$ input and $V_0$ output.

### III. DISCUSSION

Based on the information presented thus far, we can analyze the data collected from the temperature variation measurements and exposure to ionizing radiation.

#### A. Resistance behavior considering the effects of temperature

In Fig.4, we can observe the device's output voltage ($V_0$) over time, considering the temperature variation from 30°C to 60°C in intervals of 10°C and the input voltage ($V_I$). The input parameters already were highlighted previously.

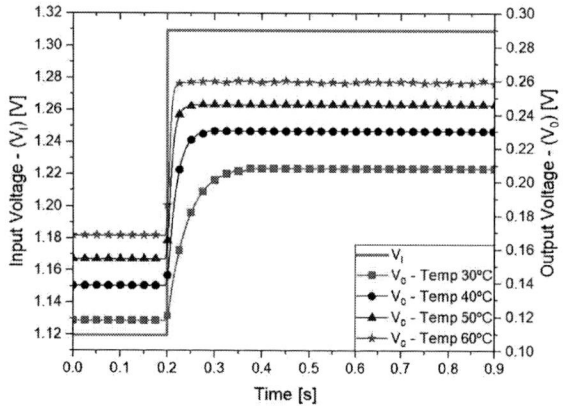

Fig.4 Input and output voltage over time, considering the influence of temperature.

As we can see in Fig.4, just by changing the temperature while keeping other parameters constant, there has already been a considerable shift in the output voltage of the device, as expected, [12]. It is also worth noting that the slope changes significantly, which means a variation in the dynamic behavior. This occurs because the increase in temperature leads to more free carriers within the semiconductor, which results in a variation in the transistor's biasing, reducing the required threshold voltage. As is known, there is also an opposite effect due to mobility reduction, but it is a minor effect for the observed range.

In Fig.5, we can observe the behavior of the equivalent resistance of the pseudo-resistor at the different analyzed temperatures. As the temperature rises, the equivalent resistance decreases.

Consequently, it decreases equivalent resistance, causing the RC time constant to vary with changing temperature. Consequently, the capacitor charges more rapidly at higher temperatures, bringing the output waveform, proportional closer to the input waveform, as evidenced by Fig.5.

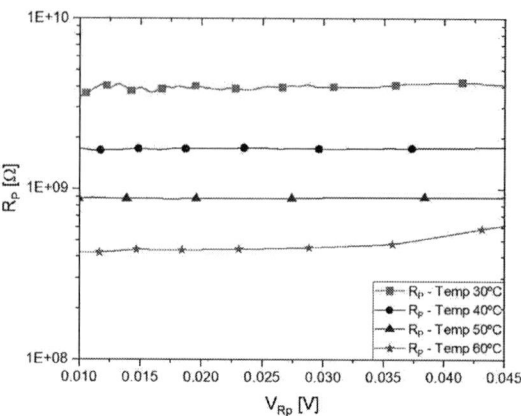

Fig.5 Equivalent resistance of the device as a function of pseudo-resistor voltage with temperature variation.

#### B. Resistance behavior considering exposure to ionizing radiation

In Fig.6, the variation in the output voltage ($V_0$) of the device over time can be observed, considering the total dose accumulated at specific moments.

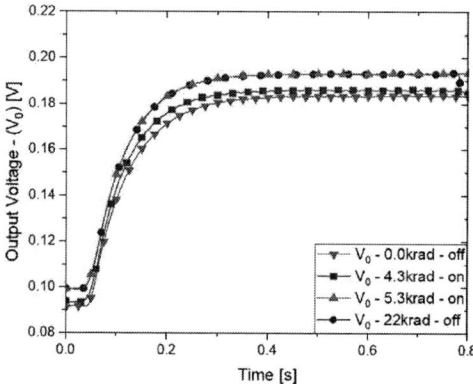

Fig.6 Output voltage over time, considering exposure to X-rays.

The curves shown in Fig.6 represent different exposure times to the effects of ionizing X-ray radiation; a first measurement was made with a diffractometer turned off; that is, there was no exposure to X-rays on the device in this first measurement. After the onset of radiation, a measurement was taken at an accumulated TID of 4.3krad, followed by another measurement at 5.3krad. Lastly, a final measurement was conducted when the device reached a total accumulated TID of 22krad.

The Fig.7 presents the resistance of the pseudo-resistor at different moments when accumulated doses of radiation were applied. A subtle decrease in resistance is observed compared to the non-irradiated system, reflecting the effect of trapped charges in the oxide regions and at the oxide-semiconductor interfaces of the transistors comprising the pseudo-resistor. This statement is valid because, as NMOS transistors are in common drain configuration, their gain is approximately unity and even if there is a change in the threshold voltage, it will not affect the resistance of the pseudo-resistor. SPICE simulations were performed to verify this.

979-8-3315-4064-7/24 $31.00 © 2024 IEEE

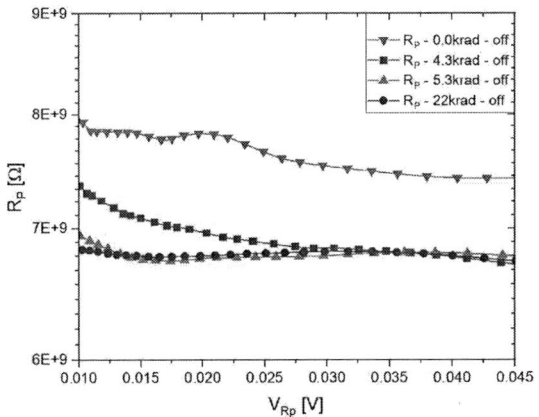

Fig.7 displays the resistance of the pseudo-resistor for the different accumulated TID values.

After the irradiation on the device was completed, 65 hours (65h) were waited before conducting a new analysis with temperature variation and another with 16 days (16d) after TID events, observing only the adopted extreme temperatures, 30°C and 60°C (T30 and T60). In Fig.8, we can observe the device's equivalent resistance behavior at these temperatures before and after the TID events.

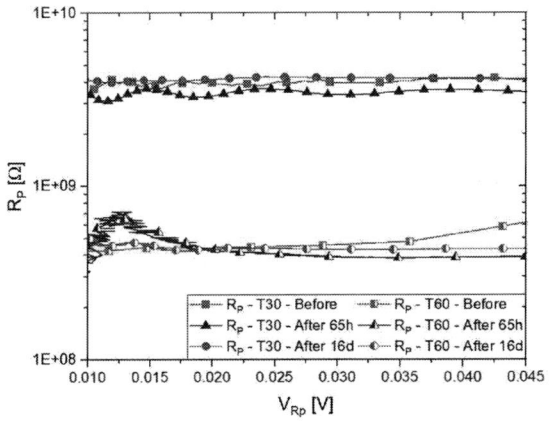

Fig.8 displays the resistance of the pseudo-resistor with temperature variation, before and after TID events.

As we can observe, the pseudo-resistor exhibited a slight reduction in its resistance value following the TID events, in the linear region, as per graph 0V to 45mV, thereby confirming the accumulation of charges at the device interfaces, but after some time and some measures involving temperature variation, the device is tending to return to its original characteristics, as observed.

## IV. CONCLUSIONS

When subjected to adverse situations, the pseudo-resistor exhibits distinct behaviors for these conditions. Considering the temperature variation in which the device is operating, the equivalent resistance of the pseudo-resistor decreases as the temperature increases. As mentioned, a proportional increase in free carriers in the semiconductor can be considered, altering the threshold voltage of the semiconductor itself by lowering it and raising the average voltage level over the pseudo-resistor. With this reduction in resistance, there is a decrease in the RC time constant, shortening the transition time and bringing the output voltage ($V_0$) closer to the input voltage waveform. On the other hand, when irradiated by X-rays, the pseudo-resistor does not exhibit a significant variation in its resistance value compared to the total dose accumulated by the device. Even after the device is no longer exposed to this type of ionizing radiation, this lack of variation is due to the accumulation of charges at the oxide-silicon interface.

Therefore, when using a pseudo-resistor, it is important to thoroughly understand the temperature variation in which it will operate, as this will directly impact its resistance value. Regarding the total accumulated radiation dose, the resistance value will decrease even after the irradiation has ceased, but after a few days of a temperature variation it tends to return to its original characteristics.

## REFERENCES

[1] F. G. H. Leite, "METODOLOGIA DE TESTES DE TOLERÂNCIA À RADIAÇÃO IONIZANTE EM COMPONENTES E SISTEMAS ELETRÔNICOS," São Bernardo do Campo.

[2] A. H. Johnston, *Reliability and radiation effects in compound semiconductors*. World Scientific, 2010.

[3] W. F. Brinkman, D. E. Haggan, and W. W. Troutman, "A history of the invention of the transistor and where it will lead us," *IEEE J Solid-State Circuits*, vol. 32, no. 12, pp. 1858–1865, 1997.

[4] T. Delbruck and C. A. Mead, "Adaptive photoreceptor with wide dynamic range," in *Proceedings of IEEE International Symposium on Circuits and Systems-ISCAS'94*, IEEE, 1994, pp. 339–342.

[5] C. F. Pereira, "PROJETO, IMPLEMENTAÇÃO E MODELAGEM COMPACTA DE TRANSISTORES MOSFET NA CONFIGURAÇÃO PSEUDORRESISTOR PARA CIRCUITOS AMPLIFICADORES DE BIOSINAIS," Centro Universitário FEI, São Bernardo do Campo, 2020, pp. 34–49.

[6] C. F. Pereira, P. L. Benko, M. Galeti, J. C. Lucchi, and R. C. Giacomini, "Experimental study and modeling of pseudo-resistor's non-linearity," in *2017 32nd Symposium on Microelectronics Technology and Devices (SBMicro)*, IEEE, 2017, pp. 1–4.

[7] C. F. Pereira, P. L. Benko, J. C. Lucchi, and R. C. Giacomini, "Transitory recovery time of bio-potential amplifiers that include CMOS pseudo-resistors," in *2014 International Caribbean Conference on Devices, Circuits and Systems (ICCDCS)*, IEEE, 2014, pp. 1–5.

[8] C. F. Pereira, P. L. Benko, J. C. Lucchi, and R. C. Giacomini, "Teraohm pseudo-resistor experimental characterization aiming at implementation of bio-amplifiers," in *2016 31st Symposium on Microelectronics Technology and Devices (SBMicro)*, IEEE, 2016, pp. 1–4.

[9] P. L. Benko, M. Galeti, C. F. Pereira, J. C. Lucchi, and R. Giacomini, "Innovative approach for electrical characterisation of pseudo-resistors," *Electron Lett*, vol. 52, no. 25, pp. 2031–2032, Dec. 2016, doi: 10.1049/el.2016.3390.

[10] P. L. Benko, M. Galeti, C. F. Pereira, J. C. Lucchi, and R. C. Giacomini, "Bio-Amplifier based on MOS bipolar Pseudo-Resistors: A New Approach using its non-linear characteristic," *Journal of Integrated Circuits and Systems*, vol. 11, no. 2, pp. 132–139, 2016.

[11] G. N. Pereira, "CARACTERIZAÇÃO PRÁTICA DO DISPOSITIVO MOS DENOMINADO 'PSEUDO-RESISTOR', NAS TECNOLOGIAS BICMOS 0,13MM GLOBAL FOUNDRIES E TSMC 0,18MM," São Bernardo do Campo, 2019.

[12] C. F. Pereira *et al.*, "Modeling and experimental evaluation of pseudo-resistor's temperature dependence," *Semicond Sci Technol*, vol. 36, no. 1, pp. 1–7, 2020.

[13] A. A. de S. Gomes, "ANÁLISE DA OPERAÇÃO DE FOTODIODOS PIN MULTIFINGER COMO DETECTORES DE RADIAÇÃO X," Centro Universitário FEI, São Bernardo do Campo, 2020, pp. 13–15.

# Smaller Bond Pad for Device Reliability

Ng Hong Seng
*Quality, Reliability Qualification*
*X-FAB Sarawak*
Kuching, Malaysia
hongseng.ng@xfab.com

Lee Kuan Fang
*Quality, Reliability Qualification*
*X-FAB Sarawak*
Kuching, Malaysia
kuanfang.lee@xfab.com

Eddie Chaim Tau Tat
*Quality, Reliability Qualification*
*X-FAB Sarawak*
Kuching, Malaysia
eddie.chiam@xfab.com

Florinna Sim
*Quality, Reliability Qualification*
*X-FAB Sarawak*
Kuching, Malaysia
florinna.sim@xfab.com

Jerald Sim Mong Joo
*Quality, Reliability Qualification*
*X-FAB Sarawak*
Kuching, Malaysia
jerald.sim@xfab.com

Deborah Debbie Anak Philip
*Quality, Reliability Qualification*
*X-FAB Sarawak*
Kuching, Malaysia
deborahdebbie.philip@xfab.com

*Abstract*—The reliability of smaller scribe line & bond pads has been evaluated in this work to improve cost-efficiency from lesser wafer space consumption. Both fully automatic probers with probe cards and semi-auto probers with micro positioners have been analyzed. Two cantilever types of probe cards have been assessed. Various reliability test method [10,11] with Wafer Level Reliability have been assessed for transistor, capacitor, and metal lines. The wafer level probing was conducted on probe chuck at room temperature, 150°C and 175°C. The bondability with smaller bond pads was collaborated with two external assembly houses using ceramic dual-line package. Wire Bond Shear [12] and Wire Bond Pull [13] tests have been carried out to check the reliability of bonding. Package Level Reliability Electromigration and Time Dependent Dielectric Breakdown tests have been executed at 220°C and 175°C to confirm the reliability impact from smaller bond pads.

*Keywords— Semiconductor, Reliability, Small Bond Pad*

## I. INTRODUCTION

Scribe lines have been created for wafer dicing and process monitoring purposes [1,2,3]. The test vehicles of Wafer Acceptance Test (WAT) known also as Process Control Monitor (PCM) [4] or E-Test or parametric test are placed at scribe lines. WAT is referred to the electrical measurement to confirm the performance at semiconductor fabrication before subsequent processes such as wafer sort test and dicing. The test parameters are normally of active devices (threshold voltage, I-V characteristics, series resistance etc.), passive devices (capacitances, breakdown, sheet resistance). Besides, Wafer Level Reliability (WLR), parameters are examined as well either fast WLR (fWLR) [5] on production PCM testers or laboratory lifetime prediction test.

The mechanical contact between the probe needle and aluminium bond pad is important to ensure good accuracy of electrical measurement. This is challenging to straight away reducing the bond pad sizes. In this project, the scribe line has been reduced from 80µm to 60µm to maximize the wafer utilizable area. Since the scribe line width was reduced, the bond pads size must be decreased too to about 36 x 36µm. The bond pads of ~36 x 36µm, denoted as SMALL for test structures have been evaluated compared with original ~50 x

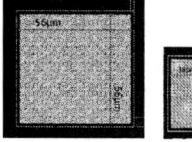

Fig.1 Graphic design system, gds layout of bond / probe pads
(left: REF ; right: SMALL)

50µm size, denoted as REF in standard 80um scribe lines (Fig.1). Prior work has been studied too on 50 x 50µm bond pads [6].

## II. EXPERIMENT

There are four parts of experiment in this project, part 1 for WLR, part 2, 3 for Packaged Level Reliability (PLR) and part 4 for fWLR. The selection of reliability test methods were based on international JEDEC guideline [7,8] the test temperature and either WLR or PLR test equipment. All other test methods can be leveraged on one or two selected test methods. This is because the impact of mechanical contact due to probing and wire bonding performance is approximate as long as temperature and mechanical force is constant.

### A. Part 1:Wafer Level Probing and Reliability Verification

The purpose is to study the performance of reliability test execution in term of probing at wafer level, WLR for fully automatic prober using probe cards and semi-auto prober (Cascade Summit 12K) using micro positioners and probe pins. Two cantilever types of probe cards are from Celadon (VC20E-16.1) [9,10] and STAr Technologies (Virgo Pro II-UT-SP) [11] have been assessed.

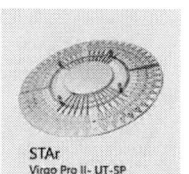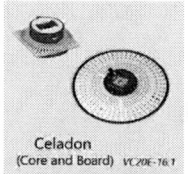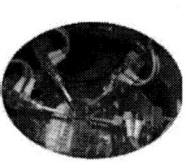

Fig.2 Photographs of STAr (left), Celadon (middle) probe cards and positioner (right)

Hot carrier injection (HCI), negative bias instability (NBTI) tests were measured on transistors, gate oxide reliability including voltage ramp oxide breakdown (VBD) and current stress charge-to-breakdown (QBD) on capacitor test structures and Standard Wafer Level Electromigration Accelerated Test (SWEAT) on Kelvin contact resistor.

HCI tests were measured to verify the impact of micro positioners probing performance on smaller bond pad. This was conducted on semi-auto prober at room temperature. SWEAT tests were performed to corroborate with data at 150°C of probe chuck of semi-auto prober. QBD distribution was plotted to compare probe cards from two suppliers at room temperature, whereas VBD on both room temperature and 175°C. The author used NBTI to evaluate both semi-auto and fully auto probers at 175°C.

979-8-3315-4064-7/24 $31.00 © 2024 IEEE

## B. Part 2:Wire Bonding Verification

The bondability with smaller bond pads was collaborated with two external assembly houses using ceramic dual-line package (CDIP). Both gold and aluminium wires, with few different diameters have been evaluated with ball and wedge bonding respectively. Table I tabulated the assembly houses, wire material, diameter, and wire type. Wire Bond Shear [12] and Wire Bond Pull [13] tests have been carried out to check the reliability of bonding according to international procedure.

TABLE I. WIRE TYPE USED TO EVALUATE BONDABILITY

| Assembly House | Material | Wire Diameter (mil) | Wire Type |
|---|---|---|---|
| AH 1 | Au | 0.7 | AW-14 |
| AH 1 | Au | 1.0 | AW-66 |
| AH 1 | Au | 0.7 | FP2 |
| AH 1 | Al | 0.7 | ALW29S |
| AH 1 | Al | 1.0 | ALW29S |
| AH 2 | Au | 0.6 | FP2 |
| AH 2 | Al | 0.6 | ALW29S |

## C. Part 3:Packaged Level Reliability Verification

This is to enable the study of the impact with smaller bond pad on reliability performance. PLR Electromigration (EM) and Time Dependent Dielectric Breakdown (TDDB) tests have been executed at 220°C and 175°C respectively to verify any differences from smaller pads on elevated temperature. Fig. 3 shows the example of CDIP used for PLR EM reliability test assessment.

## D. Part 4:Massive Production fWLR Verification

Fig.3 Example of CDIP 16 pins for PLR EM tests

The production PCM systems- fully automatic probers, TEL P8XL, and Keithley testers were used to measure fWLR using Celadon probe card. This is to study the impact with smaller bond pad on reliability performance on production level with massive probing with fWLR monitoring.

## III. RESULT AND DISCUSSION

### A. Part 1:Wafer Level Probing and Reliability Verification

1.8V N-type MOSFET has been measured for WLR HCI tests on semi-auto prober using micro positioners and probe pins at room temperature, 27°C. Fig. 4 shows the lifetime is comparable for the parameters Vt (threshold voltage), Idlin (Linear Drain Current), Idsat (Saturation Drain Current) and Gm (Maximum Transconductance). Example of parameter, Idlin is showed in Fig. 5. The experiment has verified insignificant impact of WLR probing using probe pins on reliability at room temperature. This should be caused by the sufficient contact of probe pins on bond pads using 19μm probe tips and good skill of engineer executing the HCI tests.

Semi-auto prober using micro positioners and probe pins were evaluated as well at 150°C by SWEAT test. The experiment has verified insignificant impact of WLR probing using probe pins on reliability at 150°C depicted in Fig.6. The wafer was soaked at 150°C for one hour before positioning the probe tips of 19μm to ensure good contact on ~36 x 36μm bond pads size.

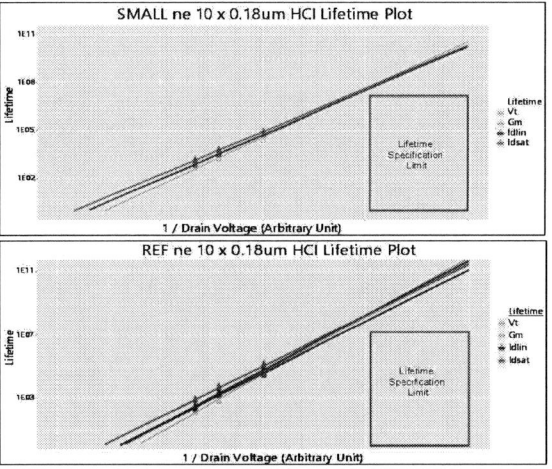

Fig.4 HCI Lifetime for ne 10x0.18

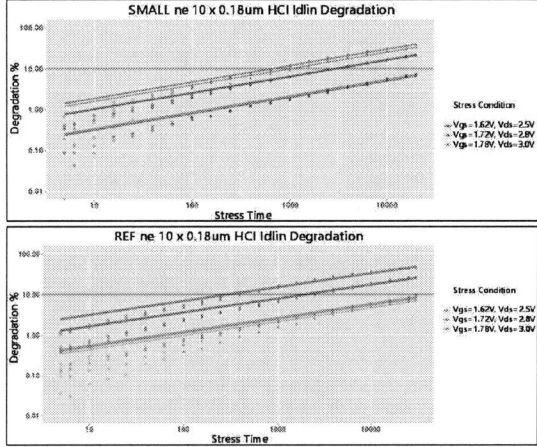

Fig.5 HCI Idlin Degradation for ne 10x0.18

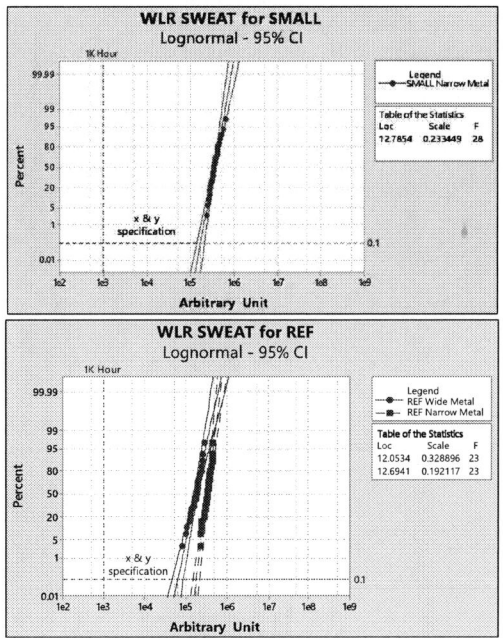

Fig.6 SWEAT Lognormal Plot for SMALL (top) and REF (bottom)

Two cantilever types of probe cards have been assessed in fully automatic prober at room temperature by QBD test. Fig. 7 shows that both Celadon and STAr probe cards contributed insignificant difference of QBD distribution with both negative and positive bias at 27°C. The evaluation has been extended to 175°C with VBD tests. The works have verified insignificant impact of WLR probing using both Celadon and STAr probe cards at both 27°C and 175°C. This is caused by the optimized overdrive and probe marks using of average of 13.2um probe tips at both temperature setting. In [9] Armendariz and team have analyzed the contact resistance and leakage current with related to scrub mark and overdrive.

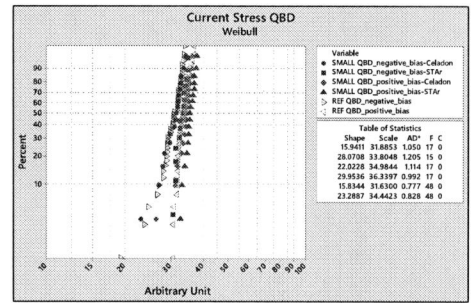

Fig.7 Current Stress QBD Comparison for 5V N-type capacitor

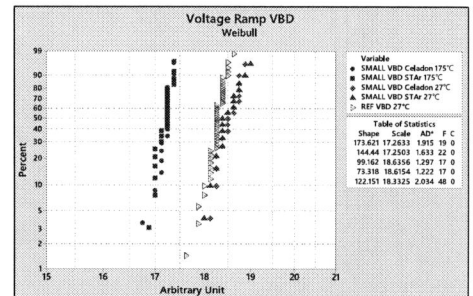

Fig.8 Voltage Ramp VBD Comparison for 5V N-type capacitor

The authors have also conducted NBTI for both semi-auto prober using micro positioners and Celadon probe card at fully auto-prober at 175°C. Result showed in Fig.9 also confirmed that insignificant difference by either semi-auto prober micro positioners pins or auto prober Celadon probe card pins. This is believed to be contributed by prober chuck leveling and planarity check and tuning if deviation during yearly equipment preventive maintenance for semi-auto and auto prober besides good contact mentioned above.

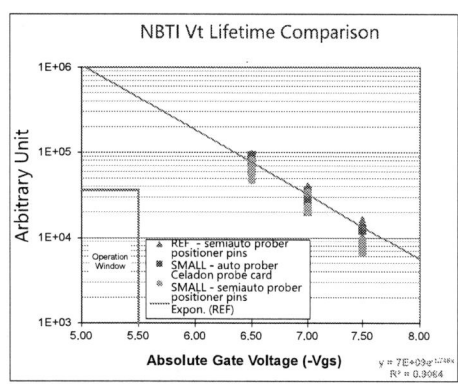

Fig.9 NBTI Lifetime of pe5 10x 0.5um

## B. Part 2: Wire Bonding Verification

Table II tabulated the result in word to elaborate the results. Fig.10 showed the example of secondary electron microscopy (SEM) images of bonding on small bond pad for gold and aluminium bonds. The condition was visualized after EM test in Fig.10b for Au 0.6 FP2 bonded by assembly house, AH2. WBS and WBP tests have been performed on the small bond pads bonded by both external packaging houses on some 0.6 and 0.7 mil wires, before and after EM tests. All passed WBP except AH1 Au 0.7 mil and Al 0.6 mil due to insufficient bonding, also AH2 Al 0.6 mil due to EM happened on bonding wires. AH1 Au 0.7 mil failed WBS for both REF and SMALL after EM test indicating weak bonding. Improvement requested to AH1 assembly house after PLR EM result for next evaluation for TDDB. Please refer to Part C for the EM and TDDB results.

TABLE II.    WIRE BOND SHEAR AND WIRE BOND PUL TEST RESULTS

| Assembly House | Wire | Test | Before EM stress | After EM stress |
|---|---|---|---|---|
| AH 1 | Au 0.7 AW-14 | WBS | Passed | Failed |
| * Failed due to low shear reading. This might indicate degradation of bonding. Lifted ball reported during failure analysis. | | | | |
| AH 1 | Au 0.7 AW-14 | WBP | Passed | Failed |
| * Failed due to low pull reading. Lifted ball reported during failure analysis. | | | | |
| AH 1 | Al 0.6 ALW29S | WBP | Marginal | Failed |
| * Lifted wedge bond might indicates insufficient bonding. | | | | |
| AH 2 | Au 0.6 FP2 | WBS | Passed | Passed |
| * Wire bond quality at bond pad (first bond) is acceptable. | | | | |
| AH 2 | Au 0.6 FP2 | WBP | Marginal | Marginal |
| * Marginal WBP result, wire break at second bond with marginal value. This indicating weak wire bond at lead finger. | | | | |
| AH 2 | Al 0.6 ALW29S | WBP | Passed | Failed |
| * Wire break at lead finger after EM stress. This may indicate degradation of wire bond due to EM stress condition. | | | | |

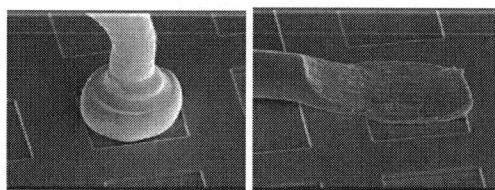

Fig.10a SEM of bond for 0.7-mil Au AW-14 (AH 1) on the left and bond for 0.7-mil Al ALW29S (AH 1) on the right

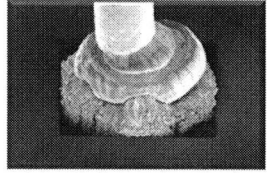

Fig.10b SEM of bond for 0.6-mil Au FP2 (AH 2) after EM test

## C. Part 3: Packaged Level Reliability Verification

Fig.11 plotted SMALL and REF bond pads with 0.6, 0.7 and 1.0 mil wires. 1.0 mil aluminium on REF bond pads is baseline reference. 0.6 mil from AH2 for both aluminium and gold wires showed comparable EM performance with baseline REF. 0.7 mil AH1 for both REF and SMALL bond pads showed slightly higher tail. This is also tallied with Table II WBS result after EM test. Further work by AH1 on 0.7 mil gold has been studied and improvement completed. TDDB at 175°C has been conducted to verify the result. Fig.12 and Fig.13 showed the comparable result.

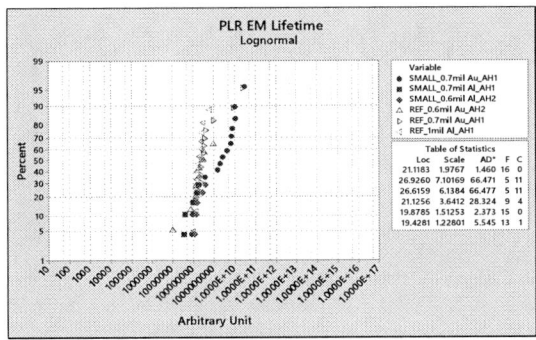

Fig.11  EM lifetime lognormal plot (tested in 220°C oven)

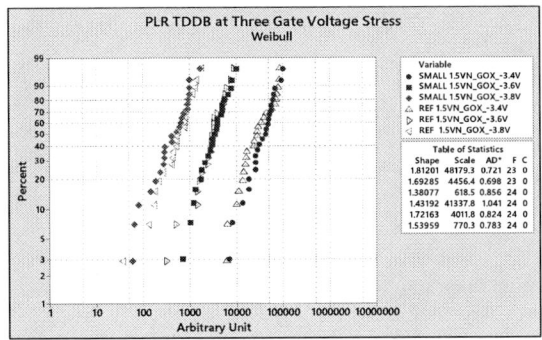

Fig.12  TDDB Weibull Plot to compare three stress condition for both SMALL and REF  (tested in 175°C oven)

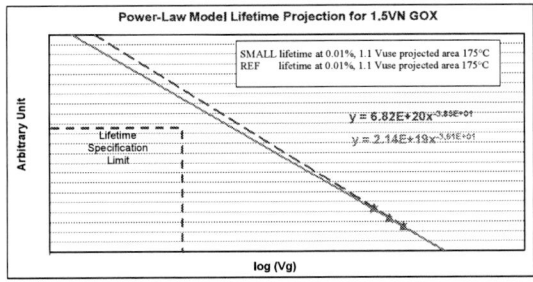

Fig.13  TDDB lifetime projection comparison for SMALL and REF

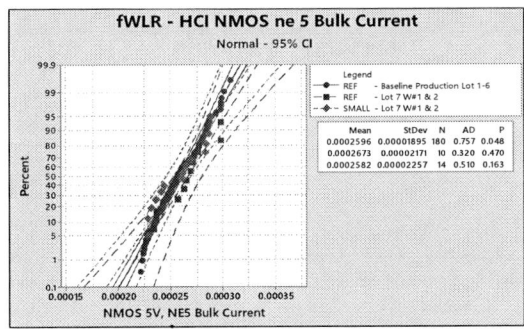

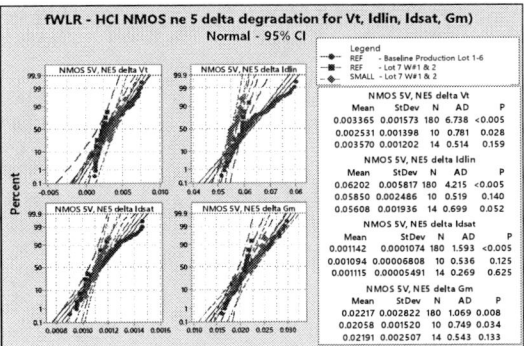

Fig.14  fWLR Comparison for HCI Bulk Current (top) and parameters degradation (bottom)

*D.  Part 4:Massive Production fWLR Verification*

Fig.14 shows some examples of fWLR parameters, such as bulk current and HCI degradation of Vt, Idlin, Idsat and Gm for ne5 (5V NMOS 10x 0.5um), comparison of small bond pads to baseline reference. With proper probing setup by well-trained production technician, the small bond pad showed within the production baseline. This is caused by the optimized overdrive and probe marks at room temperature. Regular probe marks checking procedure is implemented for production PCM.

## IV.  CONCLUSION

In conclusion, insignificant reliability impact on bondability on bond pads of less than 40μm x 40μm by two external packaging houses. Further smaller bond pads can be evaluated in future. With proper probing setup, negligible discrepancy of level 1 reliability performance from fully automatic probers with cantilever-typed probe cards. Vertical probe card will be evaluated in the next phase. With skilled technician, semi-auto probers with micro positioners showed similar performance, same go to production fWLR.

## ACKNOWLEDGMENT

The authors would like to thank the support from both Reliability colleagues and PCM group for fWLR measurement.

## REFERENCES

[1]  Fujii, US Patent  6,462,401 B2 (2002).

[2]  Wu, US Patent  8,039,367 B2 (2011).

[3]  Yoshida et al., US Patent  8,067,819 B2 (2011).

[4]  B. Choo, T. Riley, B. Schulz, and B. Singh, "Automated Process Control Monitor for 0.18-um Technology and Beyond," Proceedings of the SPIE The International Society for Optical Engineering, Vol. 3998, pp. 218-226, 2000.

[5]  A. Martin and R. Vollertsen, "An introduction to fast wafer level reliability monitoring for integrated circuit mass production," Microelectronics Reliability, Vol. 44, Issue 8, pp. 1209-1231, 2004

[6]  Y.K. Choong and J. Jaquette, "New Probing Technology Evaluation - Fine Pitch and Small Pads," Semiconductor Wafer Test Conference, 2000.

[7]  JEDEC Publication. JEP001-1A. Foundry Process Qualification Guidelines – Backend of Line. JEDEC 2018.

[8]  JEDEC Publication. JEP001-2A. Foundry Process Qualification Guidelines – Front End Transistor Level. JEDEC 2018.

[9]  K. Armendariz, J. Reynolds, G. Ttrpak, G. Bottoms and F. Huang, "Production Parametric Test – Challenges and Surprising Outcomes Running in a High-Volume Manufacturing Environment," Semiconductor Wafer Test Conference, 2015.

[10]  Celadon Systems, "VC20E™ Series Probe Card," [Online]. Available https://www.celadonsystems.com/products/[Accessed: Apr. 12, 2024].

[11]  STAr Technologies Inc., "Virgo-PRO II Keysight 4070/4080 Probe Card," [Online]. Available https://www.star-quest.com/english/info/title/Virgo-PROIIKeysight4070-4080ProbeCard . [Accessed: Apr. 12, 2024].

[12]  JEDEC Standard. JESD22-B116B. Wire Bond Shear Test Method. JEDEC 2017.

[13]  Department of Defense Test Method Standard MIL-STD-883L. Test Method Standard Mechanical Test Methods for Microcircuits, Method 2011.10 Bond Strength (Destructive Bond Pull Test), 2019

# Micromachined passive waveguide fabrication with fs laser in Ag-doped GeO$_2$–PbO glasses for photonics: straight, curved and Y shaped configurations

Thiago Vecchi Fernandes
Departamento de Engenharia de
Sistemas Eletrônicos
Escola Politécnica da USP
São Paulo, Brazil
thiago.vecchi.f@hotmail.com

Camila D. S. Bordon
Departamento de Engenharia de
Sistemas Eletrônicos
Escola Politécnica da USP
São Paulo, Brazil
camiladvieira@gmail.com

Niklaus U. Wetter
Centro de Lasers e Aplicações, Instituto
de Pesquisas Energéticas e Nucleares
IPEN-CNEN
São Paulo, Brazil
nuwetter.ipen@gmail.com

Wagner de Rossi
Centro de Lasers e Aplicações, Instituto
de Pesquisas Energéticas e Nucleares
IPEN-CNEN
São Paulo, Brazil
wderossi@gmail.com

Luciana R. P. Kassab
Departamento de Ensino Geral
Faculdade de Tecnologia de São Paulo
São Paulo, Brazil
kassablm@osite.com.br

ABSTRACT — This study aims to produce and characterize different dual waveguides using femtosecond (fs) laser irradiation on GeO$_2$-based glass samples. The work is motivated by previous results obtained with rare earth ions doped GeO$_2$ − PbO glass, with and without silver nanoparticles, in which irradiation, with fs laser was successful. The work aims to manufacture different structures such as straight, curved, and Y waveguides (using the double guide configuration) for applications in photonics (resonant rings, beam splitters, among others) in GeO$_2$ − PbO glasses with silver nanoparticles. For both, straight and S curved waveguides, better M$^2$ (beam quality factor) results were found for a distance between the guide walls of 10 µm, when compared to 25 µm. Moreover, among the two different curved guides produced it was also possible to observe better guidance when a larger radius of curvature (20 mm) was used; preliminary tests showed no guiding for 5 mm and 10 mm radius. The highest relative propagation loss was obtained for the S curved waveguide with a 25 µm distance between the guide walls whereas the lowest one was found for the Y shaped waveguide; for this configuration (opening angle of 5° and distance of 620 µm between the two arms) an output power ratio between the left and right arm of 53.9/46.1 showed promising applications for beam splitters.

**Keywords** — germanate glasses, femtosecond laser, double guides, curved guides, Y guides.

## I. INTRODUCTION

GeO$_2$-PbO glasses feature a low phonon energy (800 ~ 975 cm$^{-1}$), wide transmission window (400 ~ 4500 nm), high refractive index (2.0) associated with the high atomic mass of the elements and high polarizability, good chemical, thermal and mechanical stabilities, low melting point compared to silicate glasses, low glass transition temperature and good solubility of rare earth ions (TRs) [1, 2]. GeO$_2$-PbO glasses doped with TRs demonstrated enhanced optical properties due to the plasmonic effects of metallic NPs [3]. Moreover, these glasses with metallic

NPs exhibited significant potential for photonic applications due to their ultrafast response times and high third-order nonlinearities [4]. The first results of dual-line waveguides written by femtosecond (fs) laser and produced directly in glasses based on GeO$_2$ and TeO$_2$ were reported in [5]. More recently, we showed a new configuration of dual-line waveguides, produced in Nd$^{3+}$ doped GeO$_2$-PbO glass containing silver (Ag) nanoparticles (NPs), for applications in optical amplifiers at 1064 nm [6,7]. Inspired by these promising results, which showed improved beam quality factor and optical gain, we present for the first time the production of dual-line waveguides with diverse configurations such as, straight, curved [8, 9], and Y-shaped [10, 11], by using GeO$_2$-PbO glasses with Ag NPs. These guides were directly inscribed via femtosecond laser onto the glasses, for applications in photonic devices, such as resonant rings [12], high gain amplifiers [8], beam splitters [13], among others. We utilized a Ti:Sapphire femtosecond (fs) laser operating at 800 nm, which delivered 30 fs pulses at a 10 kHz repetition rate. We present results of beam quality factor M$^2$ (at 632 and 1064 nm), propagation loss and polarization. The present result covers a lack in the literature as there are few reports of waveguides inscribed in glasses with Ag NPs by fs laser [10]. Another contribution of the present investigation is the use of glasses for curved and Y waveguide configurations as mainly of the reports are produced in crystalline hosts.

## II. MATERIALS AND METHODS

The glass manufacturing process involved the original composition (in wt%) 40GeO$_2$-60PbO (labeled GP) with the addition of 2.0 wt% AgNO$_3$. The samples were prepared using the melt quenching technique. The process involved the melting of the reagents in high purity alumina crucible (99.999%) at 1200 °C, using a glass rod for homogenization. Subsequently, the molten material was poured into preheated brass molds, followed by annealing at 420 °C for 1 hour. This procedure is crucial to reduce internal tensions and provide greater resistance to the samples. After annealing, the samples were cooled until they reached room temperature inside the oven. Then, the samples were subjected to a final polishing process [6].

Subsequently, double waveguides were created, consisting of a pair of parallel lines each of which is produced by superimposing four laser-irradiated tracks. These tracks were manufactured using a fs laser system (Ti:sapphire, model PRO 400, Femtolasers GmbH) and a focusing lens with a focal length of 20 mm, N.A = 0.23, with the focal point positioned 0.75 mm below the surface. The laser beam was perpendicular to the polished surface of the sample, with the linear polarization inclined at 45° to the direction of movement, at a speed of 0.5 mm/s and pulse energy of 30 µJ. In this work Ag NPs are produced during the annealing; however, we cannot discard their concentration growth during fs laser irradiation, as already reported in the literature, for tungsten lead–pyrophosphate glass [6, 14]. The two straight waveguides included two parallel straight lines each, separated by distances of 10 and 25 µm, curved waveguides with a radius of curvature of 20 mm and distances of 10 and 25 µm; a Y-shaped waveguide was also produced, with an opening angle of 5° and distance of 620 µm between the arms, maintaining a distance of 10 µm between the guide walls. Curved guides with radius of 5 and 10 mm were produced but demonstrated no guiding. Figure 1, from left to right, displays two curved guides, followed by two straight guides, and the Y-shaped guide.

The structures that form the walls of the double waveguides are formed by regions with a refractive index lower than that of the original glass, and have approximate dimensions of 178 mm x 3 mm [6]. Due to these dimensions and the low contrast with the surrounding medium, these regions are very difficult to observe, and can only be seen with polarized light microscopy, and under special conditions. Therefore, any characterization of the morphology of the walls, such as their roughness, becomes an extremely difficult task and was not done in this work. The laser beam used has a Gaussian distribution, and the pulse overlap rate was quite high, which could indicate a great homogeneity in the morphology of the walls.

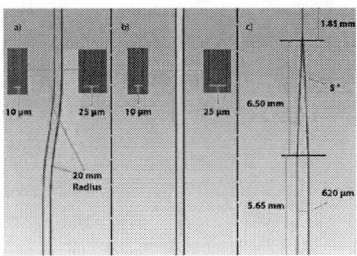

Fig. 1. Illustration of the waveguides configurations irradiated with fs laser in GeO₂PbO glass a) curved b) straight c) Y-shaped.

Using the setup shown in Fig. 2 (a), the propagation losses for all waveguide configurations were measured at 632 nm, using the straight waveguide with 10 µm separation as reference [15]; the output power distribution of the straight waveguide with 25 µm width, the two curved waveguides (with 10 and 25 µm separation between the walls) and also the two arms of the Y waveguide was obtained. Then with equation 1, it was possible to measure the relative propagation losses [15]:

$$\eta = -10 \, \log_{10}\left(\frac{\Sigma_i P_i}{P_0}\right) \qquad (1)$$

In the equation above $P_i$ is the sum of the output powers of the guide that will be compared, and $P_0$ represents the output power of the 10 µm waveguide used as a reference. This procedure is suitable as it avoids the use of the cutback method [15-17] but not adequate for samples that cannot be cut, as is the case of those with curved and Y shaped configurations. Finally, $\eta$ represents the value of the relative losses in dB.

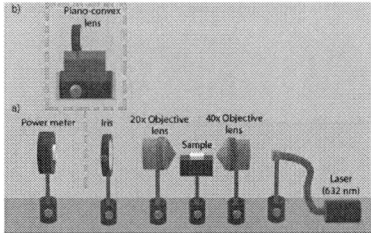

Fig. 2. Experimental setup used to measure a) propagation loss; b) M².

To determine the quality of the waveguide output beam, the M² was measured for all waveguide configurations, using standard procedures [18] and the setup illustrated in Fig. 2 (b) and replacing the power meter by a CCD camera. Using equation 2 it was possible to determine M² by measuring the beam diameter along the near and far fields, obtained for 632 nm [17]. Then the experimental values were adjusted by equation 2 where d = 2w is the diameter of the waveguide output beam (2w) measured at different focal distances (z). The diameter of the beam at the focus ($z_0$) is given by d=2 $w_0$.

$$d = d_0 \sqrt{1 + \left(\frac{(M^2 \, \lambda \, (z - z_0))}{(\pi \, d_0{}^2)}\right)^2} \qquad (2)$$

Equation 3 is used to calculate M² at 1064 nm using the values obtained with the setup of Fig. 2 (b) since $\lambda \ll w$

$$M^2 = \frac{\pi \cdot \theta_{ideal} \cdot w_{0_{ideal}}}{\lambda} \qquad (3)$$

In the above equation, $\theta_{ideal}$ is the beam divergence semi-angle measured in the far field and $w_{0_{ideal}}$ is the waist radius at the beam focus. Finally, polarization was measured using the experimental arrangement presented in Figure 3. The acquired data made it feasible to calculate a percentage relationship between the power output values of each polarization axis. These results were of importance for the assessment of polarization during the guidance process and its corresponding orientation.

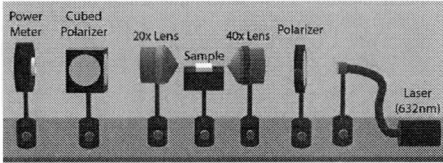

Fig. 3. Experimental setup used for measurements of polarization of dual waveguides.

Optical microscopy was performed to obtain top images of all the waveguides.

## III. RESULTS

Table 1 summarizes the values obtained for relative propagation loss (equation 1 and setup of fig. 2a) of each configuration, compared to the straight waveguide of 10 μm, which served as the reference.

TABLE 1 - PROPAGATION LOSSES CONCERNING A STRAIGHT WAVEGUIDE WITH A DISTANCE BETWEEN THE WAVEGUIDE WALLS OF 10 μm.

| Waveguide | Propagation losses (dB/cm) |
|---|---|
| Straight 25 μm | 0.74 |
| S curved 10 μm | 1.33 |
| S curved 25 μm | 2.19 |
| Y waveguide | 0.55 |

By using the arrangement of Fig. 2 (b) it was possible to measure the beam waist (d=2w) as a function of the z position. The results of the $M^2$ factor at 632 nm obtained by equation 2, for the horizontal and vertical axes, respectively, are presented in Figure 4 (for straight waveguide configuration with 10 μm distance between the walls).

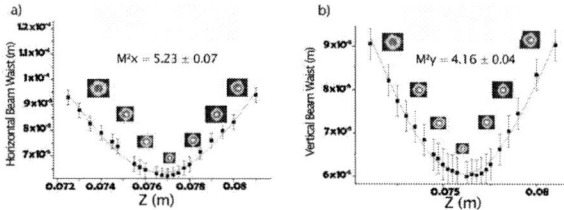

Fig. 4. GP sample results for the straight waveguide with a distance between the guide walls of 10μm a) M²x b) M²y.

The results for all waveguide configurations can be seen in Table 2.

TABLE 2- $M^2_X$ AND $M^2_Y$ VALUES AT 632 nm FOR ALL WAVEGUIDES.

| Waveguide | M²x (632 nm) | M²y (632 nm) |
|---|---|---|
| S curved 10 μm | 15.2 | 10.64 |
| S curved 25 μm | 23.5 | 28.8 |
| Straight 10 μm | 5.23 | 4.16 |
| Straight 25 μm | 6.30 | 5.31 |
| Y first arm | 5.03 | 6.99 |
| Y second arm | 5.84 | 5.46 |

Using equation 3 it was possible to obtain the results of the $M^2$ at 1064 nm, which are presented in Table 3.

TABLE 3 - $M^2_X$ AND $M^2_Y$ VALUES AT 1064 nm FOR ALL WAVEGUIDES.

| Waveguide | M²x (1064 nm) | M²y (1064 nm) |
|---|---|---|
| S curved 10 μm | 9.02 | 6.32 |
| S curved 25 μm | 13.95 | 17.09 |
| Straight 10 μm | 3.11 | 2.47 |
| Straight 25 μm | 3.74 | 3.15 |
| Y first arm | 2.99 | 4.15 |
| Y second arm | 3.46 | 3.24 |

Using the setup of Fig 2b) it was determined that the output power ratio between the left and right arms of the Y waveguide was 53.9/46.1, indicating potential photonic applications, such as beam splitters. The polarization results can be seen in Tables 4 and 5, with the input vertical and horizontal polarizations respectively. The polarization measurement was performed for the S curved and straight configurations (with the set up presented in fig. 3) with the smaller distance between the guide walls (10 μm) that presented the smaller losses. Only the S-curved (10 μm) showed horizontal polarization (12%), as presented in Table 4. Regarding the results presented in Table 5, vertical polarizations were observed for the straight waveguide

(13%) and for both Y arms (right and left) with polarizations of 10%. The polarization measurement was performed for the S curved and straight configurations (with the setup presented in Fig. 3) with distance between the guide walls of 10 μm that presented smaller losses with respect to those with 25 μm.

TABLE 4 - OUTPUT POLARIZATION PERCENTAGE WHEN INPUT LASER POLARIZATION IS VERTICAL.

| Waveguide | Vertical polarization (%) | Horizontal polarization (%) |
|---|---|---|
| S curved 10 μm | 87.9 ± 0.7 | 12.1 ± 0.7 |
| Straight 10 μm | 98 ± 2 | 2.3 ± 2.2 |
| Y first arm | 98 ± 3 | 2.3 ± 2.8 |
| Y second arm | 97 ± 2 | 3.1 ± 2.1 |

TABLE 5 - OUTPUT POLARIZATION PERCENTAGE WHEN INPUT LASER POLARIZATION IS HORIZONTAL.

| Waveguide | Vertical polarization (%) | Horizontal polarization (%) |
|---|---|---|
| S curved 10 μm | 6.7 ± 5.9 | 93 ± 6 |
| Straight 10 μm | 13 ± 2 | 87 ± 2 |
| Y first arm | 10 ± 1 | 90 ± 1 |
| Y second arm | 10 ± 4 | 90 ± 4 |

The results obtained by optical microscopy (top images) can be seen in Fig. 5a and Fig 5b for the straight waveguide with 10 and 25 μm distance between the guide walls. The results for the two curved waveguides, separated by 10 and 25 μm are presented in Fig. 6a and 6b, respectively. The corresponding output modes are also shown.

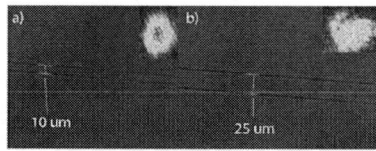

Fig. 5. Top images and the corresponding output modes of the straight waveguide with a distance between the guide walls of a) 10 μm b) 25 μm.

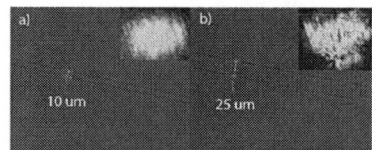

Fig. 6. Top images and the corresponding output modes of the curved waveguides with a distance between the guide walls of a) 10 μm b) 25 μm.

Fig. 7 presents the microscopy results of the Y waveguide and the simultaneous view of the output modes of the two arms.

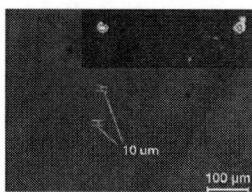

Fig. 7. Top images of the two arms of the Y waveguide and the corresponding output modes.

## IV. CONCLUSION

In short, with the analysis of the data, it was possible to observe a better result in the straight waveguide configuration with respect to propagation losses and beam quality. Other than that, the distance between the guide walls resulted in higher output powers when the distance was smaller (10 μm) for both straight and curved waveguides. Also, for both, straight and S curved waveguides, better $M^2$ results were found for a distance between the guide walls of 10 μm, when compared with 25 μm. Moreover, for curved guides it was also possible to observe better guidance when they had a larger radius of curvature, since in the initial tests with 5 mm and 10 mm radius, there was no light guidance. The highest propagation loss was obtained for the S curved guide with a 25 μm distance between the guide walls. We highlight the low propagation losses in both arms, for the Y configuration. Finally, the configuration of a Y waveguide showed promising results for applications in photonics, such as beam splitters due to the output power ratio between the left and right arms being close to 50%. In this configuration, an opening angle of 5° and 620 μm was used between the two arms. The present results show for the first time the possibility of producing passive waveguides with curved and Y configurations, using $GeO_2$-PbO glass with Ag NPs, and the configuration of double waveguides each formed by 4 superpositions. Most of the results reported in the literature for curved and Y waveguide configurations are mainly produced in crystalline hosts. The present investigation covers a lack in the literature related to curved and Y-shaped waveguides irradiated by fs laser in glasses and can be extended to different glassy hosts with Ag NPs.

## ACKNOWLEDGMENTS

This study was financed in part by the Coordenação de Aperfeiçoamento de Pessoal de Nível Superior – Brasil (CAPES) – Finance Code 001, and CAPES-PROEX 88887.670647/2022-00. We acknowledge Conselho Nacional de Desenvolvimento Científico e Tecnológico—Grants 465.763/2014 (Instituto Nacional de Ciência e Tecnologia de Fotônica), 305745/2023-9, 308526/2021-0 and also Fundação de Amparo à Pesquisa do Estado de São Paulo—Grants: 2019/06334-4, 2017/10765-5, Sisfóton-Grant 440.228/2021-2.

## REFERENCES

[1] M. Wachtler, et al. "Phonon Sidebands and Vibrational Properties of $Eu^{3+}$ Doped Lead Germanate Glasses." Journal of Non-Crystalline Solids, vol. 217, no. 1, pp. 111–114, 1997.

[2] M. Wachtler, et al. "Optical Properties of Rare-Earth Ions in Lead Germanate Glasses." Journal of the American Ceramic Society, vol. 81, no. 8, pp. 2045–2052, 2005.

[3] C.B. De Araújo and L. R. P. Kassab. "Enhanced Photoluminescence and Planar Waveguide of Rare-Earth Doped Germanium Oxide Glasses with Metallic Nanoparticles." Glass Nanocomposites, pp. 131–144, 2016.

[4] L. De Boni, et al. "Femtosecond Third-Order Nonlinear Spectra of Lead-Germanium Oxide Glasses Containing Silver Nanoparticles." Optics Express, vol. 20, no. 6, pp. 6844–6844, 2012.

[5] D.S. Da Silva, et al. "Production and Characterization of Femtosecond Laser-Written Double Line Waveguides in Heavy Metal Oxide Glasses." Optical Materials, vol. 75, pp. 267–273, 2018.

[6] C.D.S Bordon, et al. "Effect of Silver Nanoparticles on the Optical Properties of Double Line Waveguides Written by Fs Laser in $Nd^{3+}$ -Doped $GeO_2$-PbO Glasses." Nanomaterials, vol. 13, no. 4, pp. 743–743, 2023.

[7] C.D.S. Bordon, et al. "A New Double-Line Waveguide Architecture for Photonic Applications Using Fs Laser Writing in $Nd^{3+}$ Doped $GeO_2$-PbO Glasses." Optical Materials, vol. 129, pp. 112495–95, 2022.

[8] L. Li, et al. "Femtosecond-Laser-Written S-Curved Waveguide in Nd:YAP Crystal: Fabrication and Multi-Gigahertz Lasing." Journal of Lightwave Technology, vol. 38, no. 24, pp. 6845–6852, 2020.

[9] H.-D. Nguyen, et al. "Heuristic Modelling of Laser Written Mid-Infrared $LiNbO_3$ Stressed-Cladding Waveguides". Vol. 24, no. 7, pp. 7777–7791, 2016.

[10] A. Abou Khalil, et al. "Direct Laser Writing of a New Type of Waveguides in Silver Containing Glasses." Scientific Reports, vol. 7, no. 1, 2017.

[11] V. A. Amorim, et al. "Optimization of Broadband Y-Junction Splitters in Fused Silica by Femtosecond Laser Writing." IEEE Photonics Technology Letters, vol. 29, no. 7, pp. 619–622, 2017.

[12] A. Yalcin, et al. "Optical Sensing of Biomolecules Using Microring Resonators." IEEE Journal of Selected Topics in Quantum Electronics, vol. 12, no. 1, pp. 148–155, 2006.

[13] R. Dangel, and W. Lukosz. "Electro-Nanomechanically Actuated Integrated-Optical Interferometric Intensity Modulators and 2×2 Space Switches". Vol. 156, no. 1-3, pp. 63–76, 1998.

[14] J.M.P Almeida, et al. "Metallic Nanoparticles Grown in the Core of Femtosecond Laser Micromachined Waveguides." Journal of Applied Physics, vol. 115, no. 19, pp. 193507, 2014.

[15] J.R.V. De Aldana, et al. "Femtosecond Laser Direct Inscription of 3D Photonic Devices in Er/Yb-Doped Oxyfluoride Nano-Glass Ceramics." Optical Materials Express, vol. 10, no. 10, pp. 2695, 2020.

[16] D.L. Yang, et al. "Radiative Transitions and Optical Gains in $Er^{3+}$/$Yb^{3+}$ Codoped Acid-Resistant Ion Exchanged Germanate Glass Channel Waveguides." Journal of the Optical Society of America B-Optical Physics, vol. 26, no. 2, pp. 357–357, 2009.

[17] E.C. Sousa. "Otimização Da Eficiência Do Modo TEM00 Em Lasers de Nd:YLF de Alta Potência Bombeados Lateralmente". IPEN, São Paulo, vol. 89, pp. 22 – 24, 2008.

[18] D. Feise, et al. "High-Brightness 635nm Tapered Diode Lasers with Optimized Index Guiding." Proc. SPIE 7583, High-Power Diode Laser Technology and Applications VIII, vol. 75830. SPIE, 2010

# Determining Neutron-based static cross-section of a SRAM-based FPGA in a simplified setup

Julia Willow Benvenutti[1], Fábio Benevenuti[1], Lívia Streit[2], Fernanda Kastensmidt[1]

[1]Programa de Pós-Graduação em Microeletrônica (PGMICRO) - Instituto de Informática - Universidade Federal do Rio Grande do Sul.
[2]Programa de Pós-Graduação em Química (PPGQ) - Instituto de Química - Universidade Federal do Rio Grande do Sul

jwasbenvenutti@inf.ufrgs.br, fbenevenuti@inf.ufrgs.br, livia.streit@ufrgs.br, fglima@inf.ufrgs.br

*Abstract—This work presents a simplified setup composed of americium and beryllium sources that was capable of generating a neutron flux to measure a static cross-section of a SRAM-based FPGA in the Chemical Department of UFRGS. The sources are stored in a special container with an opening on top, where the board containing the SRAM-based FPGAs is placed. Data monitoring and collection occurred every three days, using a computer connected to the board. The experiment found single and multiple bit-flips, including some multiple bit-flips events whose behavior was not fully explained, requiring a more in-depth analysis. However, the results obtained were sufficient to demonstrate that a simple setup is capable of showing single, double, triple, and even quadruple events and results are comparable with related works published in the literature.*

*Keywords— Neutrons, bit-flip, upset*

## I. INTRODUCTION

FPGAs (Field-Programmable Gate Arrays) are well-known programmable devices widely used to implement many complex digital circuits. FPGAs programmed by SRAM (Static Random Access Memory) are a common category of FPGA, where the configuration of the device's logical elements is stored in SRAM memory cells, which are is a type of semiconductor memory that stores data in a static form, meaning it retains its contents as long as power is supplied to the device. It is commonly used in digital electronics for its fast access times and low power consumption compared to other types of memory like DRAM (Dynamic Random Access Memory).

The reason FPGAs need to be radiation tolerant is because radiation can interfere with the operation of electronic components, leading to temporary glitches or permanent damage. Radiation can cause various types of errors in electronic components, including single event upsets (SEUs) [1]. These are temporary errors where radiation-induced charge particles disrupt the state of a memory cell or logic element, causing incorrect data or transient faults. Although radiation causes both transient and permanent failures, for the present work only transient failures were studied, which are SEUs (also known as bit-flips). This phenomenon refers to a situation where a digital bit, typically representing a 0 or a 1 in binary representation, changes its value unintentionally due to external factors such as radiation or electrical noise.

Even on the Earth's surface, shielded by the atmosphere and urban structures, there exists the presence of neutrons and forms of radiation. Background radiation, as it is called, though small, cannot be ignored in studies involving integrated circuits. Devices such as cell phones, computers, and other household appliances can be affected by the passage of neutrons through the circuits present in these technologies.

This work aims to demonstrate how a simplified setup

can detect SEUs, allowing for a more feasible and less costly approach than alternatives commonly used in laboratories. Nevertheless, the results obtained were adequate to demonstrate that this simplified setup can detect single, double, triple, and even quadruple events, with outcomes comparable to those reported in relevant literature. This paper is organized to firstly provide an overview of the radiation interaction with integrated circuits and to present the methodology used to obtain the cross-sectional area. Finally, the obtained results are also presented, along with a comparison with existing literature.

## II. NEUTRON EFFECTS IN SRAM-BASED FPGAs

One of the causes of errors and failures in integrated circuits is radiation, especially when it comes to devices used in aerospace technologies. These interactions can be divided into four different types: single event effects, total ionizing dose, changes in spacecraft surface charging, and displacement damage dose. The present work aims to study the single event effects on a board.

Single Event Effects can be divided into two types: destructive or non-destructive. Among the non-destructive ones are SEU, SET, and MCU/MBU [2]. Among the destructive ones are SEL, SEGR, SEB, and SHE. The difference between these events is how radiation passes through the board causing inversions in the information. When only one inversion occurs in the same memory cell, it is called SEU. When there are more than one in the same memory array, it is called MCU [2]. Tests with neutrons pose similar challenges to alpha particle tests. Robust equipment, such as particle accelerators, is needed for tests with higher energy. However, results involving Single Event Effects (SEEs) with neutrons often yield positive outcomes [4]. Despite having lower energy, neutrons are capable of producing the same errors and failures as alpha particles.

This type of error can appear in more than one bit within the same device, being referred to as a multiple-event. Errors of this nature typically occur in memory cells. Therefore, the development of mechanisms to protect FPGAs and make them radiation-tolerant becomes necessary.

Static cross-section is the measure used to determine the likely collision area of a particle. In the current experiment, neutrons were used as particles, and it was also chosen to divide the cross section by the number of bits in the chip. This procedure ensures greater accuracy in determining the probability of collision. The cross section is a measure that depends on various factors. The technology used and the energy of the neutron influence this determination [5]. Therefore, it is important to emphasize that the cross section found in this article refers to the

28 nm technology and neutrons with 2 MeV of energy.

$$\sigma = \frac{Nerrors}{bits\ \Phi} \quad (1)$$

The calculation of the cross section is done according to equation (1), where $\sigma bit$ is the cross section per bit, Nerrors is the total number of observed errors, bits is the total number of bits in the device, and $\Phi$ is the particle fluence. For better analysis, the division per bit is included in this equation. The calculation of the cross-section is relevant because from it, one can calculate the susceptibility of a device to radiation-induced faults. Susceptibility (S) is calculated by multiplying the cross section ($\sigma bit$) by the total area of the device (A), according to equation (2).

$$S = \sigma bit * A \quad (2)$$

### III. PROPOSED METHODOLOGY

The present study used the setup described in Figure 1. The FPGAs used (a) were positioned above the container of the neutron-emitting source (b). Inside the container was the americium/beryllium source (c), responsible for the continuous flow of neutrons.

A set of two boards was used, details of which will be described later in this article. This set was positioned so that the channel between the americium and beryllium source was perpendicular to the circuits (figure 4). This setup was utilized for approximately three weeks.

Below are detailed the characteristics and functions of each part of the setup used in the present study. Data was collected every two days. The source of neutrons used in the present work was composed of americium (Am) and beryllium (Be). Sources composed of these elements are widely used in tests involving neutron emission, with several studies indicating how to best apply them and how to improve the flow and emission of thermal neutrons [6].

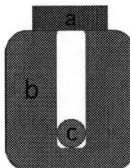

Figure 1: Setup demonstrative schematic. The board (a) on the metal contêiner (b) with the radioactive sample (c).

Figure 2: The neutron generator phenomenon.

The americium/beryllium source predominantly emits fast neutrons. This category of neutrons is preferable for tests involving microelectronic components, given the aforementioned properties [7]. Especially if the source used has good activity, since the emitted particles are in the range of 1- 10 MeV [8]. The americium and beryllium source used in the experiment had a neutron flux of $5.10^2 n.cm^{-2}s^{-1}$, with a total activity of 3000 Bq. The source was stored in a thick-walled metal container, isolating it within the radionuclide room. The only opening was at the top, to allow for a direct path to the board.

Two boards are used in the test: AES-MMP-BB2-G Avnet Mini-Module Plus baseboard 2 (motherboard), and

AES-MMP- 7K325T-G Xilinx Kintex-7 Mini-Module Plus (the design under test). The model of the tested chip is "XC7K325T," a "Kintex-7" FPGA from the 7 Series family of AMD/Xilinx, manufactured using TSMC 28 nm technology (figure 2 and 3). In total, the chip used in the experiment has 11,441,684 bytes, or 91,533,472 bits. The FPGA was placed on top of the metal container containing the source, with the only path to the board being its opening. The board was connected by cables to the notebook that would read the data. The static test lasted a total of 20 days, with data collected weekly. It was not possible to connect the notebook to the network.

### IV. RESULTS AND DISCUSSIONS

The existing literature states that the cross-section is strongly dependent on the nature of the technology in terms of thickness [9]. Therefore, it is important to bear in mind that the technology used in the present study (28 nm) was decisive for the results obtained. It is possible to find in previous articles that the cross-section is also strongly dependent on the energy of the particle. In this article, the energy of the neutrons is approximately 2 MeV.

The main data of the experiment can be found in table 1. After 22 days of static tests, the number of total counts (or total experiment data) was 88826, with one count every 22 seconds. Therefore, the entire experiment lasted around 1954172 seconds approximately.

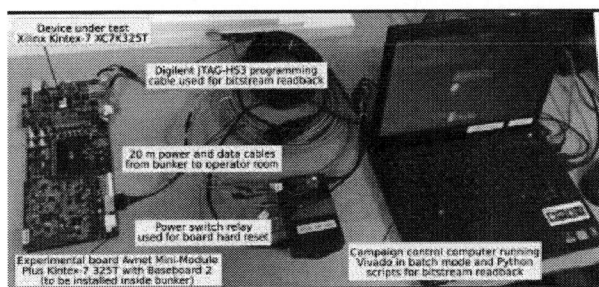

Figure 3: The board and all the equipments utilized in experiment.

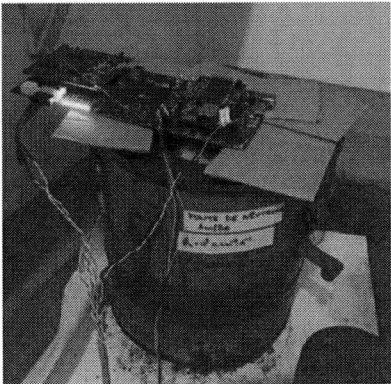

Figure 4: setup used in the experiment

Table 1: General experiment data Where $\varphi$ is the flux and t is the time in seconds.

| Total cycles | Total events | Neutron energy | Flux ($\varphi$) |
|---|---|---|---|
| 88826 | 419 | 2 MeV | $5.10^2$ n.s⁻¹cm² |

Based on the data from Table 1, the fluence of the source can be estimated using the formula:

$$\phi = \varphi t \qquad (3)$$

$$\phi = 5.10^2 \times 1.78.10^6 = 8.9.10^8 \text{ n.cm}^{-2}$$

With the fluence value, it is possible to estimate the cross section with equation (1).

$$\sigma bit = 5,14.10^{-15} \text{ cm}^{-2} \text{ bit}^{-1}$$

The cross section depends on the energy of the particles used in the experiment [10], as well as the nature of the technology employed. Based on these parameters, the result obtained by the current study is in accordance with the findings present in the literature.

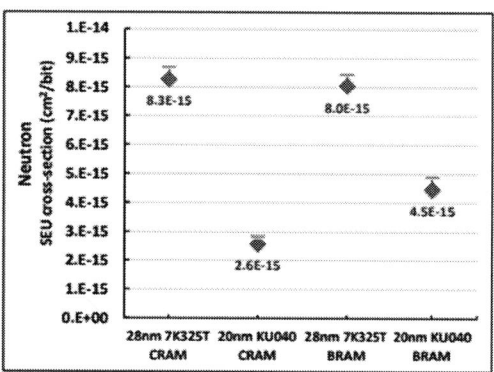

Figure 5: The cross-sectional area for Single Event Upsets (SEUs) in 20nm UltraScale Kintex CRAM and BRAM. Additionally, the SEU cross-sections for control 28nm Kintex-7 are also included for comparison [11].

Table 2: Different single event upset event

| Single | Dual | Triple | Quadruple |
|---|---|---|---|
| 315 | 102 | 1 | 1 |

Table 3: Different types of double events.

| Same column | Same line | Undefined |
|---|---|---|
| 100 | 1 | 1 |

Figure 5, published by P. Maillard et al [11], demonstrates the behavior of the cross-sectional area (y- axis) as a function of different technologies (x-axis). The results in the present study are consistent with those obtained by P. Maillard.

The results obtained by previous studies indicate that as the technology decreases in thickness, there is a decrease in the magnitude of the cross section. They also state that as the energy of the incident neutron decreases, the reduction is even more pronounced. Therefore, the results obtained in this study, in a static manner, are consistent when compared to other tests. According to Figure 5, for the type of technology used the experiment obtained events of four different types, as it is seen in table 2. Events with a bit flip make up the absolute majority, but double, triple and quadruple events were also found.

Table 3 displays the various types of double events that occurred in the experiment. Mostly, the double events were caused in bytes of the same column. This information is obtained by analyzing the numbering of the altered bytes. Since the events occurred at intervals of 100, it can be stated that they occurred in the same column. For events to be in the same row, they must occur in bytes whose difference is equal to 1, according to the board numbering.

The only unspecified event occurred in bytes with a difference greater than 101 but without presenting a plausible route. This type of event occurs when the neutron hits bytes that are adjacent but with numbering that does not follow the same logic. To accurately confirm this fact, a more detailed investigation into the board numbering would be necessary. The triple event exhibited the same peculiar phenomenon. It showed intervals of 100 between the first and second (corresponding to the interval between one row and another), but 3000 between the second and third. The nature of this phenomenon is uncertain for this study. During the experiment, a quadruple event was detected. In other words, in one of the events, a single neutron collided and caused changes in the information of four different bytes. The trajectory is illustrated in figure 5, based on the numbering obtained on the board. Unlike the triple event, this particular event did not exhibit undefined behavior; it was possible to graphically describe its trajectory.

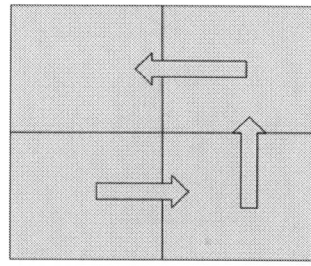

Figure 6: Graphic description for the quadruple event Its possible to calculate the event percentage.

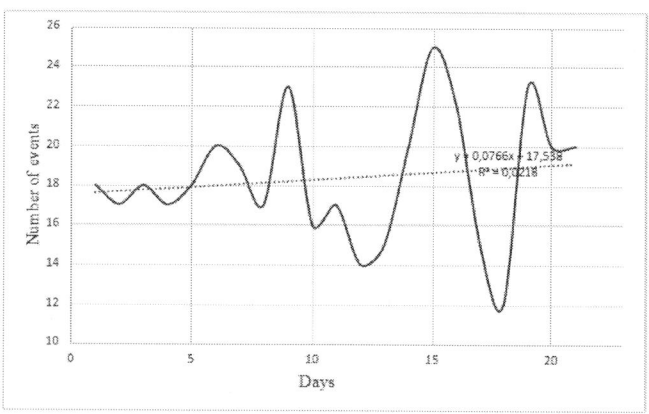

Figure 7: Events for each day.

979-8-3315-4064-7/24 $31.00 © 2024 IEEE

It is possible to calculate the event percentage with (4). Where %E is the percentage of events, E is the number of events that occurred, and Ntotal is the total number of counts.

$$E\% = \frac{E}{N_{total}} * 100 \quad (4)$$

$$\%E = \frac{419}{88826} * 100 = 0,47\%$$

The probability of an event occurring in the assembled setup is 0.47%. In other words, an event occurs every 212 cycles. Each cycle takes around 22 seconds, so there is an event approximately every 4240 seconds. To better visualize the data, it can be converted to minutes, hours, and days, resulting in: 1 event every 70.67 minutes, or every 1.18 hours, or every 0.049 days. This totals an average of approximately 20 events per day, regardless of the nature of the event.

However, this method is not appropriate for calculating this type of event. By having access to the number of events each day in figure 7, it is possible to observe that the average is not the most frequently occurring value, requiring the use of a Poisson distribution [12]. In Table 4, you can find the probabilities of each quantity of events occurring per day, according to the Poisson distribution.

## V. CONCLUSIONS

The present study demonstrates how a simplified setup is capable of detecting events such as bit flips of various natures. In addition to showcasing this capability, the experiment obtained data consistent with those found in the literature for cross-sectional areas. It was also possible to determine, using the Poisson distribution, the probabilities of events occurring each day.

### ACKNOWLEDGMENT

This study was only possible with the collaboration of an interdisciplinary team, involving the Fault Tolerance Team from the Institute of Informatics and the Isotopic Tracers Laboratory from the Institute of Chemistry at the FederalUniversity of Rio Grande do Sul.

Table 4: number of events per day and its probability events

| Events for day | X<16 | X<17 | X<18 | X<19 | X<20 |
|---|---|---|---|---|---|
| Probability (%) | 15.65 | 22.11 | 29.70 | 38.14 | 47.02 |

References

1.  R. Viégas and L. A. Guedes, "Real-time communication over 802.11e wireless network using a publish-subscribe model," 2011 IEEE Third Latin-American Conference on Communications, Belem, Brazil, 2011, pp. 1-6, doi: 10.1109/LatinCOM.2011.6107406.

2.  L'Annunziata, Michael F. Radioactivity: introduction and history, from the quantum to quarks. Elsevier, 2016.

3.  L. M. Luza, A. Bosser, V. Gupta, A. Javanainen, A. Mohammadzadeh and L. Dilillo, "Effects of Heavy Ion

and Proton Irradiation on a SLC NAND Flash Memory," 2019 IEEE International Symposium on Defect and Fault Tolerance in VLSI and Nanotechnology Systems (DFT), Noordwijk, Netherlands, 2019, pp. 1-6, doi: 10.1109/DFT.2019.8875475

4.  J. Tonfat et al., "Analyzing the influence of the angles of incidence on SEU and MBU events induced by low LET heavy ions in a 28-nm SRAM-based FPGA," 2016 16th European Conference on Radiation and Its Effects on Components and Systems (RADECS), Bremen, Germany, 2016, pp. 1-6, doi: 10.1109/RADECS.2016.8093186.

5.  Xilinx Inc., San Jose, CA, "UG116 Device Reliability Report (v10.2.1)", pp.24-25, Mar. 2015

6.  D. Kobayashi, "Scaling Trends of Digital Single-Event Effects: A Survey of SEU and SET Parameters and Comparison With Transistor Performance," in IEEE Transactions on Nuclear Science, vol. 68, no. 2, pp. 124-148, Feb. 2021, doi: 10.1109/TNS.2020.3044659.

7.  Didi, Abdessamad, et al. "Neutron activation analysis: Modelling studies to improve the neutron flux of Americium–Beryllium source." Nuclear Engineering and Technology 49.4 (2017): 787-791

8.  Sogbadji, R. B. M., et al. "The design of a multisource americium–beryllium (Am– Be) neutron irradiation facility using MCNP for the neutronic performance calculation." Applied Radiation and Isotopes 90 (2014): 192-196.

9.  X. Jin et al., "SEU in SRAMs due to 2.5 MeV, 14 MeV and reactor neutrons," 2018 18th European Conference on Radiation and Its Effects on Components and Systems (RADECS), Goteborg, Sweden, 2018, pp. 1-4, doi: 10.1109/RADECS45761.2018.9328717.

10. A. Coronetti et al., "Thermal-to-high-energy neutron SEU characterization of commercial SRAMs," 2021 IEEE Radiation Effects Data Workshop (REDW), Ottawa, ON, Canada, 2021, pp. 1-5, doi: 10.1109/NSREC45046.2021.9679344.

11. P. Maillard, M. Hart, J. Barton, P. Jain and J. Karp, "Neutron, 64 MeV Proton, Thermal Neutron and Alpha Single-Event Upset Characterization of Xilinx 20nm UltraScale Kintex FPGA," 2015 IEEE Radiation Effects Data Workshop (REDW), Boston, MA, USA, 2015, pp. 1-5, doi: 10.1109/REDW.2015.7336723.

12. Gurgel, J. T. A. "Análise estatística da distribuição de Poisson." Anais da Escola Superior de Agricultura Luiz de Queiroz 2 (1945): 299-320.

# Experimental Comparison of Junctionless and Inversion-Mode Nanowire SOI MOSFETs Down to Cryogenics Temperatures

Jefferson Almeida Matos
*Electrical Eng. Department,*
Centro Universitário FEI
São Bernardo do Campo, Brazil
jmatos@fei.edu.br

Flávio Enrico Bergamaschi
CEA-Leti
Université Grenoble Alpes
Grenoble, France

Michelly de Souza
*Electrical Eng. Department,*
Centro Universitário FEI
São Bernardo do Campo, Brazil
michelly@fei.edu.br

Sylvain Barraud
CEA-Leti
Université Grenoble Alpes
Grenoble, France

Mikael Cassé
CEA-Leti
Université Grenoble Alpe
Grenoble, France

Olivier Faynot
CEA-Leti
Université Grenoble Alpe
Grenoble, France

Marcelo Antonio Pavanello
*Electrical Eng. Department,*
Centro Univesitário FEI
São Bernardo do Campo, Brazil
pavanello@fei.edu.br

*Abstract*— **This paper evaluates the electrical characteristics of Inversion-mode (IM) and Junctionless (JNT) SOI nanowire MOSFETs across a temperature spectrum from 300K to 82K. The study examines devices with varying fin widths, utilizing experimental data to analyze crucial electrical metrics, including threshold voltage, inverse subthreshold slope, and carrier mobility throughout the temperature range. The investigation also explores how these parameters are influenced by channel length and fin width.**

*Keywords—Junctionless, SOI nanowire, cryogenic temperature, electrical characteristics.*

The ongoing reduction in MOSFET sizes has been a significant catalyst for advancements in modern electronics, in line with Moore's Law. As transistor dimensions decrease, electronic devices benefit from increased speed and performance, while also achieving reductions in overall dimensions and energy consumption. However, shrinking transistor sizes present new challenges, such as short-channel effects (SCE). Moore's Law, which forecasts exponential growth in transistor density on integrated circuits, remains a crucial guiding principle for the industry but encounters hurdle as the physical limits of miniaturization are approached [1]. To address this issue, nanowire MOSFETs have emerged due to the increased electrostatic control provided by their multi-gate structure [2]. Both devices analyzed in this paper feature a nanowire structure (or triple-gate), with the difference that the Junctionless nanowires transistors (JNTs) have the channel region doped with the same type of dopant as the source and drain regions, whereas the Inversion-Mode (IM) nanowires feature dopants of opposite types in the source and drain regions compared to the channel. Conversely, investigating MOSFET operation at low temperatures has generated significant interest due to its potential to enhance electronic device performance, including reducing the inverse subthreshold slope and mitigating short-channel effects. [3]. Such a temperature range is especially important for applications in medicine and aerospace [4]. Additionally, the growing interest in CryoCMOS for integration into quantum computing systems [5] provides a strong impetus for comprehensive low-temperature characterization.

## I. DEVICE CHARACTERISTICS

Both nanowires were produced at CEA-Leti (France), using Silicon-On-Insulator (SOI) substrate featuring with $t_{BOX}$=145nm a buried oxide thickness of [6]. Fig. 1 depicts a cross-sectional TEM image and the 3D-view representation of a nanowire transistor, showcasing its key geometrical characteristics. The silicon film remains intentionally undoped for IM nanowires and is a heavily doped n-type silicon with a concentration of $N_D$=5 ×$10^{18}$ cm$^{-3}$ for JNT The gate stack is composed of a 28 nm polysilicon layer, a 5 nm TiN metal gate, a 2.3 nm HfSiON layer, and an interfacial SiO$_2$ layer. All devices feature a fin height ($H_{FIN}$) of 9 nm and utilize a multi-fin structure with 10 parallel fins. The study examines devices with a fixed channel length (L) of 100 nm and varying fin widths ($W_{FIN}$) of 10, 15, 20, and 40 nm, as well as short-channel devices with L of 40 nm and $W_{FIN}$ of 10 nm

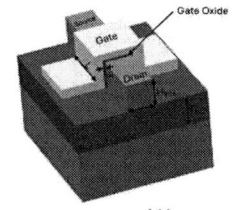 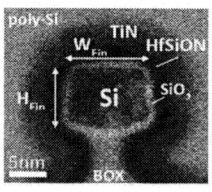

**(A)**　　　　**(B)**

**Fig. 1.** (A) Schematic 3D view of the devices and (B) cross-sectional TEM image [7].

## II. EXPERIMENTAL RESULTS AND DISCUSSION

The drain current ($I_{DS}$) *versus* gate voltage ($V_{GS}$) curves were measured with temperatures for both devices ranging from 82 K to 300 K, utilizing a Low-Temperature Microprobe (LTMP) system from MMR Technologies and a B1500A Semiconductor Parameter Analyzer. The devices were biased with a low drain voltage ($V_{DS}$) of 25mV and 40mV. Figures 2 and 3 show the experimental drain ($I_{DS}$) as a function of the gate voltage ($V_{GS}$) in linear and logarithmic scales, for IM and JNT nanowires biased with drain voltage $V_{DS}$= 25 mV in different temperatures for devices with L=100nm and $W_{FIN}$ = 10 and 40 nm, respectively.

979-8-3315-4064-7/24 $31.00 © 2024 IEEE

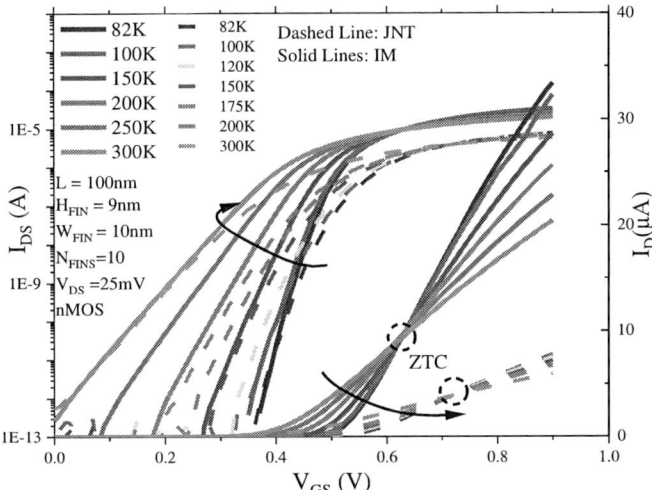

**Fig. 2** - $I_{DS}$ vs $V_{GS}$ curves as a function of temperature for JNT and IM with $W_{FIN} = 10nm$.

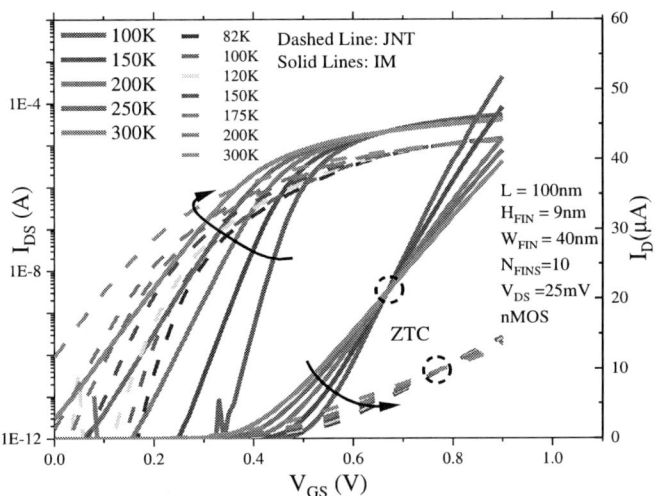

**Fig. 3** - $I_{DS}$ vs $V_{GS}$ curves as a function of temperature for JNT and IM with $W_{FIN} = 40nm$.

## A. Threshold Voltage

Using the $I_{DS}$ *versus* $V_{GS}$ curves with $V_{DS} = 25$ mV, the threshold voltage was extracted for JNT and IM nanowires using the double derivative method [8], investigating the influence of $W_{FIN}$ and L in this electrical parameter. Fig. 4a presents the comparison of the threshold voltage ($V_{TH}$) as a function of temperature between JNT and IM nanowires with L=100nm and variable $W_{FIN}$.

Both devices presented an almost linear reduction of the threshold voltage with the temperature increase. Threshold voltages calculated for JNT nanowires are higher compared with threshold voltage calculated for IM nanowires, regardless of temperature and $W_{FIN}$.

The IM nanowires exhibit a higher variation in threshold voltage with temperature, regardless of $W_{FIN}$. The narrower devices, with $W_{FIN} = 10nm$, showed approximately a 15% difference, whereas the wider devices showed 3%.

The $dV_{TH}/dT$ of these devices can be visualized in Table I, comparing JNT and IM.

TABLE I. $dV_{TH}/dT$ for different $W_{FIN}$, comparing JNT and IM nanowires.

| $dV_{TH}/dT$ of IM and JNT nanowires | | |
|---|---|---|
| $W_{FIN}$ (nm) | JNT | IM |
| 10 | -0.46 mV/K | -0.53 mV/K |
| 15 | -0.53 mV/K | -0.54 mV/K |
| 20 | -0.54 mV/K | -0.54 mV/K |
| 40 | -0.64 mV/K | -0.66 mV/K |

Comparing the $dV_{TH}/dT$ rate between the narrowest and widest devices, an increase of 24% for IM and 39% for JNT is found, suggesting a higher dependence of $W_{FIN}$ in JNT.

To gain some insight into these findings, two analytical models for the threshold voltage were employed: the model proposed by Duarte et al. [9, 10], valid for IM nanowires, and the one proposed by Trevisoli et al. [11], valid for JNT nanowires. These models are presented in Eqn. (1) and Eqn. (5):

$$V_{TH} = V_{FB} - \frac{Q_{d,n}}{C_{g,n}} + 2v_T \ln\left(\frac{N_{Si}}{n_i}\right)$$
$$-v_T \ln\left(\frac{c_{ch,n}}{c_{g,n}}\left(1 - e^{\frac{-Q_{d,n}}{v_T C_{ch,n}}}\right)\right) \quad (1)$$

In this context, $v_T$ is the thermal voltage, $N_{Si}$ indicates the channel doping concentration, q is the electron charge, $n_i$ represents the intrinsic carrier concentration, $V_{FB}$ denotes the flat band voltage, $Q_{d,n}$ is the depletion charge per unit length, $C_{g,n}$ refers to the gate oxide capacitance per unit length, and $C_{ch,n}$ represents the channel capacitance per unit length.

These terms pertain to the geometry and construction of the devices. Equations (2), (3), and (4) detail $Q_{d,n}$, $C_{g,n}$, and $C_{ch,n}$, respectively, for the transistors discussed in this work

$$Q_d = -qN_{si}H_{fin}W_{fin} \quad (2)$$

$$C_g = \frac{3.02 \times \frac{3\varepsilon_{ox}}{2}}{\ln\left(1+\frac{3t_{ox}}{2H_{fin}}\right)} - \frac{5 \times \frac{3\varepsilon_{ox}}{4}}{\ln\left(1+\frac{5t_{ox}}{4H_{fin}}\right)}$$
$$+ \frac{5 \times \frac{3\varepsilon_{ox}}{4}}{\ln\left(1+\frac{5t_{ox}}{4W_{fin}}\right)} \quad (3)$$

$$C_{ch} = W_{fin}\frac{\varepsilon_{si}}{H_{fin}} + 4H_{fin}\frac{\varepsilon_{si}}{W_{fin}} \quad (4)$$

where $\varepsilon_{si}$, $\varepsilon_{ox}$ are electrical permittivity of silicon and oxide, respectively and $t_{ox}$ is oxide thickness.

$$V_{TH} = V_{FB} - \frac{qN_D}{\varepsilon_{si}}\left(\frac{A}{P}\right)^2 - \frac{qN_D A}{C_{ox}} + \frac{\Delta E_0}{q} \quad (5)$$

where A and P represent the cross-sectional area and gate perimeter of the devices, respectively, where A= $W_{FIN} \times H_{FIN}$ and P=2 $H_{FIN}$ + $W_{FIN}$. $\Delta E_0$ is the variation of the minimum energy level in the conduction band due to quantum confinement.

The decrease in $V_{TH}$ due to the increase in $W_{FIN}$ is attributed to the terms $C_g$, $Q_d$, and $C_{ch}$ in eqn (1), establishing an inversely proportional relationship with $W_{FIN}$.

In the case of JNT, the ionization rate presents a relevant contribution to the threshold voltage variation on temperature, as the devices operate in the Mott transition [12]. Therefore, $V_{TH}$ calculation also considered the partial carrier ionization changing the effective $N_D$ over the temperature.

Both models demonstrated reduced $V_{TH}$ due to temperature increase, which is mainly associated with the reduction of Fermi potential ($\Phi_F$) (third term in Eqn. (1)) and the $\partial N_D / \partial T$ (second and third term in Eqn. (5)) [13]. Moreover, $dV_{TH}/dT$ reduction for narrower devices can be explained by the fourth term in Eqn. (1) and the second and third terms in Eqn. (5), which are directly related to $W_{FIN}$. Fig. 5 introduces the threshold voltage as a function of the temperature, comparing the influence of L in IM and JNT nanowires, with $W_{FIN}$=10nm.

The threshold voltage linearly reduced with the temperature rise, independent of the nanowire's architecture and $W_{FIN}$, as illustrated in Fig 4b. Although both architectures presented a reduction in $V_{TH}$ for the shorter nanowires, the same rate $dV_{TH}/dT$ around -0.46 mV/K for JNT, and -0.53 mV/K for IM was found. Calculating $dV_{TH}/dT$ from Eqns. (1) and (5), it was obtained $dV_{TH}/dT$ = -0.54 mV/K for IM nanowires and -0.47 mV/K for JNT.

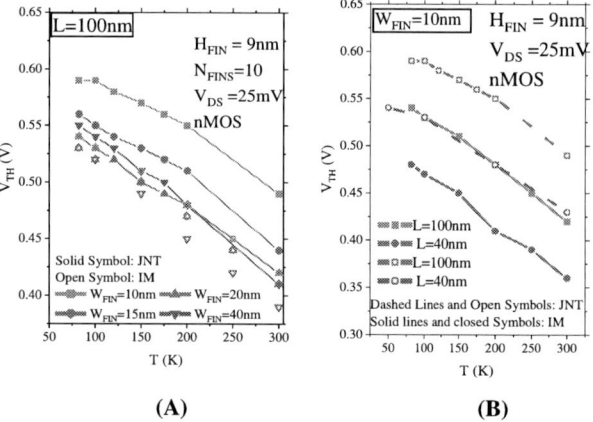

**(A)**        **(B)**

**Fig. 4.** Threshold voltage as a function of temperature for JNT and IM nanowires with (A) L of 100nm and variable WFIN and (B) WFIN of 10nm and variable L.

The threshold voltage in both devices is inversely proportional to temperature and linearly influenced by the Fermi potential ($\Phi_F$), as seen in the third term of Eqn. (1) and the first term of Eqn. (5), with the flat-band voltage ($V_{FB}$) being directly proportional to the $\Phi_F$.

### B. Subthreshold Slope

Fig. 5a presents the subthreshold slope (SS) as a function of temperature for JNT and IM nanowires with L of 100nm. In both devices, the subthreshold slope linearly rises due to

temperature increase and remains close to the theoretical limit (shown in the graph as the dashed line) for the narrower nanowires down to 150K. Bellow this temperature, a rise in interface trap density as well as the presence of band tails [14] increases the body factor and contributes to the elevation of S compared to its theoretical limit.

Increasing $W_{FIN}$ from 10nm to 40nm results in a slight degradation of the SS in the entire temperature range. Specifically, the observed degradation at room temperature transitions from 60mV/dec to 62.5mV/dec for JNT and 60mV/dec to 61.4mV/dec for IM. This decline can be attributed to decreased electrostatic control exerted by the gate.

Fig. 5b presents a comparison of SS as a function of temperature between JNT and IM nanowires with $W_{FIN}$ of 10nm and L of 40nm and 100nm. The longer devices present SS close to the theoretical limit (shown in the graph as the dashed line) down to 150K, as previously discussed. The shorter devices demonstrate the degradation of SS in a whole range of temperatures, with values ranging from 60mV/dec to 61.7mV/dec for JNT and 60mV/dec to 65.3mV/dec for IM at room temperature, due to the short-channel effect (SCE).

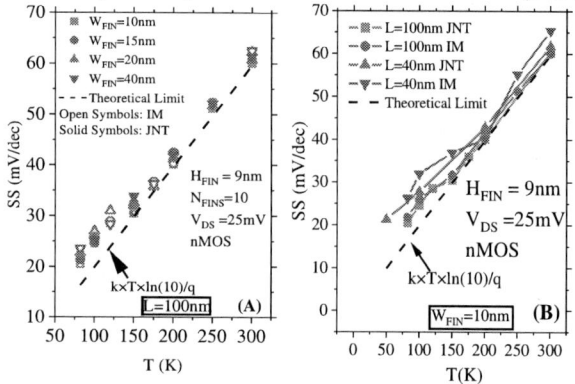

**Fig. 5.** Subthreshold slope as a function of temperature for JNT and IM nanowires with (A) L of 100nm and variable $W_{FIN}$ and (B) $W_{FIN}$ of 10nm and variable L.

### C. Low-Field Mobility

The low-field mobility was extracted using the Y-Function Method [15] in the $I_{DS} \times V_{GS}$ curves, at low drain bias. Matthiessen's rule, presented in Eqn. (6), can be applied in this situation to evaluate $\mu_0$.

$$\mu_0 = \frac{1}{\left(\frac{1}{\mu_{psii}} + \frac{1}{\mu_{cc}} + \frac{1}{\mu_{ni}}\right)} \qquad (6)$$

where $\mu_{psii}$ is the mobility degraded by phonon scattering and ionized impurity scattering, $\mu_{cc}$ is the mobility degraded by carrier-to-carrier scattering and $\mu_{ni}$ is the mobility degraded by neutral impurity scattering.

Fig. 6a shows the low-field mobility as a function of temperature for JNT and IM nanowires with a channel length of L=100 nm. Both nanowires exhibit a clear decrease in mobility with increasing temperature for the measured $W_{FIN}$

values, indicating the prevalence of phonon (lattice) scattering as the primary mechanism of mobility degradation. This scattering takes place due to the carrier's interaction with the crystal lattice. As the temperature rises, the crystal lattice vibrates more intensely, which in turn adversely affects it.

The main difference between the devices is related to mobility, which is over twice as high for IM devices compared to JNT in a whole range of temperatures measured. However, the variation of mobility with the temperature is larger for IM, independent of $W_{FIN}$. Specifically, $\mu_0$ varied from $64.7 cm^2/Vs$ at 300K to $119.7 cm^2/Vs$ at 82K, whereas variated from $169.6 cm^2/Vs$ at 300K to $389.4 cm^2/Vs$ at 82K, for devices with $W_{FIN}$ of 20nm mobility. Moreover, the variation with $W_{FIN}$ is negligible across the entire temperature range for all transistors.

In Fig. 6b, the low-field mobility is presented as a function of the temperature for JNT and IM nanowires, for devices with $W_{FIN}$ of 10nm. As occurred in the previous analysis, $\mu_0$ degrades due to a temperature increase, regardless of L. The mobility for IM is expressively higher compared to JNT's mobility and the variation of mobility with the temperature is higher for IM devices, independent of L. Furthermore, there is a clear degradation of $\mu_0$ as the transistor channel length decreases. Specifically, $\mu_0$ decreases from $186.4 cm^2/Vs$ to $123.1 cm^2/Vs$ for IM devices and from $74.7 cm^2/Vs$ to $23.8 cm^2/Vs$ for JNT devices, regardless of temperature.

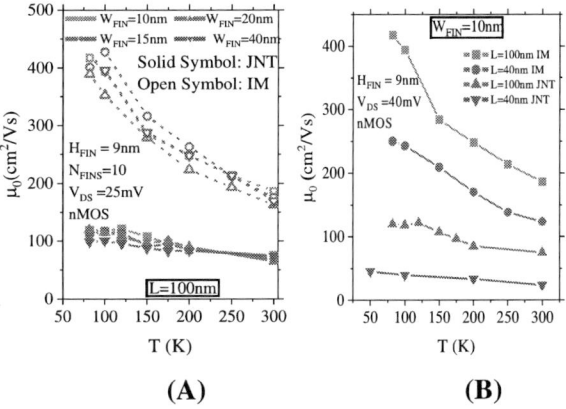

**Fig. 6.** Mobility under Low-Field Conditions as a Function of Temperature for IM and JNT Nanowires with (A) L of 100nm and variable $W_{FIN}$ (B) $W_{FIN}$ of 10nm and variable L.

Due to the significantly higher doping concentration in the channel of the JNT compared to the IM, it is well-established that the carrier mobility in the JNT is lower than the IM [16]. This relationship stems from the fact that mobility is inversely related to doping concentration.

## III. CONCLUSION

This work presented a comparative analysis of the main electrical parameters of JNTs and IMs nanowires with different fin widths and channel lengths, in a temperature range between 300 K and 82 K. For both nanowires, the temperature-dependent variation in threshold voltage is similar in both long-channel and short-channel transistors ($dV_{TH}/dT \approx$ -0.53mV/K for IM $dV_{TH}/dT \approx$ -0.46mV/K for JNT). However, the threshold voltage variation with temperature increases with the fin width. Comparing the narrower and wider devices rate, an increase of 24% and 39% is found for IM and JNT, respectively. The subthreshold slope reduction with the temperature exhibits a slight degradation for temperatures below 150K. Shorter and wider devices show deviations from the theoretical limit across a wide range of temperatures. The carrier mobility of the IM transistor is consistently higher than that of the JNT transistor, regardless of L, $W_{FIN}$, and temperature. However, a lower variation in mobility with temperature was observed for the JNTs in both geometric analyses conducted. Therefore, it is possible to establish that JNT exhibits better than IM thermal stability of mobility within the studied temperature range.

## ACKNOWLEDGMENT

The authors express their gratitude to CNPq (#2019/15500-5) and FAPESP and (#2022/14874-1) for the financial support provided during the execution of this work.

## REFERENCES

[1] B. Parvais et al., "Scaling CMOS beyond Si FinFET: an analog/RF perspective," 2018 48th European Solid-State Device Research Conference (ESSDERC), 2018.

[2] T.A.Oproglidis et al. "Effect of temperature on the performance of triple-gate junctionless transistors" IEEE Transactions on Electron Devices, Aug.2018.

[3] R Trevisoli, M. de Souza, R.T. Doria, V. Kilchtyska, D. Flandre, M.A. Pavanello. "Junctionless nanowire transistors operation at temperatures down to 4.2 K", 2016.

[4] E.A. Gutierrez, M.J. Deen, C. Claeys. Low-temperature electronics: physics, circuits, and devices, applications. Boston: Academic Press; 1991.

[5] B. Patra et al., IEEE J. Solid-State Circuits, vol. 53, no 1, p. 309–321, 2018.

[6] S. Barraud et al., "Performance of omega-shaped-gate silicon nanowire MOSFET with diameter down to 8 nm", IEEE Electron Device Letters, vol. 33, no. 11, pp. 1526–1528, Nov. 2012.

[7] Barraud S, Previtali B, Lapras V, et al. Tunability of parasitic channel in gate-all-around stacked nanosheets. In: 2018 IEEE International Electron Devices Meeting (IEDM), San Francisco, CA, USA. pp. 21.3.1–21.3.4. 2018.

[8] A. Ortiz-Conde. et al. "A review of recent MOSFET threshold voltage extraction methods". Microelectronics Reliability, 2002.

[9] J.P. Duarte, S-J Choi, D-I Moon, J-H Ahn, J-Y Kim, S. Kim, et al. "A universal core model for multiple-gate field-effect transistors. part I: charge model". IEEE TransElectron Dev 2013;60(2):840–7.

[10] J.P. Duarte, S-J Choi, D-I Moon, J-H Ahn, J-Y Kim, S. Kim, et al. "A universal core model for multiple-gate field-effect transistors. Part II: drain current model". IEEE Trans Electron Dev 2013;60(2):848–5.

[11] R. D. Trevisoli, R.T. Doria, M. de Souza, M.A. Pavanello. "Threshold voltage in junctionless nanowire transistors". semiconductor science and technology 26.10 (2011): 105009–2. Print.

[12] N. F. Mott "On The Transition To Metallic Conduction In Semiconductors". Canadian Journal of Physics. 34(12A):

[13] G. Xiao, J. Lee, JJ. Liou, A. Ortiz-Conde. "Incomplete ionization in a semiconductor and its implications to device modeling" Microelectronics Reliability, vol 39, no 8, p. 1299-1303, 1999.

[14] A. Beckers, F. Jazaeri, C. Enz "Theoretical limit of low-temperature subthreshold swing in field-effect transistors", in IEEE Electron Device Letters, vol. 41, no. 2, pp. 276-279, 2020, doi: https://doi.org/10.1109/LED.2019.2963379.

[15] G. Ghibaudo, "New method for the extraction of MOSFET parameters", IEEE Electronics Letters, vol. 24, pp. 543-545, April 1988.

[16] J. P. Colinge, et al, "Junctionless Transistors: Physics and Properties.", Semiconductor-On-Insulator Materials For Nanoelectronics Applications (pp. 187-200). (Engineering Materials). Springer-Verlag London Ltd..

# Single nanofabrication step of low series resistance nanowire-based devices for giant piezoresistance characterization

Kung Shao Chi
*Faculty of Electrical and Computer Engineering*
*Universidade Estadual de Campinas (UNICAMP)*
Campinas, Brazil
k200689@dac.unicamp.br

Lucas Barroso Spejo
*Faculty of Electrical and Computer Engineering*
*Universidade Estadual de Campinas (UNICAMP)*
Campinas, Brazil
lucas.spejo@gmail.com

Renato. A. Minamisawa
*Institut für Mathematik und Naturwissenschaften Fachhochschule Nordwestschweiz*
*CH-5210 Windisch, Switzerland*
renato.minamisawa@fhnw.ch

Marcos V. Puydinger dos Santos
*Faculty of Electrical and Computer Engineering*
*Universidade Estadual de Campinas (UNICAMP)*
Campinas, Brazil
mpuyding@unicamp.br

*Abstract*— In this work, we aim to fabricate strained silicon nanowires (sSiNWs) to study their electric mobility and giant piezoresistance. Through techniques compatible with the CMOS technology, individual nanowires (NWs) were fabricated from strained silicon-on-insulator (sSOI) thin films with 0.8% biaxial strain. Subsequently, the buried oxide (BOX) was removed from the SOI film, thus suspending the NWs, and the new boundary condition of its surface induces mechanical stress amplification, now uniaxial in the NW longitudinal direction. The proposal is to stress the NWs to levels higher than those employed in industry and, hence, fabricate prototypes using a single-step fabrication protocol, yet with an optimized contact resistance. Parameter optimization can further result in the fabrication of MOSFETs based on a single ultra-strained NW in gate-all-around (GAA) topology, as well as chemical and physical sensors for various technological applications.

*Keywords*— *strained silicon nanowires, giant piezoresistance, MOSFETs, nanofabrication, ohmic contact.*

## I. INTRODUCTION

The industry has been employing the mechanical stress of the active region of transistors since around 2004 as a way to improve the device's performance. Mechanical stress increases the carrier mobility and allows for higher clock frequencies of logic circuits, as well as overall processor performance [1]. Recent studies show that mechanical stress can change semiconductor band structure and, hence, the charge distribution on the NW surface [2], which can be beneficial for applications in chemical and physical sensors.

sSiNWs can exhibit giant piezoresistance effect [3], which consists of the modulation of the material electrical resistivity under the influence of mechanical stress. Several studies have been carried out with theoretical explanations and different approaches. However, works with sSiNWs with higher deformations (> 2%) are scarce, especially without external actuators for driving stress.

Moreover, the fabrication of sSiNWs under relatively high stress levels is generally limited with regards to fabrication yield and reproducibility [3]. Additionally, the solutions proposed in the literature usually require external actuators [4],[5] or even complex multi-step nanofabrication infrastructure. Generally, measuring piezoresistance requires relatively high doping levels and metal deposition for defining electrodes, which makes the fabrication more complex.

In this sense, this work demonstrates that sSiNWs can be fabricated in a CMOS-compatible platform and further strained at relatively high levels without external actuators, which is interesting from the technological point of view. It

represents a step forward from our previous work and the innovation here is that this platform offers the possibility to directly control the stress and measure the piezoresistance using a single-step lithographic process. The devices offer a huge potential in applications of highly sensitive pressure and force sensors, as well as in the fabrication of high-performance microelectronics with sensitivity to physical and chemical stimuli.

## II. METHODOLOGY

### A. Strained Silicon on Insulator (sSOI)

The sSOI substrate consists of an undoped 15 nm-thick silicon film with 0.8% biaxial strain (about 1.44 GPa of biaxial stress in the [110] and [-110] directions). The silicon film relies on the top of a 145 nm-thick buried silicon dioxide on a bulk silicon substrate. The crystallographic orientation of the silicon surface is (001).

### B. Device Fabrication

Initially, we spin-coat the electrosensitive resist PMMA ARP 679.04 at 4000 rpm for 30 seconds and evaporate the solvent on a hot plate at 180 ºC for 2 minutes. Then, we expose the resist using electron beam lithography system. The resist was developed with ARP 600.56 for 2 minutes. Afterward, the wafers undergo a dry etching process in an inductively-coupled-plasma (ICP) plasma system, which is used to etch the Si and define the NW. Then, an organic cleaning protocol with acetone and isopropanol was performed to remove any remaining resist. Finally, in order to suspend the sSiNWs, we use a buffered hydrofluoric acid (BHF) solution for approximately 3 minutes to remove the silicon oxide underneath the sSiNWs. Fig. 1 shows a simplified flowchart of NW fabrication steps.

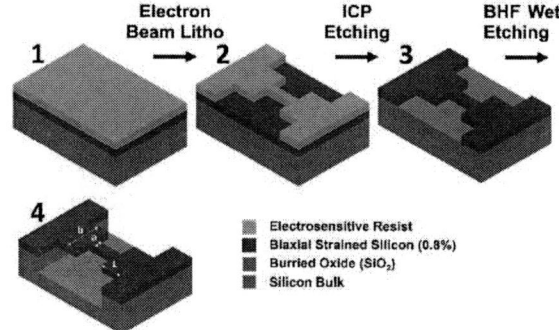

Fig. 1. Diagram of the nanofabrication steps.

## III. RESULTS AND DISCUSSIONS

### A. Suspended Nanowires

The dimensions of the suspended nanowires, such as width and length, were controlled by the lithography parameters, thus leading to distinct residual stresses in the sSiNWs. In this sense, one can obtain different piezoresistance values by varying the sSiNW width and length. The length of the sSiNW varied from 0.5 to 2.5 µm, while its width ranged from 50 to 400 nm. Fig. 2 shows an example of the fabricated suspended sSiNWs. Note that the pads are the peripheral region at the ends of the nanowires. The width of the pads also varied in this experiment between 0.5 to 1.75 µm, being wider than the width of the nanowires. The pads are partially or fully suspended, depending on their dimensions and the etching time in BHF. Simulations conducted by our group using the finite element method (FEM) in COMSOL show that before the removal of SiO₂, the nanowires exhibit predominantly uniaxial strain of around 0.8%, while the pads show a 0.8% biaxial strain. However, removing the SiO₂ underneath the nanowires and pads modifies the boundary conditions on the silicon surface, thus relaxing the pads, which amplifies the stress on the NWs [3]. The sSiNWs ultimately exhibit amplified uniaxial stress magnitude depending on their dimensions.

### B. FEM Simulation

The computational analysis employed a three-dimensional finite element method (FEM) using COMSOL Multiphysics® to accurately determine stress levels on the nanowire. This approach considered various geometric dimensions alongside parameters like corner radius and corrosion profile to enhance precision. Fig. 3 (top) displays a color map illustrating simulated stress along the X-direction ($\sigma_{xx}$ [110]), while Fig. 3 (bottom) depicts stress components along an imaginary horizontal line passing through the center of both the pads and the nanowire.

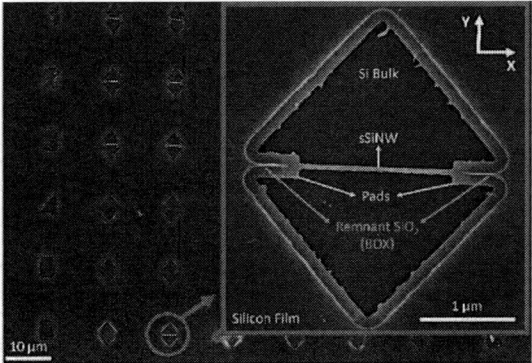

Fig. 2. sSiNW with length and width of 2.5 µm and 90 nm, respectively. Z direction is out of the plane.

As observed in the simulation for sufficiently narrow sSiNWs, they predominantly exhibit a uniform and uniaxial stress component in the X-direction. Other stress components, such as $\sigma_{YY}$ and $\sigma_{ZZ}$, respectively in the Y and Z directions, can be disregarded since $\sigma_{XX} = 2.74\ GPa$ exhibits significantly higher stress. Transitioning into the region between sSiNW and the pads, one can observe a considerable reduction in the $\sigma_{XX}$ component to values below 1 GPa, accompanied by an increase in the $\sigma_{YY}$ component to

approximately 0.5 GPa. Moving further into the PAD region, a biaxial strain of about 1.5 GPa is observed. This corresponds to the previously mentioned 0.8% biaxial strain.

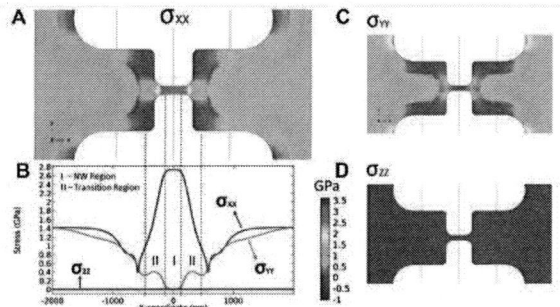

Fig. 3. A) Color map of the simulated $\sigma_{XX}$ stress component in the nanostructure; B) cut-line graph of the triaxial stress components in the structure from an imaginary horizontal line passing along the center of the pads and the nanowire; C) and D) are color maps of $\sigma_{YY}$ and $\sigma_{ZZ}$ stress components. The nanowire is 1431 nm long and 138nm nm wide. Adapted from [3].

### C. Raman characterization

Raman spectroscopy is a highly versatile technique used in this work to track the residual mechanical stress in sSiNWs. It measures the scattering of an incident light beam on the substrate. In this work, a 532 nm beam was used. The photon incident on the sample interacts inelastically with the crystal lattice, and a portion of the energy is transferred to excite phonon states of the sSiNWs, while the other portion is scattered into a photon with a new frequency, which is detected by the spectrometer. For each material, there is a specific vibrational resonance mode. For the unstrained silicon, the Raman peak position is $\omega_0 = 520.5$ cm⁻¹. On the other hand, in the strained silicon, the mechanical stress disturbs the vibrational modes of the crystal lattice and shifts the strained Raman peak ($\omega_i$) to the left (in the case of tension) [6].

The Raman shift between unstrained and strained silicon, $\Delta\omega = \omega_i - \omega_0$, allows for indirectly obtaining the stress along the [110] crystallographic direction ($\sigma_{xx}$) of the sSiNW. Equation (1) characterizes the Raman shift [3] as a function of the three-dimensional stress:

$$\Delta\omega = \omega_i - \omega_0 = \frac{1}{2\omega_0}\{[pS_{12} + q(S_{11} + S_{12})] \times (\sigma_{XX} + \sigma_{YY}) + (pS_{11} + 2qS_{12})\ \sigma_{zz}\} \quad (1)$$

$\Delta\omega$ is the Raman shift, p and q are the phonon deformation potentials (PDPs) of silicon, and $S_{11}$ and $S_{12}$ are the silicon's elastic constants (Pa⁻¹) obtained from the compliance matrix. $\sigma_{YY}$ and $\sigma_{ZZ}$ are stresses in the crystallographic directions [-110] and [001]. According to Fig. 3, the nanowire region has negligible stress in the Z direction, thus we can neglect the stress component $\sigma_{ZZ}$ and simplify (1) for the in-plane stress of the silicon thin film:

$$\Delta\omega = \frac{1}{2\omega_0}\{[pS_{12} + q(S_{11} + S_{12})] \times (\sigma_{XX} + \sigma_{YY})\} \quad (2)$$

In the case of uniaxial stress along the longitudinal direction [110], which is achieved for sufficiently narrow nanowires, we can neglect the stress in the Y direction and simplify (2):

$$\Delta\omega = \frac{1}{2\omega_0}[pS_{12} + q(S_{11} + S_{12})] \times (\sigma_{XX}) = SSC \times \sigma_{XX} \qquad (3)$$

Here, the stress shift coefficient (SSC), in units of $cm^{-1}Pa^{-1}$, linearly relates stresses and Raman shift of a tensile sample and depends on the PDPs (p and q) and their elastic constants ($S_{11}$ and $S_{12}$). Ref. [3] reports that the SSC is around $-1.9 \pm 0.1 \times 10^{-9}\ cm^{-1}Pa^{-1}$. Therefore, we ultimately estimate the nanowire stress from the Raman shift as follows:

$$\Delta\omega = -1.9 \times 10^{-9} \times \sigma_{XX} \qquad (4)$$

Fig. 4 shows a Raman spectrum of a 300 nm wide, 1.5 μm-long sSiNW showing the peaks of bulk Si (unstrained) and the sSiNW. It is worth mentioning that the diameter of the laser beam is larger than the width of the nanowires and the light penetration depth at 532 nm wavelength is around 1 μm. Therefore, both the bulk Si peak at $\omega_0 = 520.5\ cm^{-1}$ and the sSiNW peak were detected concurrently in the Raman spectrum. In our measurements, the stress in the Y direction cannot yet be neglected, as the width of the nanowires presented here does not guarantee predominantly uniaxial stress. For 1.5 μm-long sSiNWs of different widths and bridging 500 nm-wide pads, similar $\Delta\omega$ values were obtained. For 750 nm, 300 nm, and 200 nm-wide sSiNWs, $\Delta\omega$ values of $-4.4\ cm^{-1}$, $-4.6\ cm^{-1}$ and $-5.6\ cm^{-1}$, respectively, were obtained. It should be mentioned that these results refer to sSiNWs whose pads were not completely suspended. Therefore, the sSiNWs were not ensured to be uniaxially strained and at the maximum stress. This is in agreement with the FEM simulation of our previous work [3]. In fact, FEM simulation suggests that narrow nanowires with predominantly uniaxial stress should present larger $\Delta\omega$ values.

The $\Delta\omega$ of the 750 nm and 300 nm wide sSiNWs are comparable, thus suggesting residual stress in the Y direction. As shown in (2), the Y component of the stress contributes to the increase in $\Delta\omega$, thus causing the 750 nm wide biaxially strained sSiNW to have a higher $\Delta\omega$ value compared to the narrow nanowire. On the other hand, comparing the $\Delta\omega$ value for 300 nm and 200 nm wide sSiNWs, the shape anisotropy of such narrow nanowires leads to a considerable reduction of the $\sigma_{YY}$ component, which is supported by the FEM simulation. Hence, they become predominantly uniaxially stressed. Therefore, in this case, with a reduction of about 100 nm in the sSiNW width, it was already possible to observe an increase of about 1 $cm^{-1}$ in $\Delta\omega$ due to an increase in the uniaxial stress (longitudinal direction) of the sSiNW. Hence, the increase in the $\sigma_{XX}$ component yields an increase in $\Delta\omega$, as expected.

Therefore, considering that the 200 nm wide sSiNW already has predominantly uniaxial stress, we can use (4) to obtain the direct relationship between $\Delta\omega$ and $\sigma_{XX}$. In this case, the uniaxial stress obtained was around 2.9 GPa, which is higher than that used in the industry, limited to about 1 GPa.

### D. Contact resistance engineering for the piezoresistance characterization

We have developed a platform designed to strain sSiNWs. This platform features autonomous strain, thus eliminating the need for external actuators, and also being compatible with the CMOS technology. As outlined in Section Device Fabrication, this platform offers a rapid and efficient way for prototyping strained nanowires for the investigation of their electrical properties, and ultimately for the fabrication of nanowire-based devices. One of the most relevant electrical parameters of such strained nanowires is the piezoresistance, π, which consists of the variation of the nanowire's electrical resistivity, $\Delta\rho$, as a function of the applied stress, $\sigma$ [6]:

$$\pi = \Delta\rho/\sigma \qquad (5)$$

Furthermore, another relevant parameter for characterizing the stress of NWs is the longitudinal Gauge factor, $GF_l$, which is defined as [6]:

$$GF_l = \frac{dR/R}{dl/l} = \frac{dR/R}{\varepsilon_l} = \frac{d\rho/\rho}{\varepsilon_l} \qquad (6)$$

where R and $\rho$ are the nanowire's electrical resistance and resistivity and $\varepsilon_l$ represents the percentage deformation of the nanowire in the longitudinal direction. $\varepsilon_l$ is related with $\sigma$ by the Young's modulus, $E = \sigma/\varepsilon_l$ [7]. Therefore, in order the measure de Gauge factor, one simply need to measure the nanowire's resistance and normalize by the geometrical factor to obtain resistivity. The deformation is obtained by the longitudinal stress, which comes from the Raman characterization.

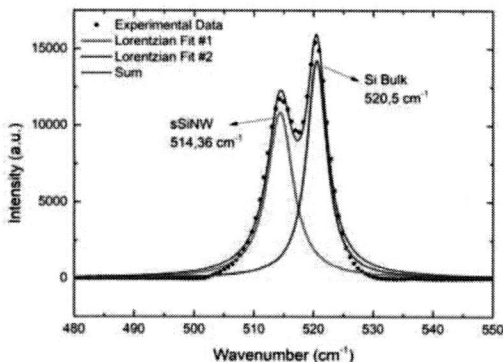

Fig. 4. The Raman peak for sSiNW was observed at 514.4 $cm^{-1}$. We used the Horiba XploRA with a 532 nm laser for the measurements.

Moreover, one of the major challenges for the piezoresistance measurements is to achieve ohmic contact between the metal electrode and the silicon pads, once the low-doped silicon has poor carrier density and, hence, one typically obtains a Schottky-like contact. In this sense, we aim here to reduce the contact resistance in order to achieve ohmic contact. Therefore, we have diffused phosphorus into the top Si layer to enhance the electron carrier density and improve the contact resistance. It's worth mentioning that the piezoresistance of sSiNWs is doping-sensitive, as previously reported by Kumar Bhaskar et al. [8]. They have demonstrated that π can decrease by a factor of 5 as the donor concentration, $N_A$, in the silicon increases from $5 \times 10^{17} cm^{-3}$ to $1 \times 10^{18} cm^{-3}$. One possible reason for such behavior is that highly doped silicon exhibits increased carrier density, resembling metal behavior with low resistance. Therefore, we chose a phosphorous doping level of around $N_D = 1 \times 10^{16}\ cm^{-3}$ to ensure an ohmic contact, yet with a significant π signal.

This doping level was carefully chosen using the semiconductor band energy theory in order to establish a metal-semiconductor (MS) ohmic contact. We have chosen a probe material with adequate workfunction, such that we could simplify further lithographic steps for alignment marks and metal contacts and, thus, directly measure the piezoresistance in a conventional probe system after a single lithographic step for nanowire fabrication. This way, metal probes can directly make ohmic contact on lightly doped Si pads. For a doping level of $N_D = 1 \times 10^{16}\ cm^{-3}$, the calculated silicon work function is [9]:

$$\phi_{Si} = \chi_{Si} + (E_g/2) - kTLn(N_D/n_i) = 4.23\ eV \quad (7)$$

where $\chi_{Si} = 4.05\ eV$ is the silicon electron affinity, $E_g = 1.12\ eV$ is the silicon bandgap, $n_i = 1 \times 10^{10} cm^{-3}$ is the intrinsic carrier density of silicon at room temperature, and $kT = 0.0258\ eV$ is the thermal energy at 300 K. Therefore, a metal workfunction close to this value is the stainless steel, $\phi_{Steel} = 4.08 \sim 4.19$ eV [10]. We used steel probe tips model SE-S from SIGNATONE Corporation in this work, featuring tip diameter of 12 µm. The MS barrier lies in the range $0.03\ eV < \phi_B = \phi_{Steel} - \chi_{Si} < 0.14\ eV$, thus the MS contact is expected to be ohmic.

Fig. 5A depicts the experimental setup for the piezoresistance measurements. The stainless-steel probes were directly placed to the doped Si for the 2-point measurement of the sSiNWs. The probes contact the silicon region near the pads in an area with $SiO_2$ underneath; thus, the probes drive no additional stress. In addition, the pads were electrically isolated by defining a 400 nm-wide trench surrounding the 500 µm wide pads between the nanowires. The trench was defined in the nanowire lithography level, hence no additional process step was necessary. Fig. 5B shows a voltage-current (V x I) curve of a 2.0 µm long, 1 µm wide sSiNW taken using a 4200 SCS Keithley analyzer. The linear relationship of the V x I curve confirms that we could achieve an ohmic MS contact, with $R_0 = 68.7\ k\Omega$ before $SiO_2$ etching.

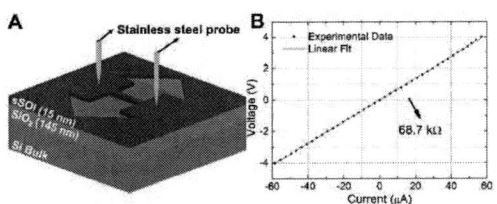

Fig. 5. A) 2-probes measurement setup; B) Ohmic contact measurement of a 2.0 µm long, 1 µm wide sSiNW.

This nanowire is not suspended yet, hence it presents a residual strain of 0.8% along the longitudinal direction. After removing the underneath $SiO_2$, one can measure the resistivity variation, $\Delta\rho$, and further calculate the piezoresistance of each sSiNW We assume that the $SiO_2$ etching yields a strain increase mostly on the nanowire, which is supported by the FEM simulations. Hence, the stress variations in the pads and, therefore, their piezoresistance are negligible compared to the nanowire. In addition, we assume that the contact resistance between the steel probe and the pads does not change. In this sense, we can measure the relative resistance variation, $\Delta R/R_0 \approx \Delta\rho/\rho_0$, of the whole device by a 2-probes setup and infer by this value the nanowire piezoresistance, $\pi$, for a given

stress level.

Therefore, this work proposes an innovative method for prototyping sSiNWs without external actuators and measuring their piezoresistance directly in a 2-probes setup without further lithographic and metal deposition steps for the electrical contacts. We were able to simplify the steps involved in the lithography fabrication, thus making possible fast prototyping and electrical characterization. Furthermore, as previously mentioned, the low doping level used in this work aims to maximize the π signal.

## IV. CONCLUSIONS

Strained silicon nanowires were fabricated using a CMOS-compatible technique. The mechanical stress was achieved without external actuators by electrical beam lithography of suspended silicon nanowires on a sSOI wafer. Raman spectroscopy was used to characterize the uniaxial mechanical stresses of around 2.9 GPa for 200 nm-wide sSiNWs. The electrical contact engineering proposed in this work makes possible an ohmic contact between metal-semiconductor for low doping levels, which also helps improving the piezoresistance signal. This also makes possible fast prototyping and piezoresistance characterization of nanowires upon a single lithography step. As a perspective for the continuation of this work, we aim to measure the giant piezoresistance for stress levels larger than 4 GPa, which requires process parameter improvements, especially in the nanowire dimensions.

## ACKNOWLEDGMENT

The authors thank CCSNano/UNICAMP staff for the experimental support, as well as the financial support from the Sao Paulo Research Foundation (FAPESP, 2022/16809-2) and The National Council for Scientific and Technological Development (CNPq: 310021/2021-9 and 406193/2022-3).

## REFERENCES

[1] I. Newsroom, "Intel 22 nm 3-D tri-gate transistor technology." Link: http://newsroom. intel. com/docs/DOC-2032, 2011.

[2] D. Nam et al., "Strain-induced pseudoheterostructure nanowires confining carriers at room temperature with nanoscale-tunable band profiles," Nano Lett, vol. 13, no. 7, pp. 3118–3123, 2013.

[3] L. B. Spejo et al., "Non-linear Raman shift-stress behavior in top-down fabricated highly strained silicon nanowires," J Appl Phys, vol. 128, no. 4, Jul. 2020, doi: 10.1063/5.0013284.

[4] F. Ureña-Begara, R. Vayrette, U. K. Bhaskar, and J.-P. Raskin, "Raman analysis of strain in p-type doped silicon nanostructures," J Appl Phys, vol. 124, no. 9, 2018.

[5] H. Ando and T. Namazu, "Influence of vacuum annealing on mechanical characteristics of focused ion beam fabricated silicon nanowires," Journal of Vacuum Science & Technology B, vol. 41, no. 6, 2023.

[6] A. S. Fiorillo, C. D. Critello, and A. S. Pullano, "Theory, technology and applications of piezoresistive sensors: A review," Sensors and Actuators, A: Physical, vol. 281. Elsevier B.V., pp. 156–175, Oct. 01, 2018. doi: 10.1016/j.sna.2018.07.006.

[7] W. D. Callister Jr and D. G. Rethwisch, Callister's materials science and engineering. John Wiley & Sons, 2020.

[8] U. Kumar Bhaskar, T. Pardoen, V. Passi, and J. P. Raskin, "Piezoresistance of nano-scale silicon up to 2 GPa in tension," Appl Phys Lett, vol. 102, no. 3, Jan. 2013, doi: 10.1063/1.4788919.

[9] R. F. Pierret, Semiconductor device fundamentals. Pearson Education India, 1996.

[10] F. Marlow, S. Josten, and S. Leiting, "Electronics with stainless steel: The work functions," J Appl Phys, vol. 133, no. 8, Feb. 2023, doi: 10.1063/5.0142185.

979-8-3315-4064-7/24 $31.00 © 2024 IEEE

# Influence of disorder on the structural analysis of a quantum Bragg mirror detector

Germano Maioli Penello
Instituto de Física
*Universidade de São Paulo*
São Paulo, Brasil
https://orcid.org/0000-0001-9067-6407

Pedro Henrique Pereira
Departamento de Engenharia Elétrica
*Universidade do Estado do Rio de Janeiro*
Rio de Janeiro, Brasil
https://orcid.org/0000-0002-1861-847X

Guilherme Monteiro Torelly
LabSem
*Pontifícia Universidade Católica do Rio de Janeiro*
Rio de Janeiro, Brasil
https://orcid.org/0000-0003-3019-2069

Rudy Massami Sakamoto Kawabata
LabSem
*Pontifícia Universidade Católica do Rio de Janeiro*
Rio de Janeiro, Brasil
https://orcid.org/0000-0001-5565-1618

Lucas Andrade Teixeira de Souza
Instituto de Física
Universidade de São Paulo
São Paulo, Brasil
https://orcid.org/0009-0009-9448-8628

Sérgio Luiz Morelhão
Instituto de Física
*Universidade de São Paulo*
São Paulo, Brasil
https://orcid.org/0000-0003-1643-0948

Alain André Quivy
Instituto de Física
*Universidade de São Paulo*
São Paulo, Brasil
https://orcid.org/0000-0002-6654-6239

**This study investigates the growth quality of a far-infrared asymmetric quantum Bragg mirror detector (QBMD) based on GaAs/Al$_{0.15}$Ga$_{0.85}$As. The sample, grown by molecular beam epitaxy (MBE), was characterized by high-resolution x-ray diffraction (HRXRD) to assess the impact of structural disorder on the QBMD's final performance. The experimental results were compared with simulated x-ray diffraction spectra, highlighting the excellent quality of the QBMD growth despite fluctuations up to two monolayers in the thickness of the quantum wells and barriers of the active region.**

***Keywords— Quantum Bragg mirror detector, far-infrared, III-V semiconductors, high-resolution x-ray diffraction***

## I. INTRODUCTION

Photodetectors are used in applications spanning from telecommunications to environmental monitoring, and from medical imaging to astrophysics. The state-of-the-art in infrared photodetection technology is driven by a quest for greater sensitivity, efficiency, and versatility. Recent advances in the field are being achieved by novel fabrication techniques [1], novel materials [2], and innovative devices [3, 4]. These improvements are supported by a deeper understanding of the fundamental principles governing light-matter interactions and electronic confinement in the semiconductor structure.

The sample's heterostructure is a key point in the development of photodetectors based on quantum wells and quantum dots, namely, quantum well infrared photodetectors (QWIPS) [5, 6], quantum dot-in-a-well (QDWELLs) [7], quantum Bragg mirror detectors (QBMDs) [4, 8], and submonolayer quantum dots (SMLQDs) [3, 9]). By controlling their heterostructure, one can control the energy levels of the electronic states inside a quantum well or quantum dot. Using different thicknesses of the quantum wells and coupling wells to form superlattices allows even better control over the electronic structure of the sample.

The QBMD is one example of recent advance in the photodetector field. The possibility of having an asymmetric heterostructure creates a preferential direction for electron extraction, thereby establishing it as a photovoltaic device (operating at null bias voltage) with minimal dark current. One direct advantage of the reduction in dark current is a higher operational temperature, which is an important aspect in order

to get rid of the cryogenic temperatures usually needed for operation of such devices.

The growth of a QBMD depends on precise manipulation of the layers' compositions and thicknesses within the sample. An accurate growth, providing a sample with a structure as close as possible to the nominal one, allows the adjustment of the electronic levels and, thus, the tuning of radiation absorption and electron extraction. Thus, the growth of high-quality semiconductor heterostructures is paramount for obtaining improved QBMD devices. Since the growth of a QBMD involves hundreds of very thin semiconductor layers, in this work we investigate the influence of the structural disorder during the growth of QBMDs by high-resolution x-ray analysis.

## II. SAMPLE DESIGN AND GROWTH

The sample was originally designed to explore an optical transition around 111 meV. The active layer consists of 5 coupled quantum wells in an asymmetric structure to use the

| Material | Thickness (nm) | |
|---|---|---|
| GaAs (n-doped) | 500 | |
| Al$_{0.15}$Ga$_{0.85}$As | 30 | |
| GaAs | 2.3 | |
| Al$_{0.15}$Ga$_{0.85}$As | 7.1 | |
| GaAs (*n*-doped) | 5.1 | |
| Al$_{0.15}$Ga$_{0.85}$As | 7.1 | |
| GaAs | 2.3 | |
| Al$_{0.15}$Ga$_{0.85}$As | 7.1 | |
| GaAs | 2.3 | |
| Al$_{0.15}$Ga$_{0.85}$As | 7.1 | |
| GaAs | 2.3 | |
| Al$_{0.15}$Ga$_{0.85}$As | 30 | |
| GaAs (n-doped) | 800 | |
| GaAs substrate | | |

(ACTIVE REGION; REPEATED 20x)

Fig. 1. The QBMD heterostructure (total of 203 layers) is grown on top of a GaAs substrate. The active region is surrounded by two *n*-doped GaAs contact layers.

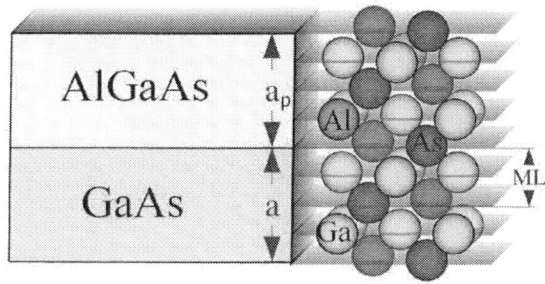

Fig. 2. On the left, pictorial representation of the thickness of one lattice parameter in comparison with the atomic representation, on the right, to explicit the planar interatomic distance during growth. The lattice constants a and $a_p$ are not exactly the same due to tetragonal lattice strain. ML stands for one monolayer (equal to two planar interatomic distances, or half lattice constant). Ga atoms are represented by light green sphere, As atoms are dark green, and Al atoms are orange.

advantage of the photovoltaic operation mode [4, 8]. The active region is repeated 20 times to improve absorption, resulting in a sample with a total of 203 layers among quantum wells, quantum barriers and top and bottom contact layers (Fig. 1).

The QBMD sample was grown by MBE on an epi-ready GaAs(001) substrate. Prior to the growth, oxide removal and outgassing were performed at 580 °C and 600 °C, respectively. Reflection high-energy electron diffraction (RHEED) was used to calibrate the fluxes of the MBE cells and guarantee an accurate growth rate for the $Al_{0.15}Ga_{0.85}As$ barriers and the GaAs quantum wells. The thin layers of the active region were grown at 580 °C with a growth rate of approximately 0.15 nm/s for every layer. The active region was surrounded by two $n$-doped contact layers ($n = 1 \times 10^{18}\ cm^{-3}$), and the main quantum well of each period (where the absorption of the infra-red radiation of interest actually occurs) was also doped with the same doping concentration as the contact layers.

Fig. 2 shows the comparison between a pictorial representation of the layers designed in Fig. 1 versus an atomistic view of a perfect interface between layers with different compositions. The planar interatomic distance and the lattice constant are also represented, showing that the lattice constant (a) is four times the planar interatomic distances along the [001] growth direction. It is also important to notice that one lattice constant of the zincblend structure consists of two monolayers of GaAs or AlGaAs layers.

Although AlAs and GaAs have a slightly different lattice parameter, 5.6622 Å and 5.6533 Å, respectively, they are close enough to allow heteroepitaxy of high-quality layers of both materials on a GaAs substrate without the introduction of any relevant structural defects. Due to the tetragonal lattice strain, the lattice constant of the $Al_{0.15}Ga_{0.85}As$ ($a_p$) is 5.6555 Å.

### III. HIGH-RESOLUTION X-RAY DIFFRACTION

The structural analysis was performed in the $2\theta/\omega$ high-resolution scan mode with 0.001-degree increments (3.6 arcsecs) and centered at $\theta = 33.028°$ (GaAs (004) Bragg diffraction peak). The diffractometer used was a D8 Discover model from Bruker Corporation with a copper anode (wavelength equal to 1.5406 Å).

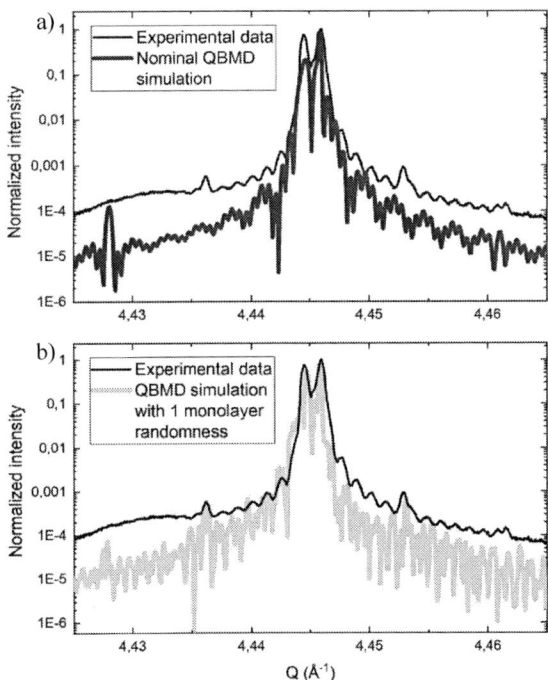

Fig. 3. Comparison between the experimental data of high-resolution x-ray diffraction and the simulation performed with software based on [10]. a) Simulation of a heterostructure with the nominal thicknesses of the layers. b) Simulation of the nominal heterostructure in which the thicknesses were allowed to randomly fluctuate by one monolayer from their nominal values.

The simulation of the experimental x-ray diffraction spectra from our QBMD were performed in two different ways. In the first one, we used an in-house software that calculates multiple x-ray rescattering between atomic layers [10, 11] that was adapted in this work to deal with GaAs/AlGaAs heterostructure. As an alternative, we also used a commercially available software designed for high-resolution measurements named Diffrac.Leptos (version 7.8), that accompanies the Bruker diffractometer.

The software based on [10] allows an intricate control of each parameter related to the x-ray diffraction process. To simulate our sample, the Al content was fixed to 15% and the heterostructure was simulated considering the nominal thicknesses of all the layers (Fig. 3a). To include a small disorder in the growth, a fluctuation of one monolayer around the nominal value was allowed for the thickness of every layer. Since there is no correlation in the thickness fluctuation in each of the 20 periods of the active region, the layers of subsequent active regions can vary independently. The results can be seen in Fig. 3bm where the simulation considers a perfect experimental setup for the x-ray measurements. No smoothing of the curve was performed to artificially include effects of beam divergence or any other artifact included by the x-ray optics during the experiment. The graphs are plotted in function of the scattering vector $Q = 4\pi \frac{sin(\theta)}{\lambda}$, where $\lambda$ is the wavelength of the copper anode. It is clear from Fig. 3b that the simple addition of one monolayer fluctuation for all nominal thicknesses improved considerably the agreement of the experimental and simulated x-ray spectra without the need to alter the compositions of the layers, confirming that there

979-8-3315-4064-7/24 $31.00 © 2024 IEEE

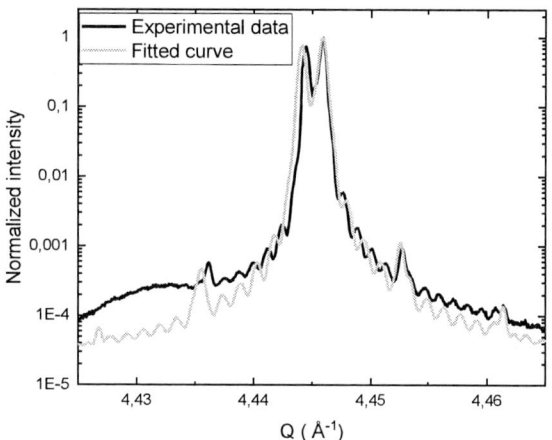

Fig. 4. Comparison between the experimental data of high-resolution x-ray diffraction and the best fit from software Diffrac.Leptos.

was no systematic growth error due to any bad calibration of the fluxes of materials.

Diffrac.Leptos allowed the use of genetic algorithm to fit the experimental data with the simulated x-ray diffraction spectrum. The result of the best fit is shown in Fig. 4. The QBMD heterostructure has several parameters to fit and depending on the restrictions involved, the values of theses parameters can vary significantly even though the fit is very good in all situations. For this reason, we decided to show only the results in which the Al content was fixed to 15% and the thicknesses of all the layers of the full active region were allowed to change independently. Thus, the genetic algorithm had 203 thicknesses to optimize independently in our sample. To reduce the search space, a restriction was imposed on the possible range for every layer. Layers with nominal values of 2.3 nm were allowed to vary around 0.3 nm (approximately one monolayer). Layers with 5.1 nm and 7.1 nm were allowed to vary approximately two monolayers (0.5 nm). The other thicker layers were allowed to vary around 2 nm. . Once again, one can see that the experimental data and simulations are in good agreement.

Both software give excellent results about the growth quality of the sample. The transport properties and the optical properties of the QBMD will be presented and this sample will be processed by regular photolithography with chemical wet etching and metallization to define the photodetector mesa.

## IV. CONCLUSION

The growth and the structural characterization of a GaAs/AlGaAs QBMD was presented. Two different software were used to analyze the experimental high-resolution x-ray diffraction data. Both software show that the quality of the growth is excellent. Variations of up to two monolayers in every layer of the active region can explain the difference from the idealized nominal x-ray diffraction from the actual measured data. Based on the results of both types of simulations, we will now be able to run band-structure calculations and predict the actual absorption energy of the device that will be compared with the nominal one (111 meV). The next step is to process the sample using photolithography, wet chemical etching, and metallization to define the photodetector mesa and measure the transport and the optical properties of the QBMD.

## ACKNOWLEDGMENT

This study was financed in part by Fundação de Amparo à Pesquisa do Estado do Rio de Janeiro (FAPERJ project E-26/210.321/2022) and Fundação de Amparo à Pesquisa do Estado de São Paulo (FAPESP, Process 2023/06016-8), and in part by Conselho Nacional de Desenvolvimento Científico e Tecnológico (CNPq, grant #309837/2021-9).

## REFERENCES

[1] Jiang Xin-Yang, Liu Wei-Wei, Li Tian-Xin, Xia Hui, Deng Wei-Jie, Yu Li, Li Yu-Ying, and Lu Wei, "Enhanced absorption of infrared light for quantum wells in coupled pillar-cavity arrays," Opt. Express 31, 7090-7102 (2023). https://doi.org/10.1364/OE.479106

[2] Giparakis, Miriam, Windischhofer, Andreas, Isceri, Stefania, Schrenk, Werner, Schwarz, Benedikt, Strasser, Gottfried and Andrews, Aaron Maxwell. "Design and performance of GaSb-based quantum cascade detectors" Nanophotonics (2024). https://doi.org/10.1515/nanoph-2023-0702

[3] A. Alzeidan, T.F. Cantalice, K.D. Vallejo, R.S.R. Gajjela, A.L. Hendriks, P.J. Simmonds, P.M. Koenraad, A.A. Quivy, "Effect of As flux on InAs submonolayer quantum dot formation for infrared photodetectors", Sensors and Actuators A: Physical, Volume 334, 113357, (2022). https://doi.org/10.1016/j.sna.2021.113357.

[4] Penello, G.M., Pereira, P.H., Torelly, G.M., Fernandes, F.M., Rushing, J., Tenorio, J.A., Simmonds, P. and Quivy, A.A., "GaAs/AlGaAs based quantum Bragg mirror detector," 2023 37th Symposium on Microelectronics Technology and Devices (SBMicro), Rio de Janeiro, Brazil, 2023, pp. 1-3, doi: 10.1109/SBMicro60499.2023.10302623.

[5] Schneider, H., & Liu, H. C. (2007). Quantum well infrared photodetectors.

[6] H. Schneider, P. Koidl, M. Walther, J. Fleissner, R. Rehm, E. Diwo, K. Schwarz, G. Weimann, Ten years of QWIP development at Fraunhofer IAF, Infrared Physics & Technology, Volume 42, Issues 3–5, Pages 283-289, (2001) https://doi.org/10.1016/S1350-4495(01)00086-X.

[7] Wolde, S., Lao, Y. F., Unil Perera, A. G., Zhang, Y. H., Wang, T. M., Kim, J. O., ... & Krishna, S. (2017). Noise, gain, and capture probability of p-type InAs-GaAs quantum-dot and quantum dot-in-well infrared photodetectors. Journal of Applied Physics, 121(24).

[8] Penello, G. M., Pereira, P. H., Torelly, G. M., Fernandes, F. M., Rushing, J., Tenorio, J. A., Simmonds, P., & Quivy, A. A. (in press). Tailoring optical transitions with GaAs/AlGaAs quantum Bragg mirror detectors (QBMDs). To be published in Journal of Integrated Circuits and Systems.

[9] Gajjela, R. S., Hendriks, A. L., Alzeidan, A., Cantalice, T. F., Quivy, A. A., & Koenraad, P. M. (2020). Cross-sectional scanning tunneling microscopy of InAs/GaAs (001) submonolayer quantum dots. Physical Review Materials, 4(11), 114601.

[10] Morelhão, S. L. (2016). Fundamentals of X-ray physics. Computer Simulation Tools for X-ray Analysis: Scattering and Diffraction Methods, 1-57.

[11] Morelhão, S. L., Fornari, C. I., Rappl, P. H., & Abramof, E. (2017). Nanoscale characterization of bismuth telluride epitaxial layers by advanced X-ray analysis. Journal of Applied Crystallography, 50(2), 399-410.

# Influence of Extraction Methods on the Threshold Voltage Variability Results in SOI Nanosheets

Vinícius Rodrigues Prates, Jaime Calçade Rodrigues, Marcelo Antonio Pavanello and Michelly de Souza
Department of Electrical Engineering, FEI University, São Bernardo do Campo, Brazil
unievprates@fei.edu.br, michelly@fei.edu.br

*Abstract* —**This work presents an experimental evaluation of the influence of the extraction method on the analysis of threshold voltage variability in nanosheet transistors, using seven drain current-based methods. SOI The experimental results of nanosheet transistors with different fin widths and channel lengths show that the choice of the extraction method might change not only the threshold voltage mean value but also its relative deviation.**

*Keywords*— *Variability, Threshold Voltage, Extraction Methods, Nanowire, Nanosheet, SOI MOSFET.*

## I. INTRODUCTION

To enhance transistor density within a chip, dimensional reduction is imperative. However, as the MOSFETs' dimensions are reduced, the standard planar structure presents effects that harm their electrical characteristics, rendering their use impractical. New structures such as Fully-Depleted (FD) SOI transistors were proposed, aiming to increase electrostatic control of the gate voltage over the channel charges, which allows for transistor further miniaturization. Presently, the pursuit of Moore's Law necessitates channel sizes so diminutive that nanowire and nanosheet structures have emerged as viable solution [1, 2, 3]. In these devices, the gate terminal is not only at the top surface of the silicon but also at the sides, configuring an $\Omega$-gate SOI nanosheet transistor. A schematic representation of the cross-section and the top view of an $\Omega$-gate SOI nanosheet transistor is shown in Fig. 1. These nanometric MOSFETs have been studied and offer several advantages compared to other methods, primarily due to their enhanced electrostatic coupling, which helps reduce the occurrence of short-channel effects [3].

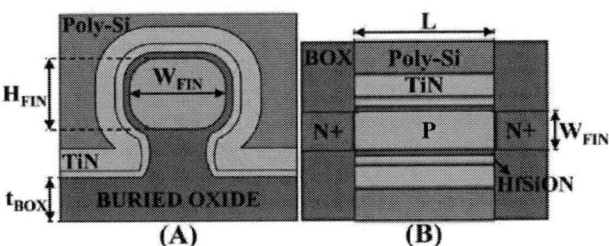

*Fig. 1     Cross-section (A) and top-view (B) representation of a $\Omega$-gate SOI nanosheet nMOSFET.*

Despite of improved short-channel characteristics, an important issue that must be taken into consideration in the scaling of a device is the variability of its electrical parameters [4, 5]. It is reported that different sources of variability, such as GER (Gate Edge Roughness), LER (Line Edge Roughness), MGG (Metal Gate Granularity), and RDD (Random Dopant Fluctuation), can play an important role in the device characteristics and may affect the performance of a circuit, mainly analog ones [6].

The threshold voltage is one of the main parameters of a transistor for circuit designs. In the literature, there are many methods to extract it, each of them presenting different results, as they are based on different premises [7]. Recent research has demonstrated that the threshold voltage of nanowire and nanosheet transistors, as well as its variation with temperature, is highly influenced by the method used for extraction [8].

Therefore, differences in the results of threshold voltage variability in nanosheet transistors can be expected, depending on the choice of the extraction method. This dependence of nanowire variability on the threshold voltage extraction method has been examined through numerical simulations, as discussed in ref. [9]. According to the presented results, the threshold voltage extraction method might play an important role in variability analysis and can even be considered an additional factor to be considered in performance comparisons. On the other hand, experimental results of FinFETs [10] suggested that the variability of the threshold voltage is independent of the extraction method, although they agree that different values of the mean threshold voltage are achieved, depending on the extraction method.

This study aims to experimentally compare how different threshold voltage extraction methods affect the variability results of $\Omega$-gate SOI nanosheet transistors. This study utilized devices manufactured at CEA-Leti with varying channel lengths and fin widths.

## II. THRESHOLD VOLTAGE EXTRACTION METHODS

This study employed seven different threshold voltage ($V_{TH}$) extraction methods, all based on drain current ($I_D$) as a function of gate voltage ($V_{GS}$) curves measured in the linear operating region [7, 11]. $I_D$ vs. $V_{GS}$ curves were obtained at low values of $V_{DS}$. The methods used include constant-current (CC), second derivative (2D), linear extrapolation (ELR), transconductance linear extrapolation (GMLE), transition (G1), second derivative logarithm (2DL), and transconductance-current method ($G_M/I_D$).

In the constant-current method (CC) the threshold voltage defined as the $V_{GS}$ value for a specific $I_D$ level, usually defined as $I_D = W/L \times 10^{-7}$ A, with $W = (2 \times H_{FIN} + W_{FIN})$ for the nanosheet devices used in this work. This is one of the most used for fast analysis, due to its simplicity, although not related to the device physics. The $V_{TH}$ is determined in the second derivative method (2D) as the $V_{GS}$ value at the peak of the $d^2I_D/dV_{GS}^2$ curve. Since this method uses derivatives of second order, it is very susceptible to noise.

In the linear extrapolation method (ELR), a linear extrapolation is made in the $I_D$ vs. $V_{GS}$ curve measured at low $V_{DS}$, in the point of maximum $G_M$. The interception between the linear extrapolation and the $V_{GS}$ axis corresponds to $V_{TH} + V_{DS}/2$. To extract the $V_{TH}$ by the transconductance linear extrapolation (GMLE), an extrapolation is made in the transconductance curve ($G_M = dI_D/dV_{GS}$) in the point of its maximum derivative. The $V_{TH}$ is the value of interception between the extrapolation and $V_{GS}$ axis. As the 2D method, the noise can impact the value of $V_{TH}$ due to the need of a high derivative order. Both methods are affected by series resistance and mobility degradation since the transconductance is dependent on these parameters.

Equation (1) is used to calculate $V_{TH}$ in the transition method (G1), where $V_{gb}$ and $V_{ga}$ are the lower and upper limits of integration. The $V_{TH}$ is given by the highest value of this curve.

$$G_1(V_g, I_D) = V_g - 2\frac{\int_{V_{gb}}^{V_{ga}} I_D(V_g)dV_g}{I_D} \qquad (1)$$

The $V_{TH}$ is defined by the second derivative logarithm method (2DL) as the $V_{GS}$ at the lowest point of the $d^2\ln(I_D)/dV_{GS}^2$ curve. This method is also susceptible to noise due to the use of derivatives. Finally, in transconductance-current method ($G_M/I_D$), $V_{TH}$ is the $V_{GS}$ at the value where $G_M/I_D$ equals half of the plateau observed in weak inversion and is based on the device's physics.

### III. DEVICES CHARACTERISTICS AND MEASUREMENTS

The n-type $\Omega$-gate SOI nanosheet transistors used in this analysis were fabricated at CEA-Leti, using Silicon-on-Insulator (SOI) wafers featuring a 145 nm-thick buried oxide [12]. The silicon film if 9 nm-high and the channel is not intentionally doped. The gate stack is composed of an interfacial $SiO_2$ layer, 2.3 nm-thick HfSiON, 5 nm-thick TiN, followed by 50 nm of polysilicon, resulting in effective oxide thickness (EOT) of around 1.4 nm. Transistors with fin width ($W_{FIN}$) of 10 nm, 15 nm, 20 nm and 40 nm, with channel lengths (L) of 40 nm and 100 nm were measured. The experimental curves of $I_D$ as function $V_{GS}$ were measured using Keysight B1500 Semiconductor Parameter Analyzer, with $V_{DS}$ equal to 40 mV. All curves were obtained at room temperature (300 K).

To allow for the statistical analysis required for the variability study, several devices with the same dimensions were extracted. Table 1 shows the number of measured devices for each dimension considered in this work. The drain current curves measured for all transistors with L = 100 nm and 40 nm, with $W_{FIN}$ = 10 and 40 nm are presented in Fig. 2 in light colors. The thick black curve in all graphs represents the obtained mean current value ($<I_D>$). It is possible to note that short-channel transistors present a higher dispersion amongst the curves. As can be seen in Fig. 3, the relative deviation of the drain current ($\sigma I_D/<I_D>$, with $\sigma I_D$ being the standard deviation of the drain current) increases both with L and $W_{FIN}$ reduction, which agrees with the theory, that states that as smaller the gate area, the larger the variability of electrical characteristics of a MOSFET [6].

### IV. EXPERIMENTAL RESULTS AND VARIABILITY ANALYSIS

The variability analysis was performed using three parameters: the mean value of $V_{TH}$ ($<V_{TH}>$), the standard deviation ($\sigma V_{TH}$), which represents the total variation of the parameter, and the relative deviation, which is defined as $\sigma V_{TH}/<V_{TH}>$, representing the percentage of variation, relatively to its mean value. This last one becomes important for this analysis, considering the different values of $V_{TH}$

obtained for different devices and with different extraction methods.

The mean values of the threshold voltage and the standard deviation obtained for all measured devices are shown in Fig. 4. It is worth mentioning that the $V_{TH}$ using the CC and $G_M/I_D$ methods could not be reliably extracted in the devices with L = 40 nm. These two methods are highly dependent on the subthreshold region and since the devices with this dimension presented some degradation due to short-channel effects (SCE), the extracted values are not reliable. Another consequence of SCE is the reduction of $V_{TH}$, which can be observed by the comparison of the results extracted for devices with L = 100 nm and 40 nm. The exception is the 2DL method, which showed larger values for shorter devices, indicating that this method might not be suitable for SCE analysis.

Table 1 - Number of Measured Devices of Each Dimension ($W_{FIN}$ and L).

| L \ $W_{FIN}$ | 10 nm | 15 nm | 20 nm | 40 nm |
|---|---|---|---|---|
| 40 nm | 35 | 32 | 31 | 37 |
| 100 nm | 29 | 28 | 27 | 26 |

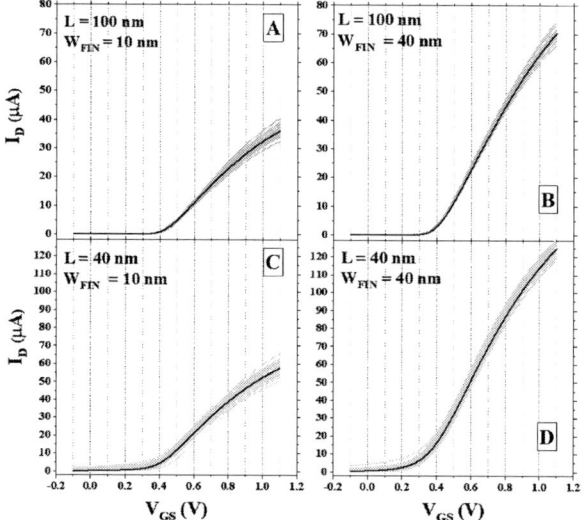

Fig. 2 Drain current vs. gate voltage for SOI nanosheet transistors with L = 100 nm and $W_{FIN}$ = 10 nm (A), L = 100 nm and $W_{FIN}$ = 40 nm (B), L = 40 nm and $W_{FIN}$ = 10 nm (C) and L = 40 nm and $W_{FIN}$ = 40 nm (D).

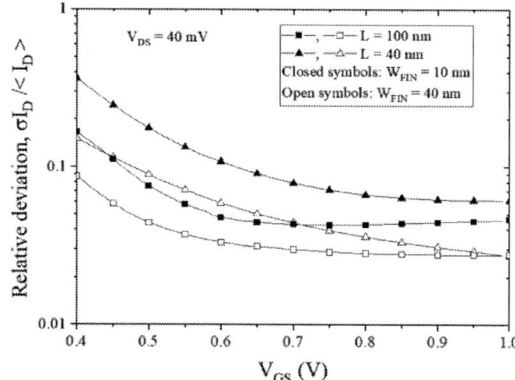

Fig. 3 Relative deviation of the drain current as a function of the gate voltage for SOI nanosheet transistors with L = 100 nm and $W_{FIN}$ = 10 nm, L = 100 nm and $W_{FIN}$ = 40 nm, L = 40 nm and $W_{FIN}$ = 10 nm, and L = $W_{FIN}$ = 40 nm.

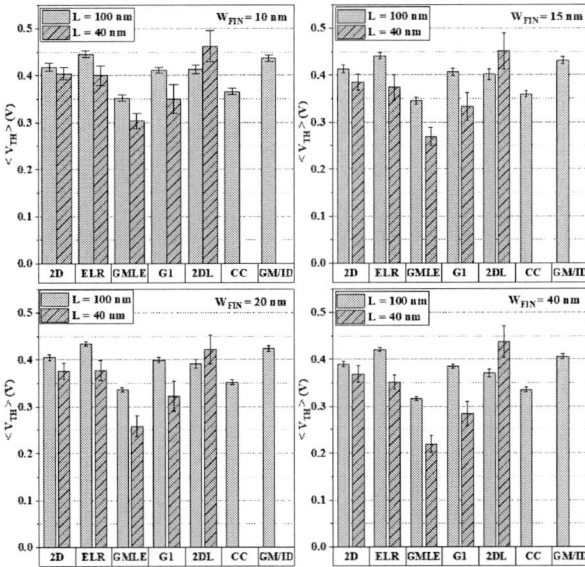

Fig. 4 Mean values of $V_{TH}$ and standard deviation for the seven methods for $L = 100$ nm in orange and $L = 40$ nm in green, for different $W_{FIN}$.

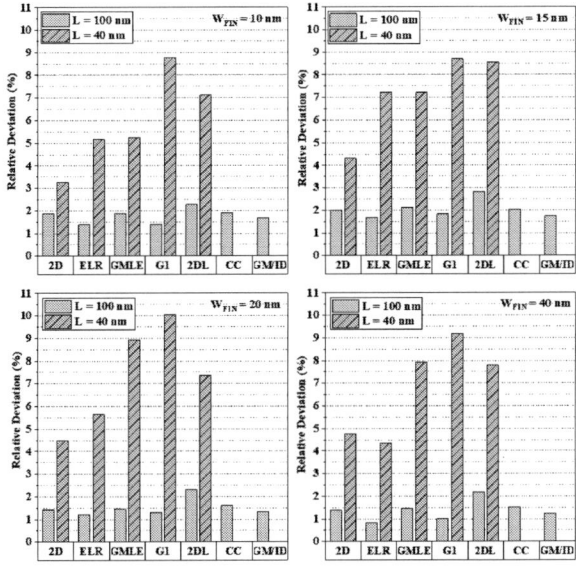

Fig. 5 Relative deviation of the threshold voltage for the seven methods for $L = 100$ nm in orange and $L = 40$ nm in green.

As expected from the inspection of the drain current curves, the standard deviation is higher for the channel length of 40 nm for all methods, with the 2DL method presenting the highest one for all $W_{FIN}$. The $\Delta V_{TH}$ (difference between the highest mean value and the minimum) is equal to 100 mV and 190 mV for long and short-channel transistors, respectively. This error is not negligible, considering that transistors with nanometric dimensions are considered for low-power applications, usually biased at a maximum of 1 V, and these values correspond to an error of 10% and 19%, respectively.

The threshold voltage relative deviation is shown in Fig. 5. Analyzing the long channel devices, it is possible to note that the obtained values for all methods are small, reaching a

maximum of 2.8% in the 2DL method. The smallest value is 0.8%, obtained when applying the ELR method in $W_{FIN} = 40$ nm, which is the device with the largest area amongst all measured transistors. For a given transistor (all samples with the same L and $W_{FIN}$) the larger ratio between maximum and minimum relative deviation in 2.75 times. With the channel length reduction, the values of relative deviation increased for all methods in comparison to $L = 100$ nm and reached up to 10.1% with G1. However, the ratio between maximum and minimum is like that obtained for long channel transistors, of approximately 2.71, obtained for $W_{FIN} = 10$ nm, suggesting that the difference of the relative deviation resulting from the different methods is more dependent on the channel length than on the method itself.

It is interesting to notice that the method with the lowest relative deviation is not the same for long and short-channel length transistors. For $L = 100$ nm, the ELR method presents the lowest values, independent of the $W_{FIN}$. Interestingly, this method is highly dependent on the series resistance $R_{SD}$ [7], whose effect would be more important with channel length reduction and could have a larger impact on the results of $V_{TH}$ with channel length reduction. Also, other methods other than ELR, such as CC and GMLE, are affected by series resistance. Therefore, Fig. 6 presents the $V_{TH}$ - $R_{SD}$ correlations for the threshold voltage extracted using the different methods for transistors with $L = 100$ nm and 40 nm. $R_{SD}$ was extracted using the method proposed in [13]. Considering $W_{FIN} = 10$ nm and 40 nm, the $G_M/I_D$ method presents a weak correlation coefficient ($\rho < 0.07$), indicating that $V_{TH}$ extracted using this method is independent of $R_{SD}$ variation. On the contrary, for ELR method, $\rho$ reaches up to 0.24 with $L = 100$ nm and exceeds the value of 0.55 when L is reduced to 40 nm.

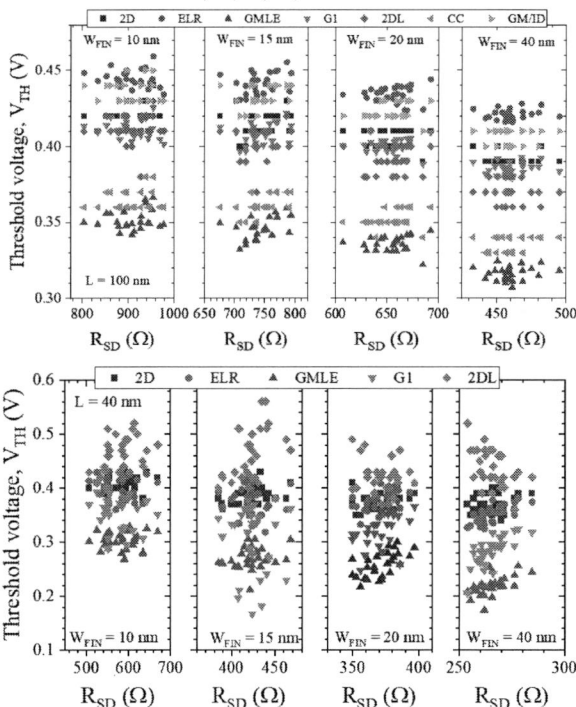

Fig. 6 Correlation between threshold voltage and series resistance plotted for all devices and extraction methods.

979-8-3315-4064-7/24 $31.00 © 2024 IEEE    113

As some of the studied methods rely on the maximum transconductance ($g_{m,max}$), and the corresponding values of $V_{GS}$ and $I_D$ at this point, Fig. 7 presents these variables as a function of the series resistance for all studied devices. The $V_{GS}$ @ maximum transconductance shows a moderate correlation with the $R_{SD}$, while the maximum transconductance and the corresponding drain current show a strong correlation with the series resistance. These values exceed those reported for FinFETs [10] and suggest that the series resistance may significantly influence the extraction of threshold voltage and the analysis of its variability, depending on the selected method.

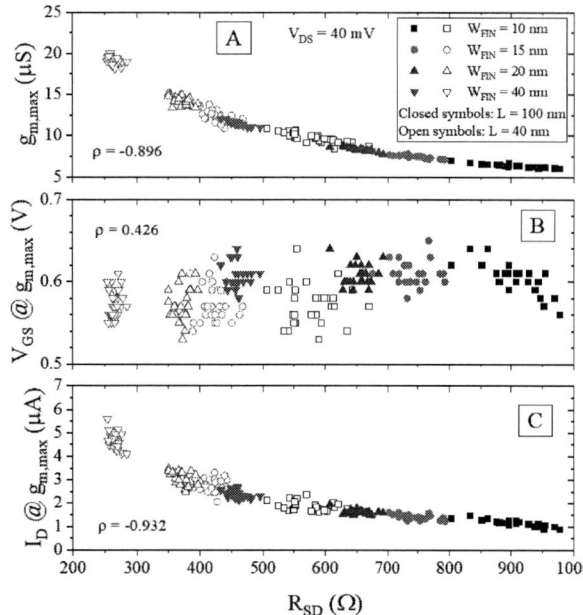

Fig. 7 Curves of (A) maximum transconductance, (B) corresponding gate voltage, and (C) drain current, as a function of the series resistance plotted for all devices biased at $V_{DS} = 40$ mV.

## V. Conclusions

This work presented an experimental evaluation of the impact of threshold voltage extraction methods in the variability results of Ω-gate SOI nanosheet transistors with different channel lengths and widths. In this analysis, seven different methods were used. It was shown that the choice of the extraction method changes the results of threshold voltage variability, and the influence becomes more pronounced for short-channel transistors, which agrees with simulated results presented for nanosheet transistors, and unlike experimental results shown for FinFETs. Also, a significant difference in the mean threshold value was seen, reaching 100 mV and 190 mV, for devices with L = 100 nm and 40 nm, respectively. The presented results also suggest that methods that are affected by series resistance tend to show a larger relative deviation when short-channel transistors are evaluated. In summary, the results demonstrated that the extraction method has a significant influence on the variability results of nanosheet transistor threshold voltage.

## Acknowledgment

The authors thank FAPESP for the financial support to this work (Grant number 2022/12622-5). This work was also partially supported by CNPq Grants 427975/2016-6, 307383/2017-2, 311421/2019-9, 311768/2022-9, 409792/2022-5. The authors acknowledge CEA-Leti for providing the devices used in this study.

## References

[1] J.-P. Colinge, J. C. Greer, "Nanowire transistors: physics of devices and materials in one dimension", Cambridge University Press, 2016. DOI: 10.1017/CBO9781107280779.

[2] R. Coquand et al., "Scaling of high-k/metal-gate Trigate SOI nanowire transistors down to 10nm width," in 13th ULIS, Grenoble, France, 2012, pp. 37–40. DOI: 10.1109/ULIS.2012.6193351.

[3] M. Saitoh et al., "Short-channel performance and mobility analysis of <110>- and <100>-oriented tri-gate nanowire MOSFETs with raised source/drain extensions," in Symposium on VLSI Technology, Honolulu, USA, 2010, pp. 169-170. DOI: 10.1109/VLSIT.2010.5556214.

[4] M. Saitoh, K. Ota, C. Tanaka, Y. Nakabayashi, K. Uchida and T. Numata, "Unified understanding of $V_{th}$ and $I_d$ variability in tri-gate nanowire MOSFETs," 2011 Symposium on VLSI Circuits, 2011, pp. 132-133.

[5] T.A. Oproglidis, D.H. Tassis, A. Tsormpatzoglou, G. Ghibaudo, C.A. Dimitriadis, "Drain current local variability analysis in nanoscale junctionless FinFETs utilizing a compact model", Solid-State Electronics, Volume 170, 2020. DOI: 10.1016/j.sse.2020.107835.

[6] M. J. M. Pelgrom, A. C. J. Duinmaijer and A. P. G. Welbers, "Matching properties of MOS transistors" in IEEE Journal of Solid-State Circuits, vol. 24, no. 5, pp. 1433-1439, Oct. 1989. DOI: 10.1109/JSSC.1989.572629.

[7] A. Ortiz-Conde et al, "Revisiting MOSFET threshold voltage extraction methods", Microelectronics Reliability, vol. 53, n. 1, pp. 90 – 104, 2013. DOI: 10.1016/j.microrel.2012.09.015.

[8] V. R. Prates, M. A. Pavanello and M. de Souza, "Experimental Comparison of Threshold Voltage Extraction Methods in SOI Nanowire Transistors," 2023 37th Symposium on Microelectronics Technology and Devices (SBMicro), Rio de Janeiro, Brazil, 2023, pp. 1-4. DOI: 10.1109/SBMicro60499.2023.10302588.

[9] G. Espiñeira, A. J. García-Loureiro and N. Seoane, "Does the Threshold Voltage Extraction Method Affect Device Variability?," in IEEE Journal of the Electron Devices Society, vol. 9, pp. 469-475, 2021. DOI: 10.1109/JEDS.2020.3046122.

[10] M. S. Bhoir, T. Chiarella, L. Å. Ragnarsson, J. Mitard, N. Horiguchi and N. R. Mohapatra, "Process-induced Vt variability in nanoscale FinFETs: Does Vt extraction methods have any impact?," 2020 4th IEEE Electron Devices Technology & Manufacturing Conference (EDTM), Penang, Malaysia, 2020, pp. 1-4. DOI: 10.1109/EDTM47692.2020.9117815

[11] Pananakakis, G., Ghibaudo, G., Cristoloveanu, S., Threshold voltage in FD-SOI MOSFETs, Solid-State Electronics (2024), doi: https://doi.org/10.1016/j.sse.2024.108947.

[12] S. Barraud et al., "Performance of omega-shaped-gate silicon nanowire MOSFET with diameter down to 8 nm", IEEE Electron Device Letters, vol. 33, no. 11, pp. 1526–1528, Nov. 2012. DOI: 10.1109/LED.2012.2212691

[13] A. Dixit, A. Kottantharayil, N. Collaert, M. Goodwin, M. Jurczak and K. De Meyer,"Analysis of the parasitic S/D resistance in multiple-gate FETs", in IEEE Trans.on Electron Devices, vol. 52, no. 6, pp. 1132-1140, June 2005. DOI: 10.1109/TED.2005.848098.

979-8-3315-4064-7/24 $31.00 © 2024 IEEE

# Optimizing broadband InGaAs/InP photodetectors for the SWIR range

1st Marcelo G. Rua
*LabSem, Engenharia Elétrica*
*Pontifícia Universidade Católica do*
*Rio de Janeiro*
Rio de Janeiro, Brazil
0009-0001-8130-7746

2nd Rudy M. S. Kawabata
*LabSem, Engenharia Elétrica*
*Pontifícia Universidade Católica do*
*Rio de Janeiro*
Rio de Janeiro, Brazil
0000-0001-5565-1618

3rd Mauricio P. Pires
*Instituto de Física*
*Universidade Federal do Rio de*
*Janeiro*
Rio de Janeiro, Brazil
0000-0001-9664-0181

4th Carlos L. Ferreira
*Subdivisão de Cursos de Graduação*
*Instituto Militar de Engenharia*
Rio de Janeiro, Brazil
0000-0002-5790-724X

5th Guilherme M. Torelly
*LabSem, Engenharia Elétrica*
*Pontifícia Universidade Católica do*
*Rio de Janeiro*
Rio de Janeiro, Brazil
0000-0003-3019-2069

6th Patrícia L. Souza
*LabSem, Engenharia Elétrica*
*Pontifícia Universidade Católica do*
*Rio de Janeiro*
Rio de Janeiro, Brazil
patricialustozadesouza@gmail.com

*Abstract* — **Two aspects of *pin* infrared photodetectors with InGaAs as the active layer for the short-wave infrared range (900 to 1700 nm) were optimized. The first one involved using a novel anti-reflective coating formed by two double layers of SiO₂/TiO₂, which presented an average 30% reduction in total reflectance compared to a device without the anti-reflective coating. The second one consisted of introducing a quaternary InGaAsP layer between the active InGaAs layer and the *n-doped* InP layer to reduce interface losses. An extraordinary increase by a factor of six in the responsivity of the device with the quaternary layer was reached.**

*Keywords* — *InGaAsP, anti-reflective, coating, InGaAs, SWIR, photodiode.*

## I. INTRODUCTION

Optoelectronic devices based on InGaAs active layers were widely used for imaging systems in the short-wave infrared range (SWIR), due to high quantum efficiency, electron mobility and absorption coefficient [1-2]. Different approaches to optimize the performance of such *pin* diodes were pursued and reported in the literature [3-5]. Optimizing the anti-reflective (AR) coating can greatly improve the performance of the device. Different materials, such as $TiO_2$, $SiO_2$ and other dielectrics have already been reported as good candidates for this AR coating. In particular, we have already proposed a bi-layer of $TiO_2$ and $SiO_2$ AR coating for the SWIR range (900 nm to 1700 nm) [6]. For avalanche photodiode, it has been reported that introducing the quaternary InGaAsP between InGaAs and InP mitigated the effects of hole trapping caused by the band offset between these materials [7-8].

In this work, we have used a $TiO_2$ and $SiO_2$ bi-layer as AR coating on the *pin* diodes and introduced an InGaAsP layer between the active InGaAs layer and the *n-type* InP buffer layer. The role played by these two novelties in improving the performance of the InGaAs infrared photodetectors was investigated.

## II. EXPERIMENTAL DETAILS

### A. Device structure

The structures of the investigated devices are shown in Figure 1. A 10 nm quaternary InGaAsP layer was incorporated between the 3000 nm InGaAs and the 400 nm InP layers. The idea was to reduce carriers trapping between InP and InGaAs and, therefore, increase the photogenerated current. The samples were grown in an MOVPE (metal-organic vapor phase epitaxy) system, model AIX-200 by Aixtron company. Our MOVPE system has an *in-situ* real-time monitoring system to guarantee the crystalline quality.

### B. Simulation of the reflectance curves

The reflectance of the AR coating was calculated using the Essential Macleod software. The AR coating was formed by two $SiO_2/TiO_2$ bi-layers with the following thicknesses: 196.9/39.5/25.2/64.9 nm. Simulations were carried out for an InGaAs/InP device with mesa structure and frontal incidence. Figure 2 presents the calculated reflectance curves (dashed

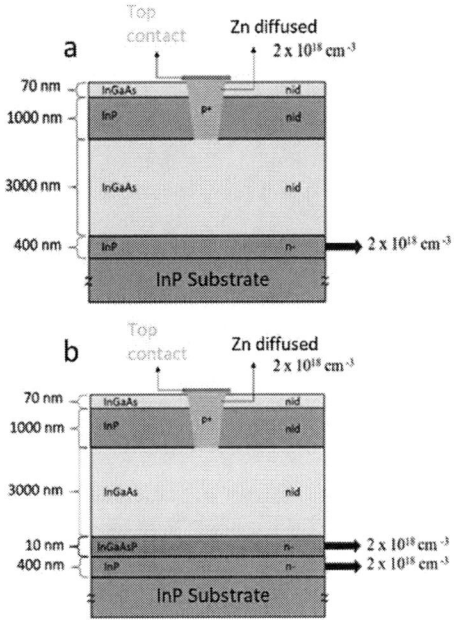

**Fig. 1**: Structures of the devices investigated in this work. (a) InGaAs *pin* diode structure without the InGaAsP quaternary layer, (b) with the InGaAsP layer in the diode. The *pn* junction was formed by Zn diffusion from the top surface. *nid* refers to not intentionally doped.

red) of the device without and with the AR coating, curves (2) and (4), respectively.

### C. AR coating deposition

The AR coatings were deposited using a sputtering system model ATC 2200-V by Aja International. The deposition process was conducted at a chamber pressure of 5 mTorr and with an Argon flow rate of 50 sccm. The sample was not heated during the process. The $SiO_2$ and $TiO_2$ targets were powered using a DC source set at 100 W, achieving a deposition rate of 0.44 Å/s and 0.37 Å/s, respectively.

### D. Spectrophotometry measurements

The reflectance spectra of the device with and without the AR coating were measured in the wavelength range between 900 and 1700 nm, using a spectrophotometer Cary 5000 UV-Vis-NIR by Agilent. The spectrophotometry technique was used to measure the reflectance of the surface of the device.

### E. Responsivity measurements

Responsivity measurements were conducted using a transimpedance amplifier model SR570 by Stanford Research Systems (SRS), a lock-in amplifier model SR530 by SRS, a chopper at 160 Hz and a blackbody at 800 K model SR-2-23 by CI.

### III. RESULTS AND DISCUSSIONS

The results of the measured and simulated reflectance of the investigated photodetectors are shown in Figure 2. The data for the measured reflectance and with AR coating is depicted by the solid dark yellow curves 1 and 3, respectively, while those for the simulated data are represented by the dashed red curves 2 (without AR coating) and 4 (with AR coating). The horizontal dotted lines represent the average reflectance of the respective curves. The oscillations observed are due to interference, given that the thicknesses of the semiconductor layers are comparable to the incident wavelength range. The difference in the periodicity of the oscillations of the experimental and simulated reflectance curves shows that the nominal semiconductor thicknesses were not exactly attained.

Theoretically, according to the data in Figure 2, an average reduction in reflectance of 28.9 ± 5.0% was observed. Experimentally, in Figure 2, the average reflectance decreased from 43.2 ± 3.7% (without AR coating, curve 1) to 13.5 ± 2.8% (with AR coating, curve 4). This represents a total average reduction of 29.7 ± 4.6% instead of 28.9 ± 5.0% and shows that the proposed AR coating successfully achieved the expected decrease in reflectance. However, the absolute value of the experimental reflectance did not exactly match the simulated value, which can be attributed to the differences in the nominal and experimental thicknesses of the top 70 nm InGaAs layer.

A figure of merit of paramount importance for photodetectors is the responsivity. We chose to use the responsivity to probe the influence of the quaternary layer on the device's performance. The responsivity measurements were performed at room temperature and in photovoltaic mode (without applied bias voltage). The data presented in Figure 3 show that the AR coating increases the responsivity of the conventional device (CD) by 20% (from 4.01 to 4.86 mA/W). However, according to Figure 2, the AR coating enhances the photon transmission by around 30%, therefore, part of the transmitted photons, approximately 1/3, does not contribute to the photogenerated current. The insertion of the quaternary layer into the device (DQ) improved the responsivity of the device by a factor of almost six. This tremendous performance improvement is attributed to a reduction in carrier trapping at the InP/InGaAs interface, due to the smoother change in bandgap [9, 10]. Such an improvement exceeded expectations. Therefore, a detailed investigation of the optical and morphological properties of the interface should be undertaken to clarify this issue.

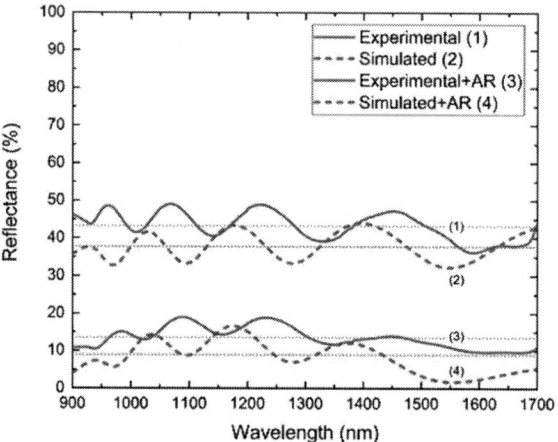

**Fig. 2**: Reflectance curves are calculated with the Essential Macleod software. In dark yellow (solid) the reflectance curves of the experimental detector without (1) and with (3) the proposed AR coating, in black (dotted) reflectance of simulated devices without (2), and in red (dashed) reflectance of simulated device without (2) and with (4) the proposed AR coating. The horizontal dotted lines represent the average of the curve with the same color, dark yellow for experimental device and red for simulated one.

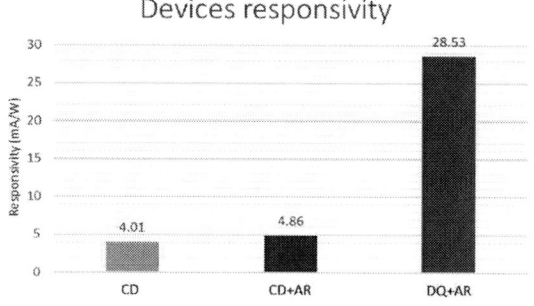

**Fig. 3**: Sample responsivity at room temperature in the photovoltaic mode. In blue are the devices without the addition of the quaternary layer (CD and CD +AR) and in red is the device with the quaternary layer and the AR coating (DQ+AR).

## IV. CONCLUSION

We proposed two approaches to improve the performance of *pin* infrared photodetectors in the SWIR wavelength range (900 to 1700 nm). The first one was the optimization of the AR coating to minimize reflectance on the surface. The introduction of two bi-layers of $SiO_2$ and $TiO_2$ as AR coating was able to attenuate the reflectance of the devices by 29.7%. This result showed that the proposed coating successfully reduced reflectance in the device, but there remains room for improvements since the experimental reflectance has not yet reached the simulated value. The second approach was the introduction of a quaternary InGaAsP layer between the active InGaAs and the n-*doped* InP layers. The responsivity increased by a factor of six, indicating that the quaternary layer played a fundamental role in reducing carriers` traps at the interface. Such an enormous improvement requires further investigation to pinpoint the real role played by this layer. These results indicate that the proposed approaches indeed considerably improve the performance of devices in the SWIR range.

## ACKNOWLEDGMENTS

The partial support by the Coordenação de Aperfeiçoamento de Pessoal de Nível Superior – Brasil (CAPES) – Finance Code 001 and other funding agencies FINEP, FAPERJ and CNPq is acknowledged. We thank CBPF for the use the facilities for the AR coating deposition.

## REFERENCES

[1] F. Rutz et al., "InGaAs infrared detector development for SWIR imaging applications," Proc. SPIE 8896, Electro-Optical and Infrared Systems: Technology and Applications X, vol. 88960, October 2013.

[2] J. Boisvert, T. Isshiki, R. Sudharsanan, P. Yuan and P. McDonald, "Performance of very low dark current SWIR PIN arrays," Proc. SPIE 6940, Infrared Technology and Applications XXXIV, vol. 69400, May 2008.

[3] B. Yang, Y. Yu, G. Zhang, X. Shao and X. Li, "Design and Fabrication of Broadband InGaAs Detectors Integrated with Nanostructures," Sensors, vol. 23, n. 14:6556, July 2023.

[4] X. Li et al., "High performance visible-SWIR flexible photodetector based on large-area InGaAs/InP PIN structure," Scientific Reports, vol. 12, n. 1, pp. 7681, May 2022.

[5] O. M. Braga et al., "Investigation of InGaAs/InP photodiode surface passivation using epitaxial regrowth of InP via photoluminescence and photocurrent," Materials Science in Semiconductor Processing, vol. 154, pp. 107200, November 2023.

[6] M. G. Rua et al., "Anti-reflection coatings for photodetectors to reduce the reflectance in the SWIR range," 2023 37th Symposium on Microelectronics Technology and Devices (SBMicro), IEEE, pp. 1-3, November 2023.

[7] W. Liang et al., "Research progress of InGaAs single photonavalanche diode arrays detector," Fourteenth International Conference on Information Optics and Photonics (*CIOP 2023*), SPIE, vol. 12935, November 2023.

[8] R. H. Hadfield et al., "Single-photon detection for long-range imaging and sensing," Optica, vol. 10, n. 9, pp. 1124-1141, August 2023.

979-8-3315-4064-7/24 $31.00 © 2024 IEEE

# Neutron-induced effects on a commercial GaN High Electron Mobility Transistor

Alexis Cristiano Vilas Bôas
Electrical Engineering Department
FEI University Center
São Bernardo do campo, Brazil
alexiscvboas@fei.edu.br

Saulo Gabriel Alberton
Physics Department
USP
São Paulo, Brazil
alberton@if.usp.br

Paulo Roberto Garcia Jr.
Electrical Engineering Department
FEI University Center
São Bernardo do campo, Brazil
uniepajunior@fei.edu.br

Nilberto H. Medina
Physics Department
USP
São Paulo, Brazil
medina@if.usp.br

Vitor Ângelo P. Aguiar
Physics Department
USP
São Paulo, Brazil
vitor_ap_aguiar@hotmail.com

Marco Antônio A. Melo
Electrical Engineering Department
FEI University Center
São Bernardo do campo, Brazil
mant@fei.edu.br

Roberto Baginski B. Santos
Physics Department
FEI University Center
São Bernardo do campo, Brazil
rsantos@fei.edu.br

Renato C. Giacomini
Electrical Engineering Department
FEI University Center
São Bernardo do campo, Brazil
renato@fei.edu.br

Tássio V. Cavalcante
Nuclear Energy Department
*IEAv - FAB*
São José dos Campos, Brazil
cavalcantetcc@fab.mil.br

Luis Eduardo Seixas Jr.
Engineering Department
CTI
Campinas, Brazil
luis.seixas@cti.gov.br

Saulo Finco
Engineering Department
CTI
Campinas, Brazil
saulo.finco@cti.gov.br

Francisco Rogelio Palomo Pinto
Electrical Engineering Department
University of Seville
Seville, Spain
fpalomo@us.es

Marcilei Guazzelli
Physics Department
FEI University Center
São Bernardo do campo, Brazil
marcilei@fei.edu.br

*Abstract*— *This work explores the effects observed in a commercial-off-the-shelf (COTS) GaN HEMT when exposed to a monoenergetic 14 MeV fast neutron source. The results emphasize the significance of both the device's technology and the neutron source set up for such conditions.*

*Keywords—GaN HEMT, Synergistic Effects, Monoenergetic Fast Neutrons*

## I. INTRODUCTION

High Electron Mobility Transistors (HEMTs) stand as a pinnacle of innovation, offering unparalleled performance across a spectrum of power applications. Their high-frequency operations, combined with unparalleled power handling capabilities, have made them indispensable in diverse fields ranging from telecommunications to space exploration [1, 2].

In contrast to traditional Metal-Oxide-Semiconductor Field-Effect Transistors (MOSFETs), which rely on charge carriers within a silicon substrate, HEMTs exploit the superior electron mobility inherent in compound semiconductor materials such as Gallium Nitride (GaN), counting with a confined bidimensional electron gas (2DEG), thus different physical mechanisms to be addressed [1- 5].

Additionally, the transistors' channel is restricted by a barrier layer and an intrinsic active substrate. Under these circumstances, GaN is deposited on a mechanical support bulk, usually, Silicon Carbide (SiC) or Silicon (Si), separated by a polycrystalline GaN layer named buffer [5- 8].

As the demand for higher speed, power, and radiation reliability continues to increase, the importance of understanding and harnessing the potential of HEMTs becomes increasingly crucial [8- 13].

Important to highlight, that our research group has conducted a comparative analysis between X-rays and gamma rays on the same device [9, 14, 15] of this work. Therefore, this work is a key sequel to the subject.

## II. METHODOLOGY

### A. Objective

This work methodology originated to analyze the effects caused by the charge collection induced by neutrons on surfaces and interfaces of a COTS HEMT based on the AlGan/GaN-on-Si heterostructure, as well as observe indications of atomic displacements. The study was performed using a 14 MeV monoenergetic fast neutron beam provided by a Deuterium-Tritium (D-T) neutron generator, with a typical neutron yield of about $10^8$ neutrons/s [16].

The energy of the fast neutron beam was accurately measured using a 100 μm thick fully depleted Si surface barrier (SSB) charged particle detector, which detected the nuclear reactions $^{28}Si(n, \alpha)$, $^{25}Mg$, and $^{28}Si(n, p)28Al$ [16].

979-8-3315-4064-7/24 $31.00 © 2024 IEEE

### B. Facilities

The tests to assess the sensitivity of the DUTs were conducted at the Instituto de Estudos Avançados (IEAv), Brazil. Details regarding the equipment and calibration procedures are provided in references [16, 17].

### C. Irradiation methodology

Throughout exposure, one device was kept unbiased (off-mode) while the other was polarized (on-mode) [14, 15]. For On-mode the device gate terminal was biased with $V_{GS} = 5.0$ V while the source and drain terminals were grounded. On the other hand, for off-mode, all terminals were grounded. The on- and off-mode DUTs were simultaneously characterized before, during, and after the exposure to a neutron beam at a flux of about $(2.25 \pm 0.06) \times 10^5$ neutrons.cm$^{-2}$.s$^{-1}$.

Measurements were taken every 30 minutes, with a total exposure time of 24 hours. Therefore, the experiment counted with a total fluence of about $(1.90 \pm 0.09) \times 10^{10}$ neutrons.cm$^{-2}$.

### D. Correlation procedure

The parameters evaluated to characterize the DUTs, under neutron exposure, were: the on-current to the off-current ratio ($I_{ON}/I_{OFF}$), which measures a transistor's performance in terms of how effectively it can switch between the "on" state (conducting) and the "off" state (non-conducting) [11, 12]; the channel resistance ($R_{DSON}$), that provides deeper insights into the behavior of the two-dimensional electron gas (2DEG) and the ohmic contact [11, 12]; and the intrinsic gain (Av), that is the gain of the transistor without any external components or loading effects influencing its performance [11, 12]. All these parameters were acquired before, during, and after the radiation exposure.

### E. Parameters's acquisition

Electrical characterization was conducted using a National Instruments PXI test platform, managed by a LabView application [14, 15, 18], in a controlled environment at 24°C. The $I_D$ vs $V_D$ characteristic curves were recorded with $V_{DS}$ ranging from 0 up to 30 mV and $V_{GS}$ set to 3.0 V. The $I_D$ vs $V_G$ curves were obtained for $V_{GS}$ ranging from 0 up to 5.0 V and drain voltage at $V_{DS} = 10$ mV. Within these voltage and current ranges, we are able to check the DUT's sensitivity to the expected effects, which manifest clearly at low currents. For additional details on the electrical characterization methodology and equipment, refer to references [14, 15].

### F. DUT

The DUT (Device-under-test) was a COTS (Component-of-the-shelf) GaN on a Si bulk HEMT, GS61008T, fabricated by Gan Systems. Other results with the same device can be seen in the ref. [9].

## III. RESULTS AND DISCUSSIONS

### A. $I_D$ vs $V_G$ curves comparison

Fig. 1 presents $I_D$ vs $V_G$ pre- and post-radiation on both on- and off-mode. The uncertainty was estimated to be around 5% based on the experimental fluctuation. T1 refers to the transistor that was irradiated in the on-mode, and T2 refers to the transistor irradiated in the off-mode.

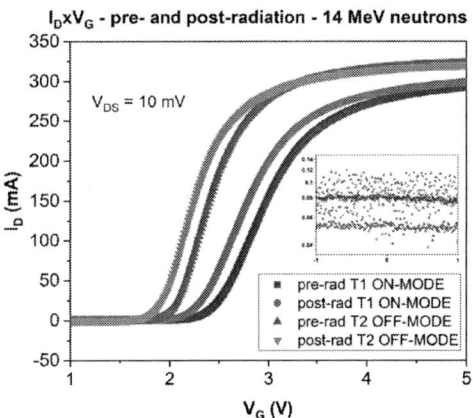

Fig. 1. $I_D$ vs $V_G$, comparing both the on- and off- modes, under exposure to a 14 MeV neutron beam.

The most prominent effect caused by fast monoenergetic neutron in a transistor is displacement damage (DD), which is the effect that reallocates some of the atomic structure of the material, this damage can cause mobility and channel resistance degradation in the device, and those effects can be permanent or just temporary [11, 12, 19-21]. Therefore, the DD can be visualized by the slope and tension shifts on the $I_D$ vs $V_G$ curve. However, it can only be quantified by analyzing those parameters.

As depicted by Fig. 1, in this case, both the on- and off-modes presented similar behavior, a left shift, of about $(192 \pm 9)$ mV, on both curves and a slight change in the slope. However, for the on-mode, we can see an increase in the on-current of about $(+6.91 \pm 0.30)$ mA, and a decrease in the off-mode of about $(-4.53 \pm 0.22)$ mA.

### B. On-current to Off-current Ratio ($I_{ON}/I_{OFF}$)

The on-current ($I_{ON}$) refers to the current flowing through the transistor when it is in the "on" state. The off-current ($I_{OFF}$) refers to the leakage current.

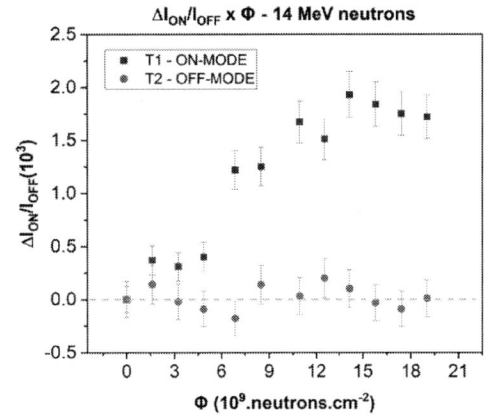

Fig. 2. $\Delta I_{ON}/I_{OFF}$ as a function of the neutron fluence ($\Phi$), comparing both the on- and off- modes, under exposure to a 14 MeV neutron beam.

The ratio $I_{ON}/I_{OFF}$ is important because it indicates how effectively the transistor can maintain a high on-current while minimizing off-current leakage. A higher $I_{ON}/I_{OFF}$ ratio indicates a higher power consumption [11, 12, 19-21].

Fig. 2 presents $\Delta I_{ON}/I_{OFF}$ as a function of the neutron fluence ($\Phi$). The reference value for the on-mode was $I_{ON}/I_{OFF}$ (T1) = (2.4 ± 0.12) x $10^3$, for the off-mode $I_{ON}/I_{OFF}$ (T2) = (3.4 ± 0.17) x $10^3$. The uncertainty was estimated to be around 5% based on the experimental $I_{ON}/I_{OFF}$ peak fluctuation.

As expected, the $I_{ON}/I_{OFF}$ parameter only degraded when the device was irradiated turned on (on-mode) [11, 12, 19-21]. All data to T2 (off-mode) showed no meaningful change since the error bar stood within the reference line (gray line in Fig. 2). Which means that this device will only increase its power consumption due to 14 MeV fast neutrons if it gets exposure when turned on. Additionally, when turned on, the device presented a maximum degradation of about (+1.93 ±0.09) x $10^3$ and a saturation tendency, with a slight decrease, afterward.

It is important to highlight, that the higher shift on this parameter occurred around the neutron fluence of $\Phi$ = 6 x $10^9$.neutrons.cm$^{-2}$, shifting from (+0.40 ± 0.02) x $10^3$ to (+1.22 ± 0.06) x $10^3$. This fluence can represent an important mark, to this experiment since it can be related to a considerable displacement event that might have occurred [11, 12, 19-21].

*C. Intrinsic Gain (Av)*

Fig. 3 presents $\Delta Av$ as a function of the neutron fluence ($\Phi$). The reference value for the on-mode was Av (T1) = (-15.53 ± 0.70) dB, for the off-mode Av (T2) = (-11.57 ± 0.50) dB. The uncertainty was estimated to be approximately 5% based on the observed fluctuation Av peak.

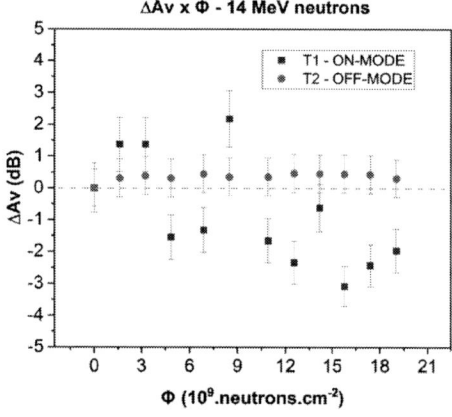

Fig. 3. $\Delta Av$ as a function of the neutron fluence ($\Phi$), comparing both the on- and off- modes, under exposure to a 14 MeV neutron beam.

The intrinsic gain (Av) quantifies the device's ability to amplify an input voltage signal. It directly influences key performance metrics such as gain-bandwidth product, linearity, and speed of operation. A higher intrinsic gain implies that a minor change in the gate-source voltage can result in a larger change in the drain current, indicating greater amplification capability [11, 12, 19- 21].

The intrinsic gain (Av) seems to show a similar behavior compared to the $I_{ON}$ to $I_{OFF}$ ratio ($I_{ON}/I_{OFF}$), since both only presented degradation in the on-mode and also a maximum parameter shift at the neutron fluence of about $\Phi$ = 6 x $10^9$.neutrons.cm$^{-2}$. For the off-mode, the decrease of

the Av represents a reduction in the device's ability to amplify an input voltage signal, a lesser gain-bandwidth product, and a smaller speed of operation [11, 12, 19- 21].

In the beginning of the irradiation, in the off-mode, we can see that the Av presents a positive shift for the amplification value, then out of the sudden it jumps to a negative shift value, showing evidence of an important displacement event that might have occurred at the described fluence.

The maximum negative curve movement took place in the off-mode, the value was about Av = (-3.1 ± 0.1) dB. Afterwards, the gain starts to recover to the reference value, due to the room temperature annealing.

*D. Channel resistance (RD$_{SON}$)*

One of the key characteristics of transistors is the drain to source on-state resistance, known as $R_{DSON}$. The channel resistance ($R_{DSON}$) can be estimated from the $I_{DS}$ vs $V_{DS}$ curve. In this case, all $R_{DS}$ were acquired with $V_{GS}$ = 3.0 V.

Fig. 4 presents $\Delta R_{DS}$ as a function of the neutron fluence ($\Phi$). The reference value for both the on and off-mode was $R_{DS\_REF}$ = (81 ± 4) m$\Omega$. The uncertainty was estimated to be approximately 5% based on the $R_{DS}$ fluctuation.

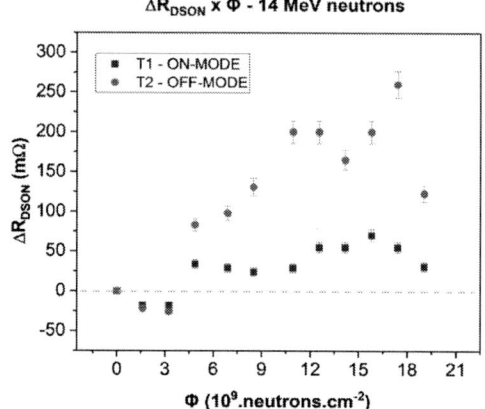

Fig. 4. $\Delta R_{DS}$ as a function of the neutron fluence ($\Phi$), comparing both the on- and off- modes, under exposure to a 14 MeV neutron beam.

In both the on- and off-modes the channel resistance ($R_{DS}$) increased, representing a degradation in the channel's mobility ($\mu_e$) and the conduction ability related to the bi-dimensional electron gas (2DEG) that characterizes the GaN-based transistor [11, 12, 19- 21].

Additionally, the off-mode showed a greater degradation compared to the on-mode. T2 had an $R_{DS}$ degradation value of about $\Delta R_{DS}$ (T2) = (+260 ± 13) m$\Omega$, on the other hand, T1 had a maximum degradation of about $\Delta R_{DS}$ (T1) = (+70 ± 3) m$\Omega$. This means that the off-mode degraded 73% more than the on-mode.

## IV. CONCLUSION

The findings illustrate the device's behavior against the impact of 14 MeV monoenergetic fast neutron exposure, which can induce both cumulative effects such as charge trapping/release and atom displacement. It is evident that the charge trapping mechanism, within defect states, varies with the irradiation mode, suggesting a considerable influence of the applied electric field in the operational state (on-mode).

Firstly, the $I_{ON}/I_{OFF}$ parameter and intrinsic gain (Av) demonstrated degradation exclusively when the device was irradiated and turned on (on-mode). Notably, this degradation reached its maximum shift jump value at a neutron fluence of $\Phi = 6 \times 10^9$ neutrons/cm$^{-2}$, marking a critical point possibly indicating substantial displacement events within the device structure.

Furthermore, the device's amplification ability, as reflected by the intrinsic gain (Av), experienced a decline in the off-mode under neutron irradiation. The channel resistance ($R_{DS}$) increased in both on- and off-modes, changing the device's functionality.

In conclusion, we can highlight the importance of the irradiation source and device technology for interpreting the resulting physical mechanism for an effective assessment of a device's robustness for applications in harsh environments, and for a deeper understanding of different devices' responses to various effects in different operating modes.

## V. FUTURE STEPS

An experimental methodology has been developed to further investigate this device's effect. A gate and channel capacitance analysis, in the form of a C versus V curve, is being prepared for pre- and post-gamma-ray, X-ray, and neutron irradiation.

### ACKNOWLEDGMENT

The authors acknowledge financial support from the funding agencies. FAPESP, Brazil 2023/16053-8, 2022/09131-0, 2018/25225-9, 2020/04867-2, 2019/07764-1; CITAR: Proc. 01.12.0224.00; INCT_FNA, Proc. 464898/2014-5; CNPq: 404054/2023-4, 408800/2021-6, 301576/2022-0, 30360/2020-9.

## REFERENCES

[1] Ma, Chao-Tsung, and Zhen-Huang Gu. 2019. "Review of GaN HEMT Applications in Power Converters over 500 W" Electronics 8, no. 12: 1401. https://doi.org/10.3390/electronics8121401

[2] L. T. Moniz and F.R. Palomo "Implementación de un SSPC com distitnas tecnologías MOSFET," in Documentacón Selena, Dpto. Ingeniería Eléctrónica,Sevilla, 2018, pp. 22–30.

[3] A. K. Visvkarma et al., "Impact of Gamma Radiations on Static, Pulsed I-V, and RF Performance Parameters of AlGaN/GaN HEMT," in IEEE Transactions on Electron Devices, doi: 10.1109/TED.2022.3161402.

[4] H. Ohta et al., "High-Electron-Mobility Transistors (HEMTs) - Devices and Circuits," Springer, 2013.

[5] A. Mishra and F. K. Knechtli, "Gallium Nitride (GaN) Based High Electron Mobility Transistors (HEMTs)," Springer, 2017.

[6] M. Shur and R. Gaska, "High Electron Mobility Transistors: Physics, Modeling, and Technology," Wiley-Interscience, 2001.

[7] T. Palacios et al., "Gallium Nitride (GaN) HEMTs: Advanced Device Design and Future Prospects," CRC Press, 2017.

[8] S. Nakamura, "GaN Growth, Devices, and Applications," Proceedings of the IEEE, vol. 101, no. 10, pp. 2161-2181, 2013.

[9] A. C. V. Bôas et al., "Ionizing Radiation Hardness Characterization of GaN HEMTs Depends on the Radiation Source," 2022 22nd European Conference on Radiation and Its Effects on Components and Systems (RADECS), Venice, Italy, 2022, pp. 1-4, doi: 10.1109/RADECS55911.2022.10412526.

[10] Yu Song at al.,"Mechanism of Synergistic Effects of Neutron- and Gamma-Ray-Radiated PNP Bipolar Transistors"ACS Appl. Electron. Mater. 2019, 1, 4, 538–547

[11] Johnston, Allan. "Reliability and radiation Effects in Compound Semiconductors." World Scientific Publishing Co. Pte. Ltd., California Institute of Technology, USA, 2010.D

[12] Attix, F.H. "Introduction to Radiological Physics and Radiation Dosimetry" 1ed. Weinheim: Germany: Wiley-VCH Verlag GmbH & Co. KGaA, 2004.D

[13] Knoll, G.F. Radiation detection and measurement. [S.l.]. Nova Jersey, EUA: Editora Wiley, 1989.D

[14] A. C. V. Bôas et al. "Ionizing radiation hardness tests of GaN HEMTs for harsh environments." Microelectronics Reliability, Vol. 116, 2021, https://doi.org/10.1016/j.microrel.2020.114000.d

[15] A.C.V. Bôas et al. "Reliability analysis of gamma- and X-ray TID effects on a commercial AlGaN/GaN based FET" Journal of Integrated Circuits and Systems, vol. 16, n. 3, 2021

[16] S G Alberton et al., "Neutron-Induced Radiation Effects in UMOS Transistor" 2022 J. Phys.: Conf. Ser. 2340 012046

[17] Goncalez, Odair L., Vaz, Rafael G., and Wirth, Gilson (2013). "A platform for TID testing of diodes and transistors." INAC 2013: international nuclear atlantic conference, Brazil

[18] Seixas, L.E., Finco, S. & Gimenez, S.P. VI-Based Measurement System Focusing on Space Applications. J Electron Test 33, 267–274 (2017). https://doi.org/10.1007/s10836-017-5651-3

[19] Sedra, A. S.; Smith, K. C. "Microeletrônica." 4. ed. São Paulo: Pearson, 2000.

[20] S.-J. Chang et al., "Comprehensive research of total ionizing dose effects in GaN-based MIS-HEMTs using extremely thin gate dielectric layer," Nanomaterials, vol. 10, no. 2175, 2020, doi: 10.3390/nano10112175.

[21] M. R. Shaneyfelt, D. M. Fleetwood, J. R. Schwank and K. L. Hughes, "Charge yield for cobalt-60 and 10-keV X-ray irradiations of MOS devices," in IEEE Transactions on Nuclear Science, vol. 38, no. 6, pp. 1187-1194, Dec. 1991, doi: 10.1109/23.124092.

# TCAD-based Performance Evaluation of Dual-Gate UTBB SOI Junctionless ISFET for pH Detection

Claudio Villela Moreira
*Electrical Engineering Department*
*Centro Universitario FEI*
São Bernardo do Campo, Brazil
https://orcid.org/0000-0002-8070-2787

Marcelo Antonio Pavanello
*Electrical Engineering Department*
*Centro Universitario FEI*
São Bernardo do Campo, Brazil
https://orcid.org/0000-0003-1361-3650

*Abstract*—This work presents a TCAD-based performance evaluation of ultra-thin body and box (UTBB) Silicon-On-Insulator (SOI) junctionless (JL) ion sensitive field effect transistor (ISFET) operating in dual-gate mode. The dual-gate operation allows to take advantage of the increased sensitivity from the SOI capacitance coupling and overcome the Nernstian limit. The threshold voltage ($V_{Th}$) and drain current ($I_{DS}$) sensitivities to the pH are analyzed. The $V_{Th}$ sensitivity showed gains up to 109.5 and $I_{DS}$ sensitivity increases up to 404.5% compared to single gate mode.

*Keywords— ISFET, Junctionless, Electrolyte, Sensor, pH*

## I. INTRODUCTION

Detecting particles, such as ions and molecules, holds significant importance in the electrochemical and biological realms [1]. The process involves measuring the potential difference between two electrodes. Bergveld developed the Ion Sensitive Field Effect Transistor (ISFET), designed to detect ions in aqueous solutions [2]. The ISFET modification involves eliminating the gate metal from a MOSFET device and submerging it in a solution alongside a reference electrode [3]. Interactions between ions in the solution and the gate oxide, serving as a sensitive layer, lead to changes in the device surface potential, rendering the ISFET capable of detecting variations in pH [4]. ISFET-based biosensors are extensively utilized for DNA, protein, and virus detection, among other applications [5]. These sensors, compatible with MOS technology, seamlessly integrate into the same substrate, simplifying the incorporation of read-out circuits [3].

Shrinking the channel length of MOSFETs posed challenges within planar technologies. In sub-20nm realms, the diminishing electrostatic control from the gate on channel charges prompted the exploration of multiple gate devices [6], some of them using Silicon-On-Insulator (SOI) substrates. Yet, even with novel inversion-mode devices, challenges persist, such as lateral dopant diffusion into the undoped channel region, particularly pronounced with high doping density gradients, prompting the search romance for novel doping techniques [7]. To relieve this, a novel MOS structure is proposed, called junctionless (JL). In JL transistors, the channel, source, and drain share identical doping types [8], effectively addressing lateral diffusion and the punchthrough effect [9]. The off state in JLs is achieved by fully depleting the silicon layer within the channel, regulated by the workfunction difference between the gate and the silicon film [8]. For an n-type JL, increasing the gate voltage reduces the depletion depth, establishing a neutral path for current flow between the source and drain ($I_{DS}$). Further increase in the gate

voltage ($V_{GS}$), surpassing the flatband voltage, introduces an additional current component associated with the accumulation layer at the interface [10].

To increase the sensitivity of an ISFET beyond the Nernstian limit, the use of dual-gate (DG) ISFET was proposed for the fully depleted (FD) SOI ISFET. In this approach, the ISFET sensitivity is extracted while biasing the device using the back gate (substrate), utilizing the enlarged oxide capacitance and reductive channel thickness, the capacitance coupling increases the sensitivity of DG ISFET. For the inversion mode FD SOI DG ISFET, the threshold voltage ($V_{Th}$) sensitivity is given by [11]:

$$\Delta V_{Th,B} = \frac{3t_{BOX}}{3t_{Ox}+t_{Si}}\Delta V_{Th,F} \qquad (1)$$

where $\Delta V_{Th,B}$ and $\Delta V_{Th,F}$ are values of the threshold voltage shift of back and top gates, respectively, and $t_{Si}$, $t_{BOX}$ and $t_{Ox}$ are the thicknesses of silicon film, the buried oxide (BOX) and effective front gate oxide, respectively [11].

The continuing reduce of device length causes reduction of electrostatic integrity, but the reduction of $t_{Ox}$ reached the limit. To continue reducing device dimensions, the silicon film thickness in FD SOI devices needs to be reduced. But to continue even further with this reduction and keep short channel effects (SCE), the drain induced barrier lowering (DIBL) and other parasitic effects controlled, the BOX thickness needs to be reduced. This thinner BOX and silicon film FDSOI transistors are called UTBB (Ultra-Thin Body and BOX) and shows advantage to high sensitive sensors [12].

In this paper, the sensitivity of UTBB JL ISFET dual-gate mode and compared with conventional mode through bidimensional simulations with TCAD Sentaurus tool.

## II. METHODOLOGY

To simulate the ISFET, the Sentaurus TCAD is used [13]. By default, the simulator does not have the necessary models to simulate ionic solutions nor the electrochemical process of ISFET operation. To overcome this, two models are implemented in TCAD using the Physical Model Interface (PMI) to calculate the electrolyte effective density of states and to consider the chemical reactions in the solution/gate dielectric interface (site-binding model). In addition to that, a Stern layer was included to compensate for the steric effects caused by the finite ionic radius, it's modelled as a thin semiconductor layer and low effective density of states, to allow the correct traps density to be placed and still behave as an dielectric, and the dielectric constant value set to

---

The authors of this work would like to thank the National Council for Scientific and Technological Development (CNPq) for the financial support.

achieve a capacitance of ~20μF/cm². These implementations have been done according to the proposed by Bandiziol [14]. Fig. 1 shows the simulated JL ISFET cross-section.

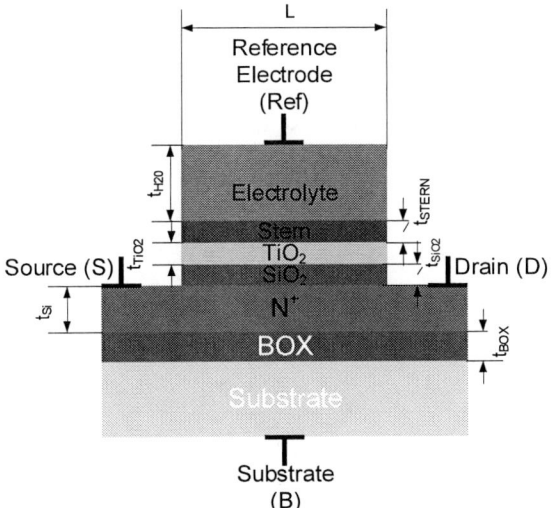

Fig. 1. Schematic view of simulated junctionless ISFET.

### III. SIMULATED DEVICE

An N⁺ silicon JL ISFET has been simulated varying doping concentration and compared the DG ISFET to traditional single-gate (SG). The dimensions have been chosen to guarantee the complete depletion of channel, a requisite to correct JL operation, while utilizing similar dimensions with MOSFETs samples at disposal to reproduce the fabrication process dimensions with the same UTBB wafer, thus no Ground Plane in the simulated device. The electrolyte thickness is chosen to optimize the simulation time, since that represents the position of reference electrode. The parameters of the electrolyte, such as salt concentration ($c_0$), are considered for NaCl.

For the carrier mobility, the IALMob was used, with SHR recombination and BandGap Narrowing in channel region. The simulation was made with $TiO_2$ sensing layer on top of $SiO_2$ as gate oxide. The choice of $TiO_2$ sensing layer is due to the great sensitivity compared to $SiO_2$ [15]. The parameters of simulation are shown in Table I.

TABLE I. SIMULATIONS PARAMETERS

| Parameter | Symbol | Value |
|---|---|---|
| Channel Length | L | 200 nm |
| Channel Height | $t_{Si}$ | 12 nm/ 10.5 nm |
| Source/Drain Length | $L_{SD}$ | 50 nm |
| Buried Oxide (BOX) Thickness | $t_{BOX}$ | 20 nm |
| $SiO_2$ Thickness | $t_{OX}$ | 5 nm/ 8.75 nm |
| $TiO_2$ Tickness | $t_{TiO2}$ | 12 nm |
| Stern Thickness | $t_{STERN}$ | 1 nm |
| Electrolyte Height | $t_{Liq}$ | 100 nm |
| Channel Doping | $N_D$ | $1 \times 10^{18}$ cm⁻³ |
| Substrate Doping | $N_{Sub}$ | $1 \times 10^{15}$ cm⁻³, P-type |
| Salt (NaCl) concentration | c0 | 1 mM |
| Drain Voltage | $V_{DS}$ | 20 mV |

The sensitivity of the ISFETs is studied using 2 methods: $\Delta V_{Th}/pH$ and $\Delta I_{DS}/pH$. Nernstian limit for the traditional ISFET is applied to $V_{Th}$ sensitivity and is about 59 mV/pH at 300K [16]. All $V_{Th}$ values in this work are extracted by the second derivative method and the sensitivities by linear extrapolation.

### IV. RESULTS

For the SG mode ISFET, the $V_{Th,F}$ sensitivity is studied by varying the $t_{ox}/t_{Si}$ rate and the thickness of the deposited $t_{TiO2}$ sensing layer. Fig 2 displays the several $I_{DS}$ versus $V_{Ref}$ curves, obtained with $V_{DS}$ of 20mV, varying the pH, in linear (solid lines) and logarithmic (dashed lines) scales. In these curves, the substrate voltage is kept at $V_{BS}=0$ V.

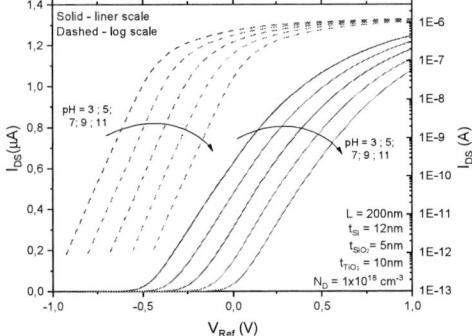

Fig. 2. $I_{DS}$ versus $V_{Ref}$ curves for various pH values

Using the $I_{DS}$ versus $V_{Ref}$ such as those from Fig. 2, the $V_{Th,F}$ has been extracted. The results of the $V_{Th,F}$ versus pH curves for two combinations of $t_{ox}$ and $t_{Si}$ thicknesses are presented in Fig 3.

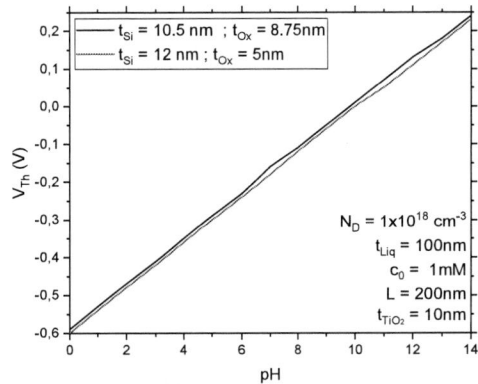

Fig. 3. $V_{Th,F}$ versus pH varying $t_{Si}$.

The $V_{Th}$ versus pH curves show a linear behavior. With the $t_{ox}$ variation, the $V_{Th,F}$ sensitivity, i. e. $\Delta V_{Th}/pH$, remains around 59.4 mV/pH and 59.2 mV/pH for $t_{Si}$ of 10.5 nm and 12 nm, respectively, also similar to inversion mode FD SOI ISFETs [17]. Fig 4 shows the $V_{Th,F}$ versus pH varying $t_{TiO2}$.

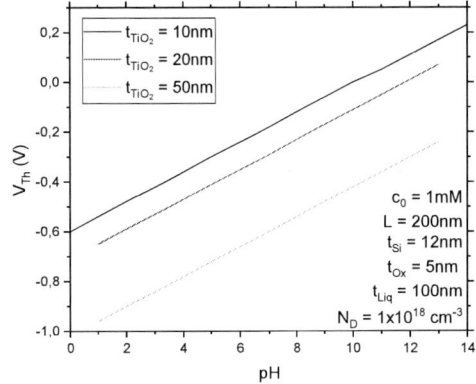

Fig. 4. $V_{Th,F}$ versus pH varying $t_{TiO2}$.

979-8-3315-4064-7/24 $31.00 © 2024 IEEE

The increase in $t_{TiO2}$ changes the $V_{Th,F}$ magnitude but does not affect the $V_{Th}$ sensitivity to pH, remaining in the order of 59.2 mV/pH for all simulated $t_{TiO2}$.

When measuring the $I_{DS}$ sensitivity varying $t_{Si}$, the better electrostatic control of the smaller gate oxide ISFET translates into better sensitivity. The extracted sensibilities for $V_{Ref}$ of 1.0V and 10nm $t_{TiO2}$ are $2.22 \times 10^{-8}$ A/pH and $1.71 \times 10^{-8}$ A/pH with linear fit correlation coefficient of r=0.980 and r=0.986 for 10.5 nm $t_{Si}$ / 8.75nm $t_{Ox}$ and 12 nm $t_{Si}$ / 5nm $t_{Ox}$ respectively. Fig 5 shows the $I_{DS}$ *versus* pH varying $t_{Si}$.

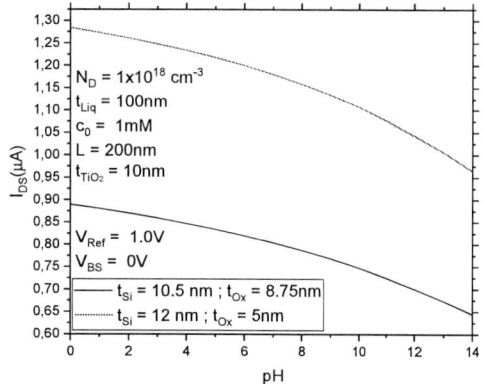

Fig. 5.  $I_{DS}$ *versus* pH varying $t_{Si}$

For the DG mode, the device has been biased using the substrate electrode and the reference voltage has been fixed at 0V. Fig. 6. shows the $I_{DS}$ *versus* $V_{BS}$ curves for various pH values and Fig. 7 shows the $V_{Th,B}$ *versus* pH varying $t_{TiO2}$.

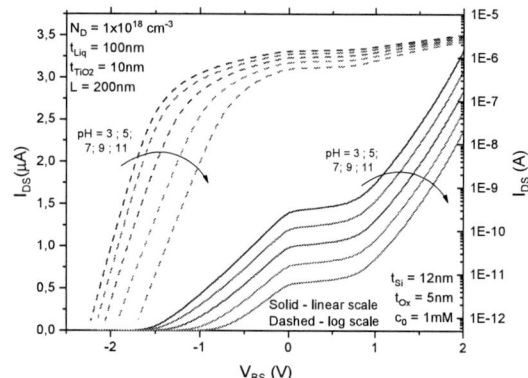

Fig. 6.  $I_{DS}$ *versus* $V_{BS}$ for various pH values

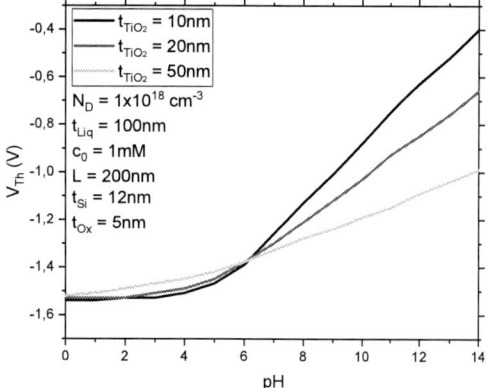

Fig. 7.  $V_{Th,B}$ *versus* pH varying $t_{TiO2}$

While in DG mode, the $V_{Th}$ sensitivity is greatly improved up to 124 mV/pH compared to 59.2 mV/pH from SG mode, but only shows the $V_{Th}$ sensitivity advantage to values of pH higher than $pH_{pzc}$ (pH for point of zero charges, about 6.2 for $TiO_2$ sensing layer).

The $V_{Th}$ sensitivity greatly increases for thinner sensing layer ISFETs, resulting in 124mV/pH for 10 nm $t_{TiO2}$/ 5 nm $t_{ox}$/12 nm $t_{Si}$ device, while it archived 59.2 mV/pH in SG mode, a 109.5% increase. But for the thickest $t_{TiO2}$, the $V_{Th}$ sensitivity is lower than SG mode, resulting in 48.3 mV/pH for 50 nm $t_{TiO2}$/ 5 nm $t_{ox}$/12 nm $t_{Si}$ device, an 18.4% reduction in sensitivity. This reduction, that is similar to eq(1), shows the thicknesses of ISFETs effect in DG mode operation, and to be advantageous needs to be thin in silicon film ($t_{Si}$) and oxide/sensing membrane ($t_{ox}$ and $t_{TiO2}$). All the extracted $V_{Th,B}$ sensitivities in DG mode are shown in Table II.

TABLE II. DG SENSITIVITY OF JL ISFET

| Sensitivity [mV/pH] | $t_{Si}$ = 12 nm $t_{ox}$ = 5nm | $t_{Si}$ = 10.5 nm $t_{ox}$ = 8.75nm |
|---|---|---|
| $t_{TiO2}$ = 10nm | 124 | 93.8 |
| $t_{TiO2}$ = 20nm | 90.2 | 70.8 |
| $t_{TiO2}$ = 50nm | 48.3 | 41.8 |

To utilize the JL ISFET below $pH_{pzc}$ in DG mode, a negative bias is applied to $V_{Ref}$. This changes the $V_{Th,B}$, allowing the DG JL ISFET to be sensible for pH values lower than $pH_{pzc}$ while maintaining the sensitivity at the linear region. Fig 8 shows the $V_{Th,B}$ sensitivity *versus* $t_{TiO2}$ varying $V_{Ref}$ for the most sensitive device ($t_{Si}$ of 12 nm, $t_{ox}$ of 5 nm and $t_{TiO2}$ of 10 nm). For this ISFET, a $V_{REF}$ = -300mV allows the usage for all pH values below $pH_{pzc}$.

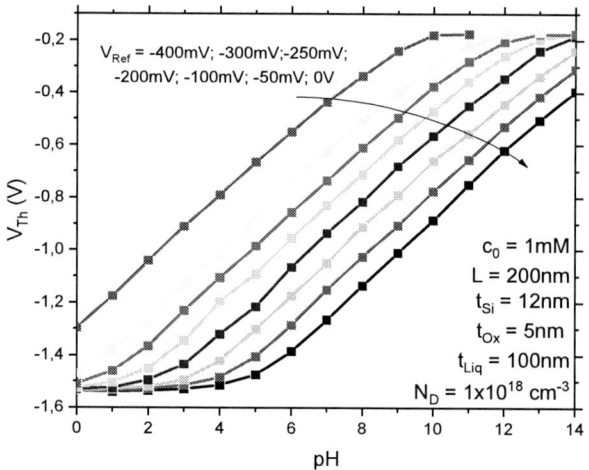

Fig. 8.  $V_{Th,B}$ sensitivity *versus* $t_{TiO2}$ varying $V_{Ref}$

For $I_{DS}$ sensitivity, the DG mode is compared to the best obtained result of SG mode ($t_{Si}$ of 12 nm, $t_{ox}$ of 5 nm and $t_{TiO2}$ of 10 nm) of $2.22 \times 10^{-8}$ A/pH at 1.0V of $V_{GS}$. Differently of $V_{Th}$ sensitivity, in this mode, the $I_{DS}$ sensitivity is linear across all the pH ranges, with a correlation coefficient of r=0.999.

For the $t_{Si}$ of 12 nm, $t_{ox}$ of 5 nm and $t_{TiO2}$ of 10 nm ISFET, extracted sensitivity is $1.12 \times 10^{-7}$ A/pH and $1.18 \times 10^{-7}$ A/pH at $V_{BS}$ of 1.0V and $V_{BS}$ of 1.5V respectively. Comparing both $V_{BS}$ show only a 5.4% increase for the higher $V_{BS}$, but when compared with SG mode the increase is 404.5% for $V_{BS}$ of

1.0V. Fig 9 show $I_{DS}$ versus pH for SG mode (dashed lines) and DG mode (solid lines).

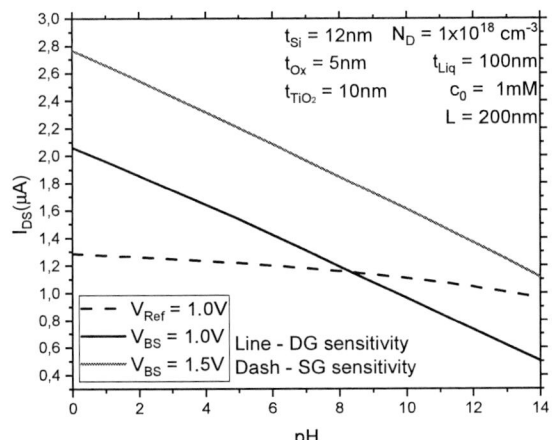

Fig. 9. $I_{DS}$ versus pH for SG mode and DG mode.

For the $t_{Si}$ of 12 nm, $t_{Ox}$ of 5 nm and $V_{BS}$ of 1.0V varying the $t_{TiO2}$, the extracted values of sensitivity $1.12 \times 10^{-7}$ A/pH, $8.04 \times 10^{-8}$ A/pH and $4.34 \times 10^{-8}$ A/pH for $t_{TiO2}$ of 10 nm, 20 nm and 50 nm respectively. These results represent an increase of 404.5%, 262.2% and 95.5% respectively when compared with the best SG mode value. Comparing the SG mode $t_{Si}$ of 12 nm, $t_{Ox}$ of 5 nm and $t_{TiO2}$ of 50 nm sensitivity, with has $2.12 \times 10^{-8}$ A/pH at $V_{GS}$ 1,0V, with the same ISFET in DG mode, the increase is increase to 104.7%. This increase is due to higher $V_{th}$ sensibility allied with higher control provided by thin BOX. Fig 10 shows $I_{DS}$ versus pH DG mode varying $t_{TiO2}$.

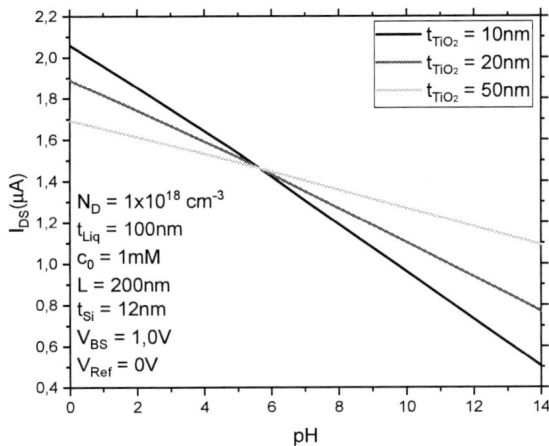

Fig. 10. $I_{DS}$ versus pH DG mode varying $t_{TiO2}$

## V. CONCLUSIONS

The DG JL ISFET shows $V_{Th}$ sensibilities that can be beyond Nernstian limit depending on device geometry for JL ISFET. In this study, this sensitivity of DG mode JL ISFET varies from 109.5% increase on 10nm $t_{TiO2}$/ 5 nm $t_{Ox}$/12 nm $t_{Si}$ device to 18.4% reduction on 50nm $t_{TiO2}$/ 5 nm $t_{Ox}$/12 nm $t_{Si}$ device compared to SG mode. But in all simulated JL ISFETs the sensitivity increases when the $I_{DS}$ is analyzed, varying from 404.5% to 104.7%. This increase in $I_{DS}$ sensitivity shows the reduction of channel control by sensing layer, with the relative increase of substrate control to those charges. With a 27nm $t_{TiO2}$ layer on 12nm $t_{Si}$/ 5nm $t_{Ox}$ ISFET, the sensing layer

and substrate have similar control of the channel charges in the 20nm $t_{BOX}$ simulated devices. To use the DG JL ISFET below the $pH_{pzc}$, a negative bias can be applied to $V_{Ref}$ to shift the $V_{Th,B}$ values, making the ISFET sensible to these pH values. The $I_{DS}$ sensitivity is also linear in all pH range, instead of only above $pH_{pzc}$ in $V_{Th}$ simulations.

REFERENCES

[1] J. S. Parmar, N. Shafi and C. Sahu, "A Novel Multi gate Junctionfree gated resistor ISFET for pH detection," in *2019 9th Annual Information Technology, Electromechanical Engineering and Microelectronics Conference (IEMECON)*, 2019.

[2] P. Bergveld, "Development of an Ion-Sensitive Solid-State Device for Neurophysiological Measurements," *IEEE Transactions on Biomedical Engineering*, p. 70–71, 1 1970.

[3] G. Verzellesi, L. Colalongo, D. Passeri, B. Margesin, M. Rudan, G. Soncini and P. Ciampolini, "Numerical analysis of ISFET and LAPS devices," *Sensors and Actuators B: Chemical*, p. 402–408, 10 1997.

[4] P. Bergveld, "Thirty years of ISFETOLOGY What happened in the past 30 years and what may happen in the next 30 years," *Sensors and Actuators B*, vol. 88, no. 1, p. 1–20, 1 2003.

[5] M. O. Noor and U. J. Krull, "Silicon nanowires as field-effect transducers for biosensor development: A review," *Analytica Chimica Acta*, vol. 825, no. 12, p. 1–25, 5 2014.

[6] J.-P. Colinge, FinFETs and Other Multi-Gate Transistors, J. -. Colinge, Ed., New York: Springer, 2008.

[7] J.-P. Colinge, A. Kranti, R. Yan, C. W. Lee, I. Ferain, R. Yu, N. D. Akhavan and P. Razavi, "Junctionless Nanowire Transistor (JNT): Properties and design guidelines," *Solid-State Electronics*, vol. 65–66, p. 33–37, 2011.

[8] J.-P. Colinge, C.-H. Lee, A. Afzalian, N. D. Akhavan, R. Yan, I. Ferain, P. Razavi, B. O'Neill, A. Blake, M. White, A.-M. Kelleher, B. McCarthy and R. Murphy, "Nanowire transistors without junctions," *Nature Nanotechnology*, vol. 5, p. 225–229, 2010.

[9] R. T. Doria, M. A. Pavanello, R. D. Trevisoli, M. de Souza, C.-L. Lee, I. Ferain, N. D. Akhavan, R. Yan, P. Razavi, R. Yu, A. Kranti and J.-P. Colinge, "Junctionless Multiple-Gate Transistors for Analog Applications," *IEEE Transactions on Electron Devices*, vol. 58, no. 8, p. 2511–2519, 2011.

[10] J. P. Colinge, C. W. Lee, N. Dehdashti Akhavan, R. Yan, I. Ferain, P. Razavi, A. Kranti and R. Yu, Semiconductor-On-Insulator Materials for Nanoelectronics Applications, Berlin, Heidelberg: Springer, 2011.

[11] J. K. Park, H. J. Jang, J. T. Park and W. J. Cho, "SOI dual-gate ISFET with variable oxide capacitance and channel thickness," *Solid-State Electronics*, vol. 97, p. 2–7, 2014.

[12] S. Monfray and T. Skotnicki, "UTBB FDSOI: Evolution and opportunities," *Solid-State Electronics*, vol. 125, p. 63–72, 2016.

[13] Synopsys, Inc., Sentaurus™ Device User Guide, 2020.

[14] A. Bandiziol, P. Palestri, F. Pittino, D. Esseni and L. Selmi, "A TCAD-Based Methodology to Model the Site-Binding Charge at ISFET/Electrolyte Interfaces," *IEEE TRANSACTIONS ON ELECTRON DEVICES*, vol. 62, no. 10, p. 3379–3386, 10 2015.

[15] W. Bunjongpru, A. Sungthong , S. Porntheeraphat, Y. Rayanasukha, A. Pankiew, W. Jeamsaksiri, A. Srisuwan, W. Chaisriratanakul, E. Chaowicharat, N. Klunngien, C. Hruanun, A. Poyai and J. Nukeaw, "Very low drift and high sensitivity of nanocrystal-TiO2 sensing membrane on pH-ISFET fabricated by CMOS compatible process," *Applied Surface Science*, vol. 267, p. 206 – 211, 2013.

[16] R. E. G. van Hal, J. C. Eijkel and P. Bergveld, "A novel description of ISFET sensitivity with the buffer capacity and double-layer capacitance as key parameters," *Sensors and Actuators B: Chemical*, vol. 24, no. 1–3, p. 201–205, 1995.

[17] A. S. M. Zain, A. M. Dinar, F. Salehuddin, H. Hazura, A. R. Hanim, S. K. Idris1 and A. M. A. Hamid., "Beyond Nernst Sensitivity of Ion Sensitive Field Effect Transistor based on Ultra-Thin Body Box FDSOI," *Journal of Physics: Conference Series*, vol. 1502, 2020.

# Proposal of [BE]SOI MOSFET source sensing region for pH monitoring applications.

Pedro H. Duarte[1], Ricardo C. Rangel[1,2], and Joao A. Martino[1], Senior Member, IEEE.

[1] LSI/PSI/USP, University of Sao Paulo, Sao Paulo, Brazil

[2] FATEC-SP, Faculdade de Tecnologia de Sao Paulo, Sao Paulo, Brazil

E-mail: phduarte@usp.br

*Abstract*— This work presents a study of the sensing regions of [BE]SOI MOSFET for pH sensing using TCAD Sentaurus simulation. A new approach was used to model the electrolyte, based on the literature, as the simulator previously lacked a model for this type of material. The simulation results show an increase in drain current levels for acidic pH values and a decrease for basic pH values. Additionally, the drain voltage demonstrates an influence on the electrolyte charges, which worsens the device sensitivity. Therefore, a new device was proposed, using only the source sensing region to avoid drain influence. This new device shows an improvement in sensitivity and optimizes the obtained results for future experimental analyses.

*Keywords*—[BE]*SOI MOSFET, pH sensing, Simulation, TCAD.*

## I. INTRODUCTION

The Back Enhanced Silicon-On-Insulator MOSFET ([BE]SOI MOSFET) is a device patented in 2015 [1], as a proposal to design a reconfigurable transistor with less complexity in the fabricating process [2]. The [BE]SOI MOSFET is a transistor that does not intentionally contain doped source and drain regions. Fabricated on a SOI (Silicon-On-Insulator) wafer, the silicon film above the buried oxide of the SOI wafer serves as the transistor channel, extending from source to drain, with only natural p-type wafer doping ($10^{15}$ carriers/cm³). Due to this low doping concentration in the film, there is no current flow between source and drain as there are no doped regions providing the necessary carriers to constitute current. However, because it features an oxide layer (buried oxide) separating the film from the substrate, it is possible to use the voltage applied to the substrate (programming gate) to induce carriers in the channel film. This method, known in the literature [3] as electrostatic doping, refers to carrier induction by an electric field. Therefore, the type and number of carriers are defined by the electric field, allowing a single transistor to function as either a N-type or P-type by simply modifying the applied potential to the substrate.

Initially, this reconfigurability feature was the main focus of research with this device and has been continuously improved upon to this day [4]-[6]. As research progressed, alternative applications were discovered, such as in digital circuits [7] and as certain types of sensors [8][9]. Another significant characteristic of the [BE]SOI MOSFET is the presence of Underlap regions, which are spaces between the gate contact and the drain/source contact. These regions are used to evaluate the electrical behavior of the device when subjected to external events, such as light incidence, to function as a light sensor, and with glucose liquid solutions to serve as a biosensor. The application as a biosensor is highly important for contributing to technological development in healthcare, especially in disease monitoring such as diabetes [10], to enhance people's quality of life.

The studies of the [BE]SOI MOSFET as a biosensor were initially evaluated through computational simulation environments to verify the feasibility of the idea before experimental implementation. The semiconductor simulator TCAD SENTAURUS was employed for all virtual research, utilizing numerical methods to simulate the electrical effects of the device. TCAD SENTAURUS does not encompass mathematical and physical models for all types of materials, such as liquids, as it is not the focus of this type of simulator. Hence, the simulations were initially adapted in a simplified manner to represent biological material in the sensing regions, like using an oxide with the dielectric constant of water ($k_{water}=80\varepsilon0$) with different amounts of effective charges [11] to assess the influence of charges in these areas, and an approach to evaluate various dielectric constants with values corresponding to certain types of biological materials (including glucose) [12].

These initial simulations greatly contributed to understanding the influences of biological material in these regions on the electrical behavior of the device, which aligned with experimental results showing similar trends to those simulated. As this was a simplistic but effective approach, efforts are underway to improve aspects of the simulation to make them more realistic. Therefore, this work presents a study of the sensing regions of the [BE]SOI MOSFET in a simulation environment with an alternative approach to representing the electrolyte in a semiconductor simulation.

## II. SIMULATION CHARACTERISTICS

### A. Electrolyte Model

A simulation study was conducted using the TCAD SENTAURUS tool to assess charge distribution in the electrolyte. The simulation was carried out based on a methodology previously outlined in relevant literature [13]. Since TCAD lacks an appropriate model for electrolytes/solutions, the adopted approach involved a definition of the solution as an intrinsic semiconductor with a relative dielectric constant equivalent to that of water ($k_{water}=80\varepsilon0$).

Considering pure water at 25°C, the dissociation of water molecules tends to maintain the ionization product in equilibrium ($K_w = 10^{-14}$). If the pH of water is 7, the pOH will also be 7, meaning both the concentrations of H+ and OH- ions will be equal to $10^{-7}$ mol/L. If the pH changes, the ion concentrations are adjusted to always keep the ionization product constant. This characteristic is very similar to the properties of intrinsic silicon in thermal equilibrium, where the product of the number of carriers in the semiconductor, holes (p), and electrons (n), is always constant ($p.n = n_i^2$). Therefore, it is possible to approach adapting the electrolyte model for the simulator.

To calculate the quantity of carriers in the electrolyte to feed into the simulator, it is assumed that the concentration of H⁺ ions, [H⁺], is equal to the concentration of holes in the semiconductor, and the concentration of OH⁻ ions, [OH⁻], is equal to the concentration of electrons in the semiconductor. With this, it is possible to use Avogadro's number ($6.022 \times 10^{23}$) to determine the number of ions per cm³ in a pH solution and estimate how many carriers need to exist in each case for the semiconductor to function as an electrolyte. The simulator uses standard equations to determine the quantity of carriers in silicon, equations that depend on the density of states, requiring the modification of model parameters for each pH value.

### B. Device Characteristic

The $^{BE}$SOI MOSFET was simulated using a Silicon-On-Insulator (SOI) substrate with intrinsic doping of $1\times10^{15}$ carriers per cm³. From the transistor dimensions, the gate length (L) was defined as 1 μm, and for a 2D simulation, the gate width (W) was also set to 1 μm. The thicknesses of the gate oxide ($t_{ox}$), channel silicon layer ($t_{Si}$), buried oxide ($t_{BOX}$), the electrolyte and the substrate are, respectively, 10nm, 20nm, 200nm, 500nm and 1 μm. The sensitive regions have a length of 500 nm, and it is in this region that the defined model will be implemented in this work. Fig. 1A shows a 3D schematic drawing to provide an idea of how the device would appear if fabricated, while Fig. 1B depicts a schematic cross-sectional profile of the device, which was used for the 2D simulation.

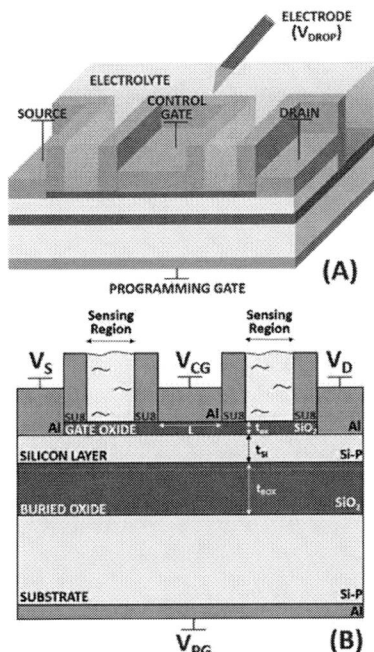

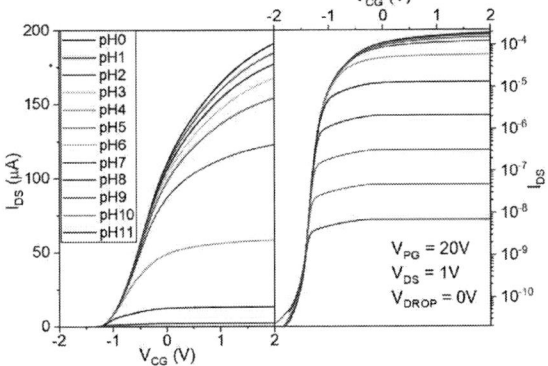

Fig. 1. Device schematic drawing on 3D (A) and 2D (B) view of $^{BE}$SOI MOSFET showing the sensing regions, where $V_S$ is the voltage represents the source voltage; $V_D$ is the drain voltage; $V_{DROP}$ is the electrolyte voltage; $V_{CG}$ is the control gate voltage and $V_{PG}$ is the programming gate voltage.

## III. RESULTS AND ANALYSIS

To commence the study of the electrical behavior of the device, the $I_{DS} \times V_{GS}$ curve was obtained, as shown in Fig. 2.

Fig. 2. Drain current as a function of the control gate voltage for different pH values on a linear (left) and logarithmic (right) scale.

In the curve, it is possible to observe different levels of current for various pH values. For more acidic values, the current increases, whereas for more basic values, the current tends to decrease. This phenomenon occurs due to the geometry of the device. Yojo [12] modeled the sensitive area as transistors connected in series along with a main transistor controlled by the gate. In other words, there is an arrangement of three transistors in series: one main transistor controlled by the gate and two secondary transistors controlled by the voltage applied to the electrode of the liquid in the sensitive area, as shown in Fig. 3.

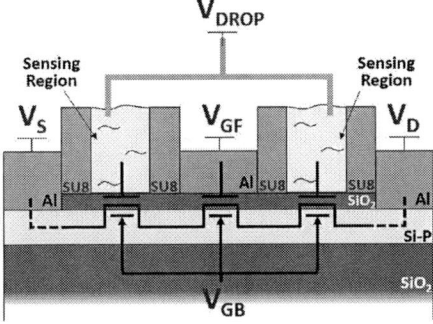

Fig. 3. $^{BE}$SOI MOSFET device electric model

In addition to extracting the current level as a function of the gate voltage, other investigations were conducted. It was observed that the voltage applied to the electrode, $V_{DROP}$, could contribute to increasing or decreasing the device current. Fig. 4 shows the electron density for pH2 (very acidic) and pH10 (very basic), with electrode applied voltages of 0.0V and 0.9V in the sensitive region near the Drain. This analysis was carried out to aid in understanding the obtained results and to examine the effect of potential in the electrolyte

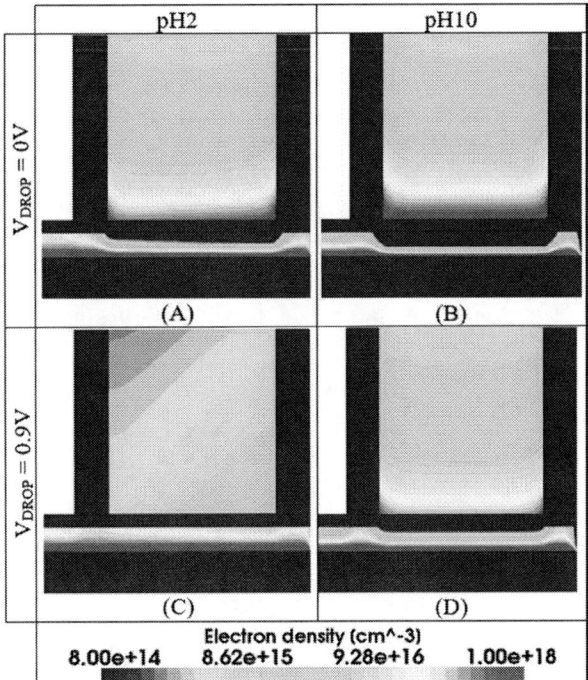

Fig. 4 Electron density analyzes for acidic pH (pH2), applying 0V (A) and 0.9V (B) to the electrolyte (VDROP) and basic pH (pH10) for the same voltage values, 0V (C) and 0.9V (D).

From the images presented in the table, it is evident that for a more acidic pH (pH2), there is a lower electron density compared to a more basic pH (pH10), which was expected based

979-8-3315-4064-7/24 $31.00 © 2024 IEEE

on the used model. The attraction of these carriers to the electrolyte/oxide interface is due to the distributed channel potential up to the electrode applying voltage to the electrolyte, causing a departure of the majority carriers (in this case, electrons) from the channel at the oxide/channel interface, thereby impeding current flow through that region. This leads to an increase in series resistance, resulting in decreased current for higher pH values. If the applied voltage to the electrolyte increases, the potential difference between the electrolyte and the channel decreases, attracting fewer electrons to the electrolyte/oxide interface, reducing the series resistance, and increasing the current. Fig. 5 illustrates the potential distribution from the substrate to the electrode in the active region near the drain.

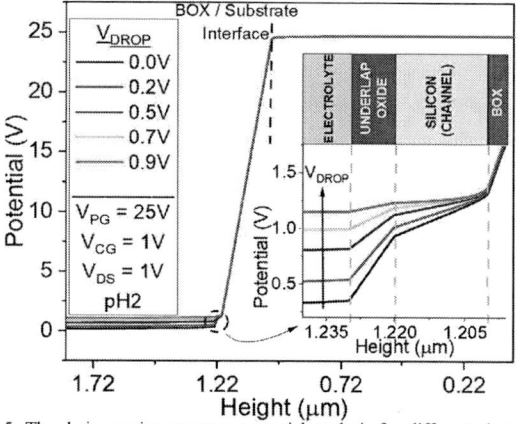

Fig. 5. The drain sensing structure potential analysis for different electrolyte applied voltage ($V_{DROP}$).

As mentioned earlier, the table was compiled using data obtained from the sensitive area near the Drain. This was done because it was noticed that for acidic pH values and higher $V_{DROP}$ voltages, the potential difference between the electrolyte and the channel near the drain was small but non-uniform. In the table, it can be observed that for pH2 with $V_{DROP}$ of 0.9V, there is a higher concentration of electrons moving towards the drain contact. This electron concentration leads to an increase in series resistance, which worsens the current level for acidic pH values, as the electron density is much higher for more basic pH values and the influence of the drain is minimal.

Based on this, the study of the $^{BE}$SOI MOSFET with only one sensitive area, on the Source side, was proposed, as shown in the schematic diagram in Fig. 6. With this area near the Source, the charges in the electrolyte will only be influenced by the voltage applied to the Source, which will be at 0V.

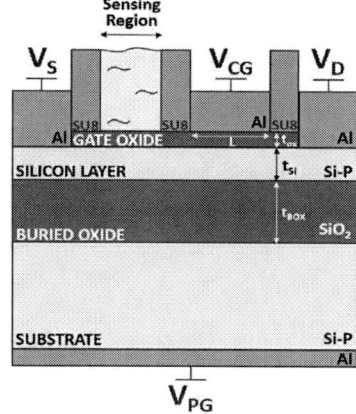

Fig. 6. 2D schematic drawing of the new device proposal, the $^{BE}$SOI MOSFET source sensing region.

Similarly to the evaluation conducted on the device with two sensitive regions, the current value ($I_{DS}$) was extracted as a function of the control gate voltage ($V_{CG}$), as depicted in the graph in Fig. 7.

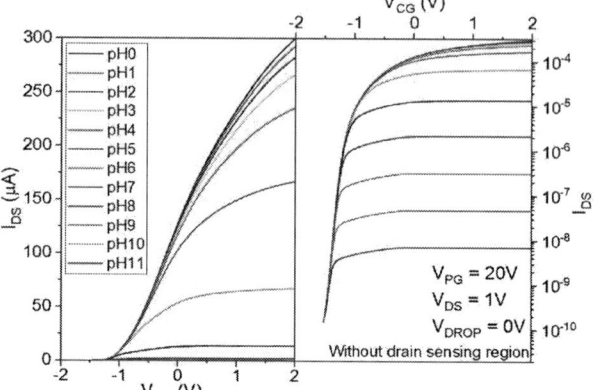

Fig. 7. Drain current as a function of the control gate voltage for different pH values for the new device proposal.

In the figure, it can be observed that the device exhibits the same characteristic as the first device but with higher current levels. This is due to the absence of the sensitive region near the drain, eliminating one of the secondary transistors that can limit the current and, consequently, reducing the series resistance and increasing the current level.

To compare the devices, a sensitivity metric (S) was defined. To calculate this sensitivity was set a different manner than in conventional ways that uses the threshold voltage variation as a function of pH, due the electrical behavior of the device. A gate voltage value of 1.5V was chosen, and the corresponding current level was verified. The reference current chosen is for pH7, which represents the neutral state of the electrolyte. Therefore, the value of S for each pH is obtained using (1): the absolute difference between the current at a pH value ($I_{DS_{pH}}$) and the reference current ($I_{DS_{ref}}$), divided by the reference current.

$$|S| = \left|\frac{(I_{DS_{pH}} - I_{DS_{ref}})}{I_{DS_{ref}}}\right| \qquad (1)$$

Therefore, the comparison between devices is presented in Fig. 8 for different values of applied programming gate voltage, along with the comparison of the sensitivities achieved in each case shown in the graph inset.

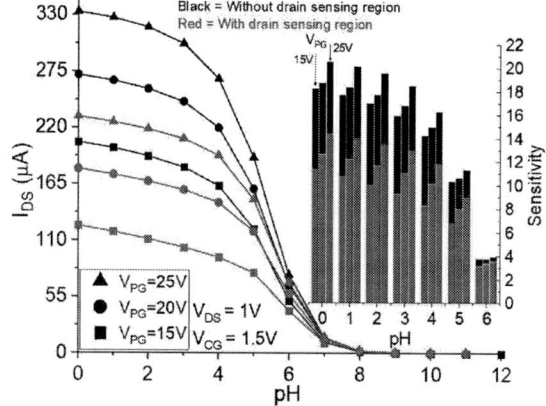

Fig. 8. Drain current as a function of pH value at 1.5V for control gate voltage, and sensitivity analyses for which device.

In the graph presented, it can be observed that for pH values greater than 7, regardless of the device used, the current levels are very low, resulting in sensitivities that are close to 1, which is the lowest possible level. For pH values greater than 7, the sensitivity achieves more noticeable values, increasing as the applied programming gate voltage changes. This occurs due to the electrostatic coupling of the transistor, where increasing this voltage will increase the number of carriers available in the channel, thereby reducing the series resistance and increasing sensitivity. Additionally, the red curves demonstrate that sensitivity is lower when the transistor has the sensitive region near the drain, as it is influenced by the voltage applied to the drain. Removing this region, indicated by the black lines, eliminates the drain's influence on the charges, removing a parameter that affects sensitivity.

After determining the optimal geometric configuration of the device, specifically without the drain region, the influence of the voltage applied to the electrolyte was investigated, as shown in Fig. 9.

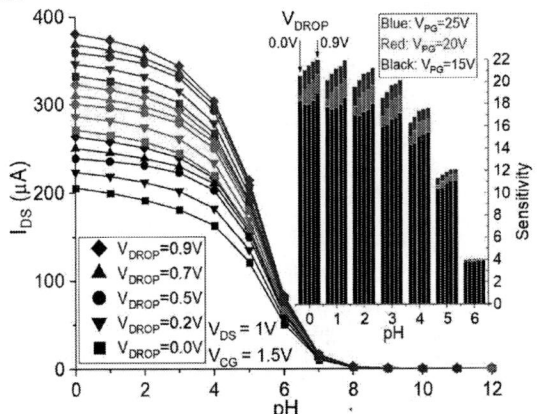

Fig. 9. Drain current as a function of pH for the device without drain sensing region, and sensitivity analyses for which electrolyte voltage.

The graphs show that increasing the voltage applied to the electrolyte leads to an increase in current and sensitivity. As shown in Fig. 5, the voltage applied to the electrolyte contributes to reducing the series resistance of the transistor, further enhancing sensitivity. This aligns with the results presented in Fig. 9.

## IV. CONCLUSIONS

This study presented the implementation of a model for electrolytes via simulation using TCAD Sentaurus. This model was used to optimize the investigation into the electrical behavior of the BESOI MOSFET as a biosensor. The sensitive Source and Drain regions were analyzed with simulated pH values obtained from the model. Variation in pH values showed an increase in current for more acidic conditions and a decrease for more basic conditions. It was observed that the Drain voltage interferes with the interaction of charges in the electrolyte with charges in the channel. Therefore, a new device was proposed without a sensitive region near the Drain. This device demonstrated increased sensitivity in all cases, indicating that the sensitive region near the Drain can indeed negatively influence the interaction of the electrolyte with the device.

Based on the results presented in this study, it will be possible to define optimal working points for conducting experimental work using the BESOI MOSFET as an optimized biosensor.

## ACKNOWLEDGMENT

The authors acknowledge CNPq (149904/2022-3), São Paulo Research Foundation - FAPESP (under grant #2020/04867-2) and Coordenação de Aperfeiçoamento de Pessoal de Nível Superior - Brasil (CAPES) - Finance Code 001 for the financial support.

## REFERENCES

[1] J. A. Martino and R. C. Rangel, "Método de Fabricação de Transistor," patent number BR102015020974-6, 2015.

[2] R. C. Rangel and J. A. Martino, "Back Enhanced (BE) SOI pMOSFET," in *2015 30th Symposium on Microelectronics Technology and Devices (SBMicro)*, Salvador, Brazil: IEEE, Aug. 2015, pp. 1–4. doi: 10.1109/SBMicro.2015.7298121.

[3] G. Gupta, B. Rajasekharan, and R. J. E. Hueting, "Electrostatic Doping in Semiconductor Devices," *IEEE Trans. Electron Devices*, vol. 64, no. 8, pp. 3044–3055, Aug. 2017, doi: 10.1109/TED.2017.2712761.

[4] D. A. Ramos, K. R. A. Sasaki, R. C. Rangel, P. H. Duarte, and J. A. Martino, "Influence of the source/drain doping region on the reconfigurability of BE SOI MOSFET," in *2023 37th Symposium on Microelectronics Technology and Devices (SBMicro)*, Rio de Janeiro, Brazil: IEEE, Aug. 2023, pp. 1–4. doi: 10.1109/SBMicro60499.2023.10302644.

[5] H. L. Carvalho, R. C. Rangel, K. R. A. Sasaki, P. G. D. Agopian, L. S. Yojo, and J. A. Martino, "Al Source-Drain Schottky contact enabling N-type (Back Enhanced) BE SOI MOSFET," in *2022 36th Symposium on Microelectronics Technology (SBMICRO)*, Porto Alegre, Brazil: IEEE, Aug. 2022, pp. 1–4. doi: 10.1109/SBMICRO55822.2022.9880960.

[6] R. A. Sasaki, R. C. Rangel, D. A. Ramos, L. S. Yojo, and J. A. Martino, "Improved Back Enhanced SOI ( BE SOI) MOSFET by adding n-doped regions," in *2021 35th Symposium on Microelectronics Technology and Devices (SBMicro)*, Campinas, Brazil: IEEE, Aug. 2021, pp. 1–4. doi: 10.1109/SBMicro50945.2021.9585735.

[7] L. Yojo, R. C. Rangel, K. R. A. Sasaki, and J. A. Martino, "Reconfigurable back enhanced (BE) SOI MOSFET used to build a logic inverter," in *2017 32nd Symposium on Microelectronics Technology and Devices (SBMicro)*, Fortaleza: IEEE, Aug. 2017, pp. 1–4. doi: 10.1109/SBMicro.2017.8112987.

[8] J. A. Padovese, R. C. Rangel, K. R. A. Sasaki, and J. A. Martino, "Thin Si channel Back Enhanced (BE) SOI pMOSFET Photodetector under different bias conditions," in *2019 Joint International EUROSOI Workshop and International Conference on Ultimate Integration on Silicon (EUROSOI-ULIS)*, Grenoble, France: IEEE, Apr. 2019, pp. 1–4. doi: 10.1109/EUROSOI-ULIS45800.2019.9041870.

[9] L. S. Yojo, R. C. Rangel, K. R. A. Sasaki, and J. A. Martino, "Study of BE SOI MOSFET Reconfigurable Transistor for Biosensing Application," *ECS J. Solid State Sci. Technol.*, vol. 10, no. 2, p. 027004, Feb. 2021, doi: 10.1149/2162-8777/abe3cc.

[10] A. Heller and B. Feldman, "Electrochemical Glucose Sensors and Their Applications in Diabetes Management," *Chem. Rev.*, vol. 108, no. 7, pp. 2482–2505, Jul. 2008, doi: 10.1021/cr068069y.

[11] L. S. Yojo, R. C. Rangel, K. R. A. Sasaki, C. A. Mori, and J. A. Martino, "Optimization of the Back Enhanced BE SOI MOSFET working as a charge-based BioFET sensor," in *2019 IEEE SOI-3D-Subthreshold Microelectronics Technology Unified Conference (S3S)*, San Jose, CA, USA: IEEE, Oct. 2019, pp. 1–3. doi: 10.1109/S3S46989.2019.9320714.

[12] L. S. Yojo, R. C. Rangel, K. R. A. Sasaki, and J. A. Martino, "Optimization of the permittivity-based BE SOI biosensor," in *2018 IEEE SOI-3D-Subthreshold Microelectronics Technology Unified Conference (S3S)*, Burlingame, CA, USA: IEEE, Oct. 2018, pp. 1–3. doi: 10.1109/S3S.2018.8640139.

[13] R. Narang, M. Saxena, and M. Gupta, "Analytical Model of pH sensing Characteristics of Junctionless Silicon on Insulator ISFET," *IEEE Trans. Electron Devices*, vol. 64, no. 4, pp. 1742–1750, Apr. 2017, doi: 10.1109/TED.2017.2668520.

# Zero Temperature Coefficient Study Regarding the Half-Diamond Layout Style for MOSFETs

M. A. P. Peixoto
COORDELT
CEFET/RJ
Rio de Janeiro, Brazil
marco.peixoto@cefet-rj.br

M. P. Braga de Lima
COPPE
UFRJ
Rio de Janeiro, Brazil
marcos.braga@coppe.ufrj.br

E. H. S. Galembeck
Electrical Engineering Department
FEI University Center
São Bernardo do Campo, Brazil
egon@fei.edu.br

M. M. Correia
Electrical Engineering Department
FEI University Center
São Bernardo do Campo, Brazil
mmarcelino.c@fei.edu.br

L. M. Camillo
COORDELT
CEFET/RJ
Rio de Janeiro, Brazil
luciano.camillo@cefet-rj.br

S. P. Gimenez
Electrical Engineering Department
FEI University Center
São Bernardo do Campo, Brazil
sgimenez@fei.edu.br

*Abstract*—This paper studies the behavior of the Zero Temperature Coefficient (ZTC) of the first element of the second generation of the gate layout styles for MOSFETs, the Half-Diamond MOSFET (HDM). This work was performed based on three-dimensional (3D) numerical simulations (3D-NS) initially fitted by experimental data at room temperature. Camillo-Martino ZTC analytical model (CM-Model) application demonstrated a good agreement (maximum error of 5.9% above 400K) with the simulations in predicting the ZTC point behavior of a Bulk HDM for high temperatures.

*Keywords—Zero Temperature Coefficient, Camillo-Martino ZTC Model, Half-Diamond MOSFET, Gate Layout Style*

## I. INTRODUCTION

More and more, semiconductor and Integrated Circuits (ICs) companies are challenged to develop innovative, robust, and more complex electronic Systems-on-a-Chip (SoCs) to meet market needs for different productive sectors linked to the areas of healthcare, automotive, avionics, space, electronic warfare, robotics, entertainment, communications, *etc*. To reach these needs, there is a lot of Research and Development to improve the ICs Complementary Metal-Oxide-Semiconductor (CMOS) manufacturing process [1], developments of new materials [2], new planar and three-dimensional (3D) devices' structures [3], new concepts of transistor operation [4], new gate layout styles for the Metal-Oxide-Semiconductor Field Effect Transistors [5]-[9], *etc*. Regarding the studies already performed until now on the innovative gate layout styles for Metal-Oxide-Semiconductor Field-Effect-Transistors (MOSFETs) with the elements of the first-generation (Diamond, Octo and Ellipsoidal) operating at room Temperature (T) [5]-[8], at high T [9]-[11], and at the ionizing radiation environments (x-rays [12], protons [13], and $^{60}$Co [14]), we could observe how the Longitudinal Corner Effect (LCE), Parallel Connections of Different MOSFET with Different Channel Lengths Effect (PAMDLE) and Deactivation of Parasitic MOSFETs in the Bird-Beaks Regions Effect (DEPAMBBRE) behaves to significantly boost the electrical performance and the ionizing radiation tolerance of the MOSFETs.

Because of these previous studies, we were motivated to propose the second generation of these gate layout styles, composed of Half-Diamond, Half-Octo and Half-Ellipsoidal

to enhance further the electrical performances and the ionizing radiation tolerances of the planar MOSFETs. These new gate layout styles were conceived further to reduce the effective channel length ($L_{eff}$) and to keep the LCE, PAMDLE and DEPAMBBRE effects [8],[9] of the first generation and therefore, the designers will be able to design analog and radiofrequency (RF) CMOS ICs still smaller than those implemented on the elements of the first generation of gate layout style for MOSFETs [9],[15]-[17]. Incipient studies done with Half-Diamond MOSFET (HDM) indicated that it can present a higher electrical performance than the equivalent Rectangular MOSFET (RM), regarding the same gate area ($A_G$) and bias conditions [15],[16]. We have observed that drain-to-source current ($I_{DS}$) normalized by the aspect ratio (W/L) in the saturation region [$I_{DS\_SAT}$/(W/L)] of HDM was 35% higher than the one found in RM counterpart, thanks to LCE and PAMDLE effects (better current driver). Afterwards, it could diminish the dissipated electrical power by approximately 62% compared to the RM counterpart (better energetic efficiency). Besides, it has also presented a unit voltage gain frequency ($f_T$) normalized by the aspect ratio [$f_T$/(W/L)] 20% greater than the one of the RM counterpart, an alternative to be used in radiofrequency (RF) CMOS ICs applications. Respectively, Fig. 1 (c), (a) and (b) illustrates the 3D structures of the HDM and of its counterparts, i.e., the conventional RM and the Diamond MOSFET (DM) of the first-generation gate layout style, where W and L are the channel width and length, respectively, α is the angle that defines the triangular shape of the interface composed of the drain/source and gate regions for the DM, and by the interface composed of the drain and gate regions for the HDM, and b and B are the smallest and largest channel lengths of the DM and HDM, respectively. It is important to emphasize that the designers can further boost the electrical performances and ionizing radiation tolerances, besides reducing the chips' total areas of analog and RF CMOS ICs by using the HDMs instead of the DMs [17].

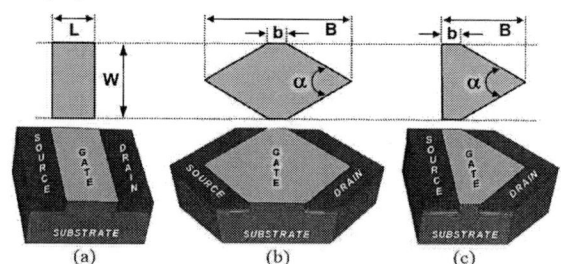

Fig. 1. Examples of the 3D structures of the RM (a), DM (b) and HDM (c)

---

M. A. P. Peixoto thanks CAPES (PROSUC Program) and CNPq (DOUTORADO-GD) for the financial support in 2023 and 2024, respectively, and to "Padre Sabóia de Medeiros" Ignatian Educational Foundation for the doctorate scholarship provided with FEI University Center. S. P. Gimenez thanks CNPq (grant number 304427/2022-5) and FAPESP (grant number 2020/09375-0) for the financial support.

979-8-3315-4064-7/24 $31.00 © 2024 IEEE

## II. ZERO TEMPERATURE COEFFICIENT BEHAVIOR PREDICTION FOR NON-CONVENTIONAL MOSFETS

The MOSFET operation is strongly dependent on T [18] and, therefore, the knowledge about a specific bias condition point, *i.e.,* the Zero Temperature Coefficient (ZTC) point of the MOSFET, which keeps its operation as insensitive to T variations as possible urge as still a critical factor for CMOS ICs design [19]-[25]. The theoretical definition of the MOSFET's ZTC point corresponds to the specific value ($V_{ZTC}$) of the gate-to-source voltage ($V_{GS}$), which ensures that the value ($I_{ZTC}$) of drain-to-source current ($I_{DS}$) remains constant despite the T variations. The design of an analog CMOS IC for thermal volubility applications must bias their transistors with their $I_{ZTC}$ values at the drain terminals, which corresponds to $V_{ZTC}$ values between gate to source terminals [24]. Thus, considering a given CMOS ICs manufacturing process means fixing the magnitude of $I_{ZTC}$ for each MOSFETs and then adjusting its corresponding W/L in order to obtain the desired values of $V_{ZTC}$ [23].

A search in the literature for the ZTC investigation applied for unconventional gate layout styles MOSFETs results that just some (*i.e.,* Ellipsoidal and DM) of the first-generation structures had received attention [10],[26]-[28]. The occurrence of the ZTC point for the triode and the saturation regions was verified in both structures by 3D numerical simulations (3D-NS) and its prediction based on the application of the Camillo-Martino ZTC analytical model (CM-Model) [23] demonstrated good agreement.

The CM-Model was developed for RMs implemented with the Silicon-On-Insulator (SOI) technology. Despite that, its application for the Bulk HDM ZTC is relatively simple order to predict $V_{ZTC}$ in both operational regions of MOSFETs (triode and saturation), as already verified on other non-conventional transistors [26]-[28]. Following this successful approach, a study of the ZTC behavior for the Bulk HDM over a wide range of T and a comparative analysis with its RM counterpart can be done, similar to those described in [26]-[28].

## III. RESULTS AND DISCUSSION

The 3D-NS data were obtained using the Atlas Device Technology Simulation applied to the devices' structures and meshes generated on the DevEdit3D, both Silvaco's tools [29]. The 3D-NS were initially fitted by experimental data at room T for all devices, after that others 3D-NS were performed at higher Ts to study the device's ZTC points. The fit got success setting up the simulator using mobility models Lombardi Model (CVT) with Klaassen Model (KLA) and Watt Model (WATT) for varying operational T (300-600 K). The dimensional features for HDM and RM, both n-channel, are: I- the thin gate oxide thickness ($t_{ox}$) equal to 3nm; II- acceptor dopants' concentration ($N_A$) of the channel region equal to $1.10^{17}$ cm$^{-3}$; III- donor dopants' concentrations of the drain and source regions ($N_D$) is equal to $1 \times 10^{19}$ cm$^{-3}$.

Fig. 2 (a) illustrates the Threshold Voltages ($V_{TH}$) as a function of T in triode (linear) region with drain-to-source voltage ($V_{DS}$) equals to 50mV for the RM and HDMs (with different values of $\alpha$), while Fig. 2 (b) illustrates in saturation region with $V_{DS}$ equals to 1.5V. In the triode region, the $V_{TH}$ was obtained through the second derivative method ($\partial^2 I_{DS}/\partial V^2_{GS}$) [30], and in the saturation region, the $V_{TH}$ was obtained as a function of the transconductance (gm) using the "Ratio Method" [$V_{GS} = I_{DS}/(gm)^{1/2}$] [25]. Fig. 2 shows that the

MOSFETs' $V_{TH}$s decrease as the T increases (350-600K), as expected, mainly due to the Fermi potential reduction [31].

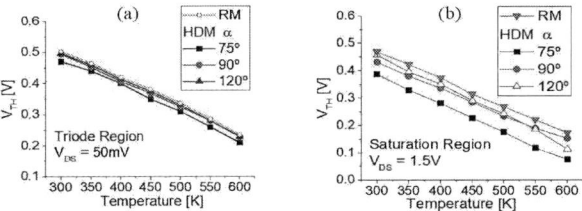

Fig. 2. $V_{TH}$ as a function of the T in triode (a) and saturation (b) regions

The HDM and RM maximum gm (gm$_{max}$) as a function of T [21] for $V_{DS}$ equal to 50mV are illustrated in Fig. 3. We observed that gm$_{max}$ degradation can be explained due to the mobile charger carriers' mobilities lowering due to the phonon scattering as T increases [21],[32],[33].

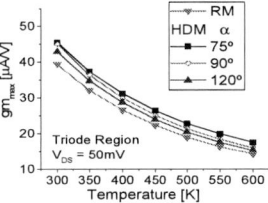

Fig. 3. gm$_{max}$ as a function of T with $V_{DS}$=50mV (triode region)

The mobile charger carriers' mobilities degradation factor ($c$), which is directly related to the gm, as a function of the T can be obtained from [28] with [34], since the MOSFET gm can be considered proportional to $c$, concerning the first-order analytical model [22].

In that way, by using the gm$_{max}$ as a function of T, regarding the triode region, or considering the same gate overdrive voltage ($V_{GT}=V_{GS}-V_{TH}$), taking into account the saturation region [35], $c$ can be calculated as in (1) under the condition of two different Ts ($T_1$ and $T_2$, with $T_1 < T_2$), respectively, where $\mu_1$ and $\mu_2$ are the mobile charger carriers' mobilities and gm$_{max1}$ and gm$_{max2}$ are the gm$_{max}$ of the MOSFETs regarding these two different T:

$$\frac{gm_{max2}}{gm_{max1}} \; \alpha \; \frac{\mu_2}{\mu_1} = \left(\frac{T_1}{T_2}\right)^c \therefore c = \frac{\log(gm_{max2}) - \log(gm_{max1})}{\log(T_1) - \log(T_2)} \quad (1)$$

Fig. 4 (a) and (b) illustrates the devices $c$ Factor as a function of the T (from 300K to 600K), respectively, in triode and saturation operations regions. We notice that the $c$ Factor increases linearly, in the first approximation, with T in the triode region, as expected, due to the dependence of the gm with the T. However, in the saturation region, the c Factor tends to be constant, because, we see that the increase of the $V_{TH}$ is compensated by the reduction of the mobile charge carriers' mobility as T increases [36].

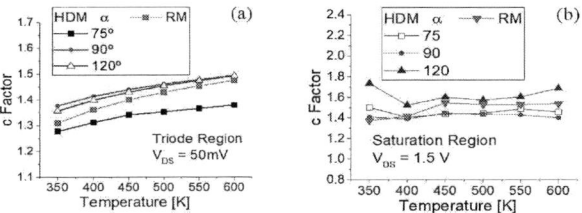

Fig. 4. $c$ factor as a function of T for triode (a) and saturation (b) regions

979-8-3315-4064-7/24 $31.00 © 2024 IEEE    131

Fig. 5 (a) and (b) illustrates the $I_{DS}$ as a function of $V_{GS}$ for RM, respectively, in the triode and saturation operational regions for different Ts (from 300K to 600 K).

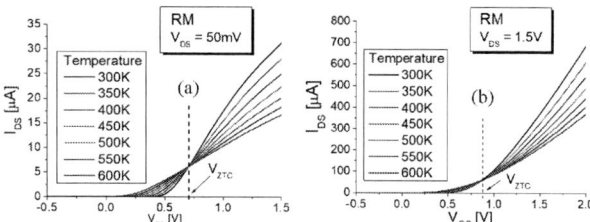

Fig. 5. RM's $I_{DS}$ as a function of $V_{GS}$ in triode (a) and saturation (b) regions for different Ts

Fig. 6 (a) and (b) illustrates the $I_{DS}$ as a function of $V_{GS}$ for HDM with α equals to 90°, respectively, in the triode and saturation operational regions for different Ts (from 300K to 600 K).

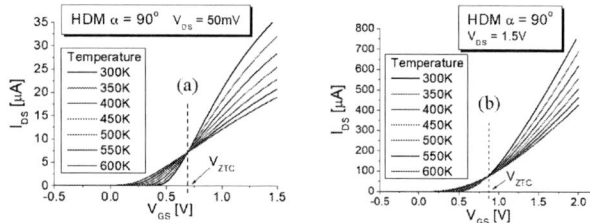

Fig. 6. 90° HDM's $I_{DS}$ as a function of $V_{GS}$ in triode (a) and saturation (b) regions for different Ts

The devices $V_{ZTC}$ obtained of Fig. 5 and Fig. 6 are summarized in Table I. We observe that the $V_{ZTC}$ of both devices are approximately the same, verifying maximum variations smaller than 2% and 4% for the triode and saturation regions, respectively, from 300K to 600K.

TABLE I: HDMs AND RM $V_{ZTC}$s (TRIODE AND SATURATION REGIONS)

| Device | | $V_{ZTC}$ [V] | |
| --- | --- | --- | --- |
| | | Triode | Saturation |
| HDM | α = 75° | 0.688 | 0.860 |
| | α = 90° | 0.690 | 0.876 |
| | α = 120° | 0.694 | 0.890 |
| RM | - | 0.700 | 0.890 |

Therefore, as the $V_{ZTC}$ are practically the same for the HDM and RM, regarding the same CMOS ICs manufacturing process, we verify that the $V_{TH}$ and the mobile charge carriers' mobilities present practically the same behaviors with the T [37],[38].

Fig. 7 (a), (b) and (c) illustrates, respectively regarding α angles of 75°, 90°, and 120°, the HDM $V_{ZTC}$ as a function of the T obtained from 3D-NS and from CM-Model.

Table II resumes the maximum errors found between the 3D-NS and the CM-Model. We can see that the maximum error found between the 3D-NS and CM-Model is equal to 2.6 % for the HDM with α equal 120° in the triode region and 13.4% in the saturation region, considering all MOSFETs studied. Furthermore, regarding the T higher than 400K, the CM -Model is capable of predicting $V_{ZTC}$s with maximum errors smaller than 6 % and smaller than 13.5% for T equal 350 K for HDM with α angle equal to 120°.

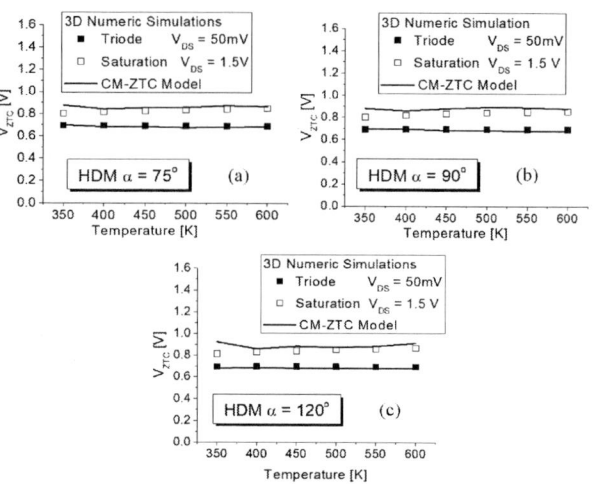

Fig. 7. $V_{ZTC}$ as a function of T for HDM α of 75° (a), 90° (b) and 120° (c) in triode and saturation regions obtained from 3D-NS and CM-Model

TABLE II: THE $V_{ZTC}$ MAXIMUM ERRORS OBTAINED FROM COMPARISONS BETWEEN THE 3D-NS AND CM-MODEL IN PERCENTAGE.

| Maximum Error Between 3D-NS and CM-Model Results for HDMs | | | | | |
| --- | --- | --- | --- | --- | --- |
| | α=75° | | α=90° | | α=120° | |
| T | Triode | Saturation | Triode | Saturation | Triode | Saturation |
| 350K | 0.7 | 9.2 | 0.2 | 9.6 | 2.3 | 13.4 |
| 400K | 1.8 | 3.3 | 0.0 | 4.8 | 1.5 | 3.7 |
| 450K | 1.3 | 3.3 | 1.8 | 5.5 | 2.6 | 4.8 |
| 500K | 2.3 | 2.5 | 1.9 | 5.9 | 2.6 | 2.4 |
| 550K | 1.2 | 3.7 | 1.9 | 4.7 | 2.0 | 2.6 |
| 600K | 0.7 | 2.0 | 1.7 | 2.6 | 1.9 | 5.2 |

## IV. CONCLUSION

This paper studies the applicability of the $V_{ZTC}$ CM-Model for the MOSFETs implemented with the Half-Diamond layout style (HDM) in comparison to the three-dimensional numerical simulations data. We conclude that the CM-Model is able to predict the $V_{ZTC}$ point behavior of the HDM, for different α angles, with a maximum error equal or smaller than 2.6% and 13.4%, considering the HDM operating in the triode and saturation regions, respectively.

## ACKNOWLEDGMENT

The authors thanks to Electrical Engineering Department of the FEI University Center for the infrastructure to carry out this work. In addition, L. M. Camillo and M. A. P. Peixoto also thanks to Electronics Technical Academic Coordination (COORDELT) of the Brazilian Federal Center for Technological Education in Rio de Janeiro (CEFET/RJ) for had given the opportunity to accomplish this work.

## REFERENCES

[1] S. Wang *et al.*, "TID effect of MOSFETs in SOI BCD process and its hardening technique," in *IEEE Transac. on Nuclear Science*, vol. 70, no. 8, pp. 1995-2001, Aug. 2023, doi: 10.1109/TNS.2023.3281597.

[2] B. -Y. Tsui, Y. -L. Chen and S. -H. Lai, "Metal contact on P-Type 4H-SiC with low specific contact resistance and micrometer-scale contact area," in *IEEE Electron Device Letters*, vol. 44, no. 9, pp. 1539-1542, Sept. 2023, doi: 10.1109/LED.2023.3299688.

[3] R. K. Maurya, R. G. Debnath, R. Saha and B. Bhowmick, "Enhanced magnetic field sensing with MAGNC-FinFET: a current mode hall effect approach," in *IEEE Transactions on Nanotechnology*, vol. 23, pp. 250-256, 2024, doi: 10.1109/TNANO.2024.3373035.

[4] M. I. B. Chowdhury, "Ion-Sensitive vertical tunnel Field-Effect Transistor for highly sensitive, low power, low pH-resolution pH

sensing," in *IEEE Sensors Letters*, vol. 8, no. 1, pp. 1-4, Jan. 2024, Art no. 1500404, doi: 10.1109/LSENS.2023.3341887.

[5] L. E. Seixas *et al.*, "Improving MOSFETs' TID tolerance through diamond layout style," in *IEEE Transactions on Device and Materials Reliability*, vol. 17, no. 3, pp. 593-595, Sept. 2017, doi: 10.1109/TDMR.2017.2719959.

[6] V. V. Peruzzi, W. S. d. Cruz, G. A. d. Silva, R. C. Teixeira, L. E. S. Junior and S. P. Gimenez, "boosting the ionizing radiation tolerance in the mosfets matching by using diamond layout style," *2019 34th Symposium on Microelectronics Technology and Devices (SBMicro)*, Sao Paulo, Brazil, 2019, pp. 1-4, doi: 10.1109/SBMicro.2019.8919344.

[7] S. P. Gimenez, M. M. Correia, E. D. Neto and C. R. Silva, "An innovative ellipsoidal layout style to further boost the electrical performance of MOSFETs," in *IEEE Electron Device Letters*, vol. 36, no. 7, pp. 705-707, July 2015, doi: 10.1109/LED.2015.2437716.

[8] S. P. Gimenez, *Layout Techniques for MOSFETS*, in Synthesis Lectures On Emerging Engineering Technologies, vol. 7, USA: Morgan & Claypool Publishers, 2016, doi: 10.1007/978-3-031-02031-5.

[9] S. P. Gimenez and E. H. S. Galembeck, *"Differentiated Layout Styles for MOSFETs: Electrical Behavior in Harsh Environments,"* Kindle Edition, Springer, 2023, doi: 10.1007/978-3-031-29086-2.

[10] E. H. Salerno Galembeck and S. P. Gimenez, "LCE and PAMDLE effects from diamond layout for MOSFETs at high-temperature ranges," in *IEEE Transactions on Electron Devices*, vol. 68, no. 8, pp. 3914-3922, Aug. 2021, doi: 10.1109/TED.2021.3086076.

[11] E. H. S. Galembeck, C. Renaux, D. Flandre, S. Finco and S. P. Gimenez, "Boosting the SOI MOSFET electrical performance by using the octagonal layout style in high temperature environment," in *IEEE Transactions on Device and Materials Reliability*, vol. 17, no. 1, pp. 221-228, March 2017, doi: 10.1109/TDMR.2017.2652729.

[12] S. P. Gimenez and D. M. Alati, "Electrical behavior of the diamond layout style for MOSFETs in X-rays ionizing radiation environments," in *Microelectronic Engineering*, vol. 148, issue 1, 2015, doi: 10.1016/j.mee.2015.09.001.

[13] L. E. Seixas Jr., S. Finco1, M. A. G. Silveira, N. H. Medina and S. P. Gimenez, "Study of proton radiation effects among diamond and rectangular gate MOSFET layouts," in *Materials Research Express*, Vol: 4, Number 1, 2017, doi: 10.1088/2053-1591/4/1/015901.

[14] L. E. Seixas Jr., O. L. Gonçalez, R. G. Vaz, A. C. C. Telles, S. Finco and S. P. Gimenez, "Minimizing the TID effects due to gamma rays by using diamond layout for MOSFETs", in *Journal of Materials Science: Materials in Electronics*, vol. 30, issue 1, 2019, doi: 10.1007/s10854-019-00747-w.

[15] E. H. S. Galembeck and S. P. Gimenez, "New hybrid generation of layout styles to boost the electrical, energy, and frequency response performances of analog MOSFETs," *IEEE Trans. on Electron Devices*, vol. 69, no.6, pp. 3310-3318, Jun. 2022, doi: 10.1109/TED.2022.3167944.

[16] G. A. da Silva and S. P. Gimenez, "The second generation of layout styles to further boost the electrical performance of analog MOSFETs," *Journal of Integrated Circuits and Systems*, vol. 17, n. 2, pp. 1-5, Oct. 2022, doi: 10.29292/jics.v17i2.586.

[17] J. R. Banin Jr., A. L. Moreto, G. A. Silva ; C. E. Thomaz and S. P. Gimenez, "Methodology to optimize and reduce the total gate area of robust operational transconductance amplifiers by using diamond layout style for MOSFETs", *Analog Integrated Circuits and Systems Signal Processing*, v. 106, p. 293-396, 2021, doi: 10.1007/s10470-020-01750-6.

[18] S. M. Sze, *Physics of Semiconductor Devices*, 2nd. ed. Singapore: John Wiley & Sons, Inc., 1981.

[19] F. S. Shoucair, "Analytical and experimental methods for zero-temperature-coefficient biasing of MOS transistors," *Electron Lett.*, vol. 25, no. 17, pp. 1196-1198, 1989, doi: 10.1049/el:19890802.

[20] Z. D. Prijic, S. S. Dimitrijev and N. D. Stojadinovic, "the determination of zero temperature coefficient point in CMOS transistors", *Microelectron. Reliab.*, vol. 32, no. 6, pp. 769-773, 1992, doi: 10.1016/0026-2714(92)90041-I.

[21] A. A. Osman, M. A. Osman, N. S. Dogan and M. A. Imam, "Zero-temperature-coefficient biasing point of partially depleted SOI MOSFET's," in *IEEE Transactions on Electron Devices*, vol. 42, no. 9, pp. 1709-1711, Sept. 1995, doi: 10.1109/16.405293.

[22] L. M. Camillo, J. A. Martino, E. Simoen and C. Claeys, "The temperature mobility degradation influence on the zero temperature coefficient of partially and fully depleted SOI MOSFETs," in *Microelectronics Journal*, vol. 37, no. 9, pp. 952-957, Sept. 2006, doi: 10.1016/j.mejo.2006.01.008.

[23] M. Bellodi, L. M. Camillo, J. A. Martino, E. Simoen and C. Claeys, "Simple analytical model to study the ztc bias point in FinFETs," in *ECS Transactions*, vol. 6, no. 4, pp. 205-209, 2007, doi: 10.1149/1.2728862.

[24] L. M. Camillo, J. A. Martino, E. Simoen and C. Claeys, "Influence of the drain bias and gate length of partially depleted SOI MOSFETs on the ZTC biasing point," in *ECS Transactions*, vol. 14, no. 1, pp. 243-252, 2008, doi: 10.1149/1.2956038.

[25] I. M. Filanovsky and A. Allam, "Mutual compensation of mobility and threshold voltage temperature effects with applications in CMOS circuits," in *IEEE Transactions on Circuits and Systems I: Fundamental Theory and Applications*, vol. 48, no. 7, pp. 876-884, July 2001, doi: 10.1109/81.933328.

[26] L. M. Camillo, M. P. Braga de Lima, M. A. P. Peixoto, M. M. Correa and S. P. Gimenez, "Zero temperature coefficient behavior for ellipsoidal MOSFET," in *Journal of Integrated Circuits and Systems*, v. 15, no. 2, pp.1-5., 2020, doi: 10.29292/jics.v15i2.166.

[27] J. L. B. A. Jorge, R. S. Alves Jr., L. M. Camillo, M. A. P. Peixoto, M. M. Correa and S. P. Gimenez, "Zero temperature coefficient behavior for diamond MOSFET," in *Chip in The Fields - 20th Microelectronics Students Forum*, vol. 20, no. 1, pp. 1-4, 2020.

[28] M. P. Braga de Lima, M. A. P. Peixoto, M. M. Correa, E. H. S. Galembeck, S. P. Gimenez and L. M. Camillo, "Impact of temperature effects in the zero temperature coefficient of the ellipsoidal MOSFET," in *36th Symposium on Microelectronics Technology (SBMICRO)*, pp.1-4, 2022.

[29] SILVACO, Atlas User's Manual Device Simulation Software, USA: Silvaco, Inc., 2016.

[30] J. Liou *et al.*, *Analysis and design of MOSFETs*. USA: Kluwer, 1998.

[31] J. P. Colinge, *Silicon-On-Insulator Technology: Materials to VLSI*, 3th. ed., USA: Kluwer Academic Publishers, 2004.

[32] D. S. Jeon and D. E. Burk, "MOSFET electron inversion layer mobilities-a physically based semi-empirical model for a wide temperature range," in *IEEE Transactions on Electron Devices*, vol. 36, no. 8, pp. 1456-1463, Aug. 1989, doi: 10.1109/16.30959.

[33] J. Pretet, A. Vandooren and S. Cristoloveanu, "Temperature Operation of FDSOI Devices with Metal Gate (TaSiN) and High-k Dielectric," *Proc. ESSDERC*, p. 573, 2003, doi 10.1063/1.2759632.

[34] C. N. Macambira, V. T. Itocazu, L. M. Almeida, J. A. Martino, E. Simoen and C. Claeys, "Ground plane influence on zero-temperature-coefficient in SOI UTBB MOSFETs with different silicon film thicknesses," *2016 31st Symposium on Microelectronics Technology and Devices (SBMicro)*, Belo Horizonte, Brazil, 2016, pp. 1-4, doi: 10.1109/SBMicro.2016.7731326.

[35] L. M. Almeida, J. A. Martino, E. Simoen and C. Claeys, "Improved Analytical Model for ZTC Bias Point for Strained Tri-gates FinFETs," *ECS Transactions*, vol. 31, no. 1, pp. 385, 2010, doi: 10.1149/1.3474183.

[36] A. A. Osman, M. A. Osman, Investigation of high temperature effects on MOSFET transconductance (gm), in: *1998 Fourth International High Temperature Electronics Conference*, HITEC (Cat. No.98EX145), Albuquerque, NM, USA, 1998, pp. 301–304, doi: 10.1109/HITEC.1998.676808.

[37] J. P. Colinge and C. A. Colinge, Physics of Semiconductor Devices, 2nd ed. Hoboken, NJ, USA: Wiley, 2002.

[38] E. A. Gutiérrez-D., M. J. Deen, and C. Claeys, Low Temperature Electronics: Physics, Devices, Circuits, and Applications. Cambridge, MA, USA: Academic, 2001.

# Characterization of AlGaN/GaN HEMTs with Different Manufacturing Characteristics

E. C. Panzo
Institute of Science and Technology
São Paulo State University
Sorocaba - SP, Brazil
eduardo-canga.panzo@unesp.br

J. Candido
Institute of Science and Technology
São Paulo State University
Sorocaba - SP, Brazil
josue.candido@unesp.br

N. Graziano Júnior
Institute of Science and Technology
São Paulo State University
Sorocaba - SP, Brazil
n.graziano@unesp.br

E. Simoen
Ghent University
Ghent, Belgium
also imec Leuven, Belgium
eddy.simoen@UGent.be

M. G. C. Andrade
Institute of Science and Technology
São Paulo State University
Sorocaba - SP, Brazil
gloria.andrade@unesp.br

*Abstract*— This work presents the characterization of high mobility transistors (AlGaN/GaN HEMTs) produced with different techniques in the manufacturing process. The study performs a comparative analysis of the respective HEMTs. Here it is demonstrated that efficient HEMTs can be developed using 2 active implants without the need for specific heat treatment processes to create ohmic contacts or activate the device. The devices thus produced had higher drain current ($I_d$), output conductance ($g_d$), transconductance ($g_m$), field effect mobility ($\mu_{eff}$), effective mobility ($\mu_{FE}$) and low field mobility ($\mu_0$), in addition to lowest series resistance ($R_{SD}$), threshold voltage ($V_T$) and subthreshold slope ($S$). These results indicate that the adopted methodology not only simplifies the manufacturing process but also significantly improves the performance of HEMTs.

**Keywords — HEMT, Manufacturing process, Gate metal, Series resistance, Ohmic contacts, GaN channel.**

## I. INTRODUCTION SECTION

In recent years, high electron mobility transistors are devices typically developed for appreciable applications in high-power, high-frequency circuits [1].

The HEMTs, developed by Fujitsu in 1980, is a voltage-controlled current source. Minor fluctuations in the gate voltage can lead to significant changes in the drain current, due to the high carrier density in Two-dimensional electron gas (2 DEG) channel. The (2 DEG) is a thin, highly electron-conducting layer between two semiconductors of different energy bands, providing superior electron mobility [2]. This results in excellent electronic transport, with less interplay between charge carriers and ions (donor). The performance of HEMTs is influenced by parameters including shape, dimensions, doping levels, and more [3].

Early these devices were made of GaAs and InP due to the high electron mobility in these materials doped with donor atoms, resulting in an excess of electrons that diffuse into the conduction band due to the availability of states with lower energies [1]. This diffusion generates an electric field that moves electrons back to the starting region, similar to a PN junction [2].

Undoped conduction band material has excess majority charges, resulting in high overload speeds. Same without doping in the lower conduction band semiconductor, there is high mobility, as there is no scattering caused by donor atoms [2-3]. It is interesting to emphasize that the GaN HEMT, a device under study, does not require doping in the AlGaN layer. Charge transfer occurs due to the spontaneous polarization and piezoelectric effects of GaN and AlGaN nitrides [3-4]

## II. TESTED SAMPLE

The investigated HEMT, depicted in Fig. 1, is built on a <111> silicon substrate with high resistivity. It includes a 2μm thick buffer layer made of GaN/AlGaN to minimize crystal imperfections. Above this, there is a 300nm gallium nitride channel layer situated under a 1nm aluminum nitride spacer, which enhances charge carrier concentration, mobility, and the drain current. The structure also features a 15nm Al$_{0.25}$Ga$_{0.75}$N barrier, where the mole fractions of aluminum and gallium are 0.25 and 0.75, respectively. Additionally, a silicon nitride (Si$_3$N$_4$) cap is incorporated, and a 140nm layer of silicon dioxide (SiO) separates the 2D electron gas from the gate metal [5]. The fabrication, as well as the respective measurements of the devices, were carried out at IMEC (Interuniversity) in Belgium.

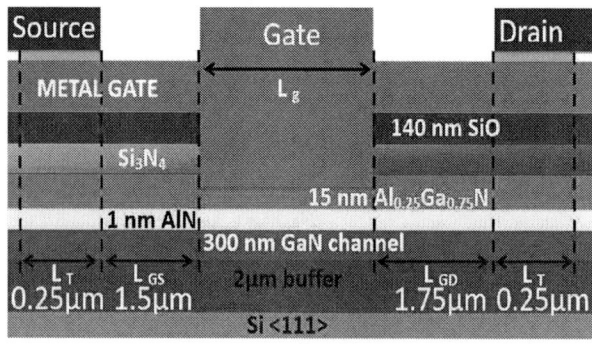

Fig. 1. Modified cross-sectional view of the AlGaN/GaN HEMT structure [5].

Table 1 presents the specific characteristics of each HEMT. The names α, β and γ were adopted to facilitate the description of each device. The general characteristics of the HEMT are shown in Fig. 1.

TABLE 1. HEMT MANUFACTURE CHARACTERISTICS

| HEMT | Characteristics |
|---|---|
| α | **Division:** 40 nm ionically implanted titanium nitride / 250 nm aluminum-copper / 60 nm titanium nitride anti-reflective layer. **Ohmic annealing/active annealing:** 525 ºC 90s. |
| β | **Division:** Post-active annealing (heat treatment to adjust doping and crystal structure after ion implantation). **Ohmic annealing/active annealing:** 700ºC 90s; 550ºC 90s (Two steps of ohmic annealing to participate as doped). |
| γ | **Division:** 2 active implants (advanced ion implantation technique to create precise doping layers). **Ohmic Annealing/Active Annealing:** NA (There is no specific heat treatment process to create ohmic contacts or activate the device). |

## III. EXPERIMENTAL ANALYSIS AND DISCUSSION

### A. Drain current ($I_d$)

The drain current can be influenced by factors such as voltage applied to the gate or by the characteristics of the device. In certain instances, this current can be modulated by altering the drain voltage, which consequently impacts the source-drain series resistance. The maximum source-drain current intensity of the HEMT can be calculated using equation 1 [2].

$$I_{d,max} = \mu_n \frac{\varepsilon_d}{\omega} \frac{W}{L} \frac{(V_g - V_T)^2}{2} \qquad (1)$$

Where $\mu_n$ - carrier mobility, $\varepsilon_d$ is the permittivity, $\omega$ −average width of the depletion layer in the channel, $\frac{W}{L}$ − aspect ratio and ( $V_g - V_T$ ), overdrive voltage or effective voltage.

Initially taken and curves $(I_d x V_g)$ were obtained for all devices, as shown in Fig. 2 (HEMT γ), to extract the average of the curves for each transistor. The same procedure was performed with the curves and $(I_d x V_d)$.

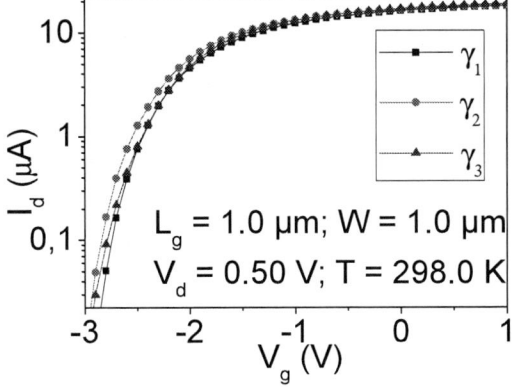

Fig. 2. $I_d x V_g$ curves for 3 HEMTs from the same wafer ($\gamma_1$, $\gamma_2$, $\gamma_3$).

The behavior of the drain current can be analyzed through the $I_d x V_g$ curves, as illustrated in Fig. 3. It is possible to observe that the current intensity is greater in HEMT γ.

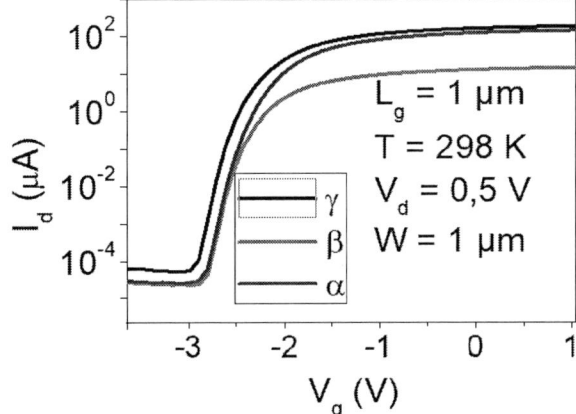

Fig. 3. $I_d x V_g$ for $V_d = 0.5$ V, varying in devices α, β and γ.

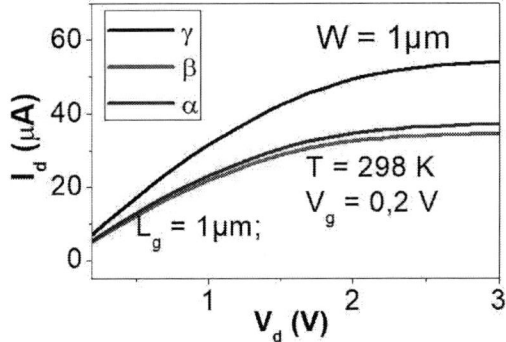

Fig. 4. $I_d x V_d$ curves for $V_g = 0.2$ V, varying in devices α, β and γ.

From the curves $I_d x V_g$ (Fig. 3) and $I_d x V_d$ (Fig. 4) the $I_d$, $g_m$, $V_T$, $g_d$, S, and $R_{SD}$ were successfully measured, analyzed, and utilized for the calculation of $\mu_{eff}$, $\mu_{FE}$ and $\mu_0$.

### B. Transconductance ($g_m$)

The transconductance evaluates the variation of the drain current according to the gate voltage, especially at low drain voltage values, crucial for determining the field effect mobility and the threshold voltage. Fig. 5 shows that $g_m$ is larger in γ devices.

Equation 2 shows the dependence on transconductance, which was extracted through the curve $I_d x V_g$ [6].

$$g_m = \frac{\partial I_d}{\partial V_g} \qquad (2)$$

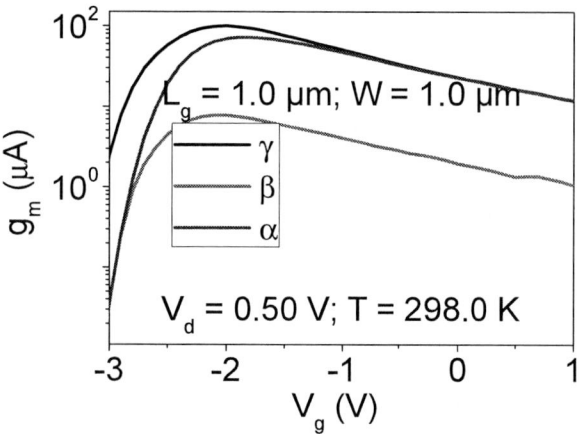

Fig. 5. $g_m \times V_g$ curves for $V_d = 0.5$ V, and $V_g$ varying in devices α, β and γ.

## C. Output conductance ( $g_d$ )

The output conductance describes the ease of current flowing at the device output, relating drain current and drain voltage (Equation 3). Expressed in Siemens (S), it is determined by the physical characteristics and design of the device, influenced by the internal structure, channel geometry, semiconductor doping and manufacturing parameters [6].

$$g_d = \frac{\partial I_d}{\partial V_d} \tag{3}$$

A $g_d$ also is larger in γ devices, as Fig. 6 indicates.

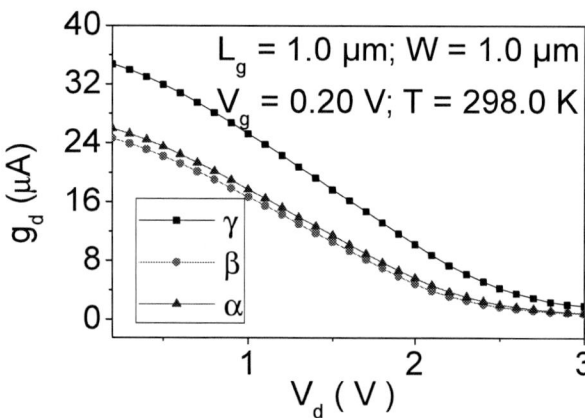

Fig. 6. $g_d \times V_d$ curves in devices α, β and γ.

## D. Series Resistance ($R_{SD}$)

It is directly proportional to the length of the path traveled by the electric current which is known as channel length (L), it is inversely proportional to the cross-sectional area (A) of that path [7]. Therefore, the greater the series resistance, the greater the voltage drop along the current path causing energy losses, heat dissipation and limitations on the maximum current that can pass through the circuit providing a decrease in the overall efficiency of the device [6].

Considering that the total resistance $R_T$ is the result of the sum of the series resistance $R_{SD}$ and the channel resistance $R_{Ch}$ in equation 4 and neglecting the resistance of ($R_{Ch} = 0$), Because the series resistance is significantly greater than the channel resistance, in addition to the respective variation in channel resistance being minimal enough to be disregarded due to the operating conditions. Which allows focusing on the series resistance, which is often dominant and of greater interest for certain analyzes and optimizations. The equation 5 was used to obtain $R_{SD}$, polarized with $V_d = 0.5 V$ and $V_g = 0.2 V$ [6]. The extraction of $R_{SD}$ is shown in table 2, and it is observed that it is greater in β - HEMTs.

$$R_T = R_{SD} + R_{Ch} \tag{4}$$

$$R_T = R_{SD} = \frac{V_d}{I_d} \tag{5}$$

## E. Threshold voltage ( $V_T$ )

It is the minimum voltage necessary for the device to begin conducting electrical current. This parameter is of great relevance, as it determines the point from which the current flow begins to operate the transistor [8].

One of the widely used methods for HEMTs is the extrapolation method according to equation 6, which allows obtaining $V_T$ through the maximum transconductance, where $g_m(máx)$ is the maximum transconductance [8] The $V_T$ results for each device can be seen in table 2.

$$V_T = V_{g(g_m(máx))} - \left(\frac{I_{d(g_m(máx))}}{g_m(máx)}\right) - \frac{V_d}{2} \tag{6}$$

## F. Subthreshold slope ( S )

The subthreshold swing describes how the drain current of a semiconductor device changes in response to varying gate voltage below the conduction threshold $V_T$. It is determined by the physical properties, transistor design, and underlying semiconductor material. Expressed in millivolts per decade (mV/dec), the lower it is (up to 60 mV/dec at room temperature), the better the control of current flow and the lower the current loss when the device is turned off [7-8]. In this research, HEMT γ recorded the lowest subthreshold slope (S), as shown in table 2.

$$S = \frac{dV_g}{d(log I_d)} \tag{7}$$

TABLE 2. RESULTS FROM $R_{SD}$ , $V_T$ AND $S$ FOR ALPHA, BETA AND GAMMA DEVICES.

| Device description | Calculation parameters | | |
|---|---|---|---|
| | $R_{SD}$ 10[Ω] | $V_T$ [V] | S [mV/dec] |
| α | 0.38 | -2.7 | 87.30 |
| β | 3.73 | -2.6 | 96.44 |
| γ | 0.28 | -2.8 | 73.90 |

Table 2, was observed that the γ device tends to present lower values of series resistance, more negative threshold voltage and smaller subthreshold slope compared to α and β. These results are consistent with the expectation that advanced doping techniques can lead to devices with relatively better electrical performance, such as lower resistance and higher operational sensitivity [2].

### G. Effective mobility ($\mu_{eff}$)

The stepwise channel approximation method, which relies on the separation of the two-dimensional electron gas (2DEG) between the source and drain regions, can be used to derive the effective mobility. In equation 8 the expression for this calculation is presented [6].

$$\mu_{eff} = \frac{n_G}{C_d(\frac{1}{g_d}-(R_{S-0}n_S+R_{S-0}n_D+R_{S-C}n_C))(V_g-R_{S-0}n_S I_d-V_T)} \quad (8)$$

Where: $C_d$ Represents the capacitance per unit surface area; $R_{S-0}$ - Uniform sheet resistance; $R_{S-C}$ - Resistance of the contact layer; $n_C = 2\frac{L_T}{W}$ , $n_D = \frac{L_{GD}}{W}$, $n_S = \frac{L_{GS}}{W}$ and $n_G = \frac{L_g}{W}$ . Represent effective square values for the ohmic contact, drain, source, and gate respectively. $L_T$ - The length over which current flows from the metal into the semiconductor; $L_{GS}$ - The distance separating the gate from the source and $L_{GD}$ - The distance from the gate to the drain.

### H. Field Effect Mobility ($\mu_{FE}$)

In order to cancel the influence of the $V_T$ on mobility, the $g_m$ is measured and then inserted into equation 9 to derive $\mu_{FE}$ [6].

$$\mu_{FE} = \frac{n_G V_d}{C_d(\frac{1}{g_m}-R_{S-0}n_S)(V_d-(R_{S-0}n_S+R_{S-0}n_D+R_{S-C}n_C)I_d)^2} \quad (9)$$

### I. Low field mobility ($\mu_0$)

The low-filed mobility ($\mu_0$) extraction method uses equation 10, approximating the drain current in relation to $V_g$ [7].

$$\mu_0 = S(1 - R_{SD}V_d S)^{-1}(C_d W/L_{eff})^{-1} \quad (10)$$

For the effective mobility ($\mu_{eff}$) which was studied by characterizing the flow of electrons in a substance exposed to a non-uniform electric field, polarizing with gate voltage ($V_g$) between -3.2 V to 0 V and the gate voltage drain ($V_d$) from 4 V to 8 V; The calculation of the field effect mobility ($\mu_{FE}$) (which is examined when the transistor is exposed to a strong, consistent electric field), was done with gate voltage polarizations ($V_g$) between -1 V to -0.5 V and drain ($V_d$) from 8 V to 10 V; And in the end, the low field mobility ($\mu_0$) (This is for tiny electric field), extracted with polarized devices with gate voltage ($V_g$) from -0.5 V to 0 V and drain voltage ($V_d$) from 1 V to 5 V [9]. Fig. 7 shows that $\mu_{eff}$ and $\mu_{FE}$ are $\mu_0$ larger in γ devices. Studies carried out in [6], show similar behavior in relation to $\mu_{eff} > \mu_{FE}$.

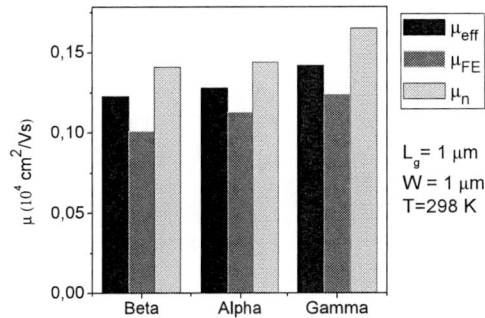

Fig. 7. Extraction of $\mu_{eff}$, $\mu_{FE}$ and $\mu_0$ in α, β and γ devices.

## IV. Conclusions

The γ devices generally in terms of performance have higher drain current conduction; transconductance; output conductance (although for better analog performance, reduced $g_d$ would be necessary) and mobilities ($\mu_{eff}$, $\mu_{FE}$ and $\mu_0$). In addition to having lower values of $R_{SD}$, $S$ and $V_T$, resulting in lower energy consumption and greater efficiency in changing the operating region. Therefore, manufacturing processes play a crucial role in HEMTs, as different processes significantly influence the behavior of each device.

### Acknowledgment

The authors extend their gratitude to the Imec; Appreciation is also given to the Brazilian research funding agency FAPESP (Process No.: 2023/00123-7), as well as CAADI and CAPES for their contributions.

### References

[1] M.A. Romero, R. Ragi and J.E. Manzoli, "High electronic mobility transistors HEMTs: Operating principles and electronic characteristics", Brazilian Journal of Physics Teaching, 2015.

[2] J. Ajayan and D. Nirmal, "Handbook for III-V High Electron Mobility Transistor Technologies," CRC Press, Taylor & Francis Group. Book standard international number 13: 978-1-138-62527-3 (hardcover), 2019.

[3] U.K. Mishra, P. Parikh and Yi-Feng Wu, "AlGaN / GaN HEMTs – an overview of device operation and applications," Proceedings of the IEEE, Volume: 90, Page(s): 1022 – 1031, 2002

[4] S.J. Farlow, "Partial Differential Equations for Scientists and Engineers," New York: Dover Publications, 1993.

[5] U. Peralagu, A. Alián, V. Putcha, A. Khaled, R. Rodríguez, A. Sibaja - Hernandez, S. Chang, E. Simão, S.E. Zhao, B. De Jaeger, D.M. Fleetwood, P. Wambacq, M. Zhao, B. Parvais, N.Waldron, N. Collaert, "CMOS- compatible GaN-based devices on 200 mm-Si for RF applications: integration and performance", International Device Meeting Electronics IEEE (Iedm), 2019.

[6] A. Aminbeidokhti, S. Dimitrijev, J. Han, X. Xu, C. Wang, S. Qu, H. A. Moghadam, P. Tanner, D. Massoubre and G. Walker, "A method for removing electron mobility in HEMTs from power," Journal of Supergrids and Microstructures. Volume 85, p. 543-550, September 2015.

[7] D. Pradeep and D.S. Rawal, "Comparison of two DC extraction methods for mobility and parasitic resistances in a HEMT", IEEE Transactions on Electron Devices, April/2017.

[8] E.C. Panzo, N.G. Junior and M.G.C. Andrade, "Mobility Extraction Methods in AlGaN/ GaN HEMTs", 37th Symposium on Microelectronics Technology and Devices, 2023.

[9] S. M. SZE and K. K. NG, "Physics of Semiconductor Devices", John Wiley & Sons, Inc., Hoboken, New Jersey, 3ª Edition, 2007.

# Proposal for an Ion-Sensitive Floating-Gate MOSFET with Tunable Sensitivity and Memory Properties

Henrique L. Carvalho[1], Ricardo C. Rangel[1,2], Joao A. Martino[1], Senior Member, IEEE

[1]LSI/PSI/USP, University of Sao Paulo, Sao Paulo, Brazil.

[2]FATEC-SP, Faculdade de Tecnologia de São Paulo,São Paulo, Brazil

E-mail:hen.lanfredi@usp.br

*Abstract* — **This work proposes a new Ion-Sensitive Floating Gate MOSFET, using ultra-thin body and box (UTBB) Silicon-On-Insulator (SOI) structure, to detect charges at the front oxide/ion concentration solution interface in the range of $5.10^{11}$ and $5.10^{12}$ cm$^{-2}$. The device operates based on a floating gate charge effect for detection and retains the charge, providing a memory effect. The device's operation and characteristics were investigated through simulations, revealing that the average electron concentration in the floating gate depends on the oxide/solution interface charge, varying from $1.10^{19}$ to $1.7.10^{19}$ cm$^{-3}$. Additionally, the write time and voltage have an impact on the memory window. Increasing the write times increases the memory window, from 1.42 to 2 V, for charge times from 2 to 40 seconds. The device's sensitivity can be controlled by adjusting the write time, offering a means to detect ions in solution. Moreover, the memory window read with charges at the oxide/solution interface can be used to sense interface charges from a reference memory window without charges. Overall, the proposed device shows promise for applications in ion sensing and detection, offering tunable sensitivity and memory properties.**

*Keywords—Ion-Sensitive; Silicion-on-Insulator; Ultra-Thin Body and Box (UTBB), Floating Gate; Memory effect*

## I. INTRODUCTION

The Moore's Law has led to rapid advancements in micro and nanoelectronics, resulting in decades of size reduction in semiconductor device, increased density, causing significant societal changes. Furthermore, there has been a trend towards decreasing energy consumption and manufacturing costs [1, 2]. In the last decades many electronic devices have been proposed, to overcome technology challenges, using FinFET, NWFET, RFET, TFET, NCFET, NCTFET and [3-6]. However, decades have led to expansion in various sectors, catalyzing progress in fields ranging from telecommunications to biomedical engineering including the field of III-V materials, integrated memory devices and sensors [7-11].

Furthermore, integrated memory devices have revolutionized data storage and processing capabilities, enabling the creation of faster and more energy-efficient computing systems. Meanwhile, advancements in sensor technologies have opened up new possibilities for real-time monitoring and analysis in critical areas such aerospace and health, using UV-light, biological, ionic, radiation sensors [12-19].

The ion-sensitive field effect transistor (ISFET) sensor has become an attractive topic in last year's [8,9,17-20], combining the function of material recognizer (electrical charge of the material) through electrical signals with microfabrication technologies, resulting in some benefits for reaction monitoring, presenting ionic and biological sensitivity. Firstly, the ISFET is based on complementary metal-oxide-semiconductor technique, can be mass produced and integrated into analysis circuits [19]. Second, ISFET requires only one reference electrode (RE), giving advantage in size. Finally, the ISFET has an easily operate and similar to a metal-oxide-semiconductor field effect transistor (MOSFET), where the metal gate voltage controls current between drain and source contacts, but in ISFET the ionic solution on the top oxide gate can modify their behavior.

The SOI group in Integrable Systems Laboratory at University of Sao Paulo (LSI/USP) has studied and fabricated ions sensitive field effect transistors and biological integrated biological sensors [13,14,16]. However, the usual architectures of ISFET work near at solution, thus built-in logic circuit from sensitive transistor and metallic lines interacting with analyte solution [17,19]. The floating gate (FG) or extended gate is a promising architecture for ISFET to separate the analyte solution and sensitive surface with the logic circuits and metallic lines [17,19,20]. Also, the floating gate architecture are present in modern non-volatile memory devices, used in USB drives, SSDs (Solid State Drives) and other storage devices, that has a function of storage charge. The floating gate metal oxide semiconductor field effect transistor (FGMOS) has compatibility with MOS manufacturing processes, integration into mixed-signal integrated circuit designs, and its impact as a high-density, high-speed, and low-power non-volatile memory [10,11,21,22].

In this work, a new ion-sensitive floating gate metal oxide semiconductor field-effect transistor is proposed, using a ultra-thin body and buried silicon-on-insulator structure (UTBB-SOI) to obtain the thickness of the top oxide (tunneling oxide), silicon film (floating gate), and buried oxide (gate oxide) for future fabrication, based on the memory effect of floating gate charging.

## II. FUNDAMENTALS OF FLOATING GATE MOSFET

The Floating-Gate MOSFET has a three operations states, write, erase and read. The Fig. 1 shows the operation of a Floating Gate MOSFET, initially, there is no charge stored on the floating gate $Q_{FG} \approx 0$ (Fig. 1 (a)), as the applied voltage increased on the control gate ($V_{CG} \geq V_{write}$), the Fowler-Nordheim tunneling mechanism transports charge from substrate to the floating gate, making $Q_{FG} \neq 0$ [10, 23]. When the applied voltage is removed, the charge is trapped in the floating gate, this mode of operation is defined as the "write" operation (Fig. 2 (b)) [10, 21, 23].The read operation with and without charges in floating gate (Fig. 1 (a) (c)) is similar to MOSFET, applying voltage to the control gate ($V_{CG} \geq V_{Th}$), electrons move to the interface bottom oxide and substrate, creating an electron channel enabling conduction current between drain and source contacts for $V_{DS} > 0$. The erase operation has the goal, to remove the charge stored, when negative voltage is applied to the control gate ($V_{CG} \leq V_{Erase}$), the electrons in floating gate transported to the substrate (Fig. 1 (d)) [10, 23].

The Fig. 2 (a) show the charge in the floating gate in function of time (s), for write ($V_{CG} = 12$ V), retention ($V_{CG} = 0$ V) and erase ($V_{CG} = -12$ V). The transistor's threshold voltage depends of the charges storages in the floating gate after write ($Q_{FG} < 0$), and erase ($Q_{FG} \approx 0$) operations and the difference between the threshold voltage written ($V_{Th}$) and erased ($V_{Th0}$) is called memory windows and given by equation (1), where $C_{CG}$ is the capacitance between floating gate and control gate [10]. The Fig. 2 (b) show the drain current ($I_{DS}$) in function of control gate voltage ($V_{CG}$) in read operation, with (programmed; $Q_{FG} < 0$) and without charges storages (erased; $Q_{FG} \approx 0$) in floating gate.

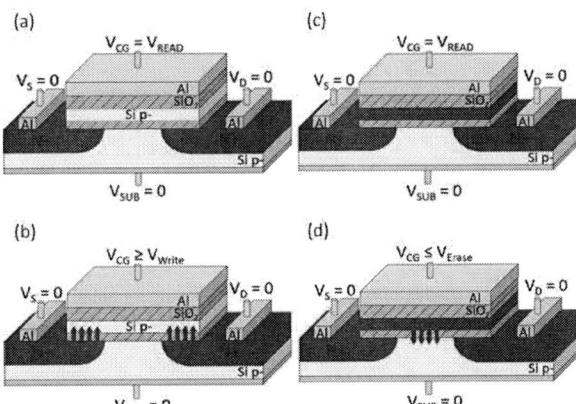

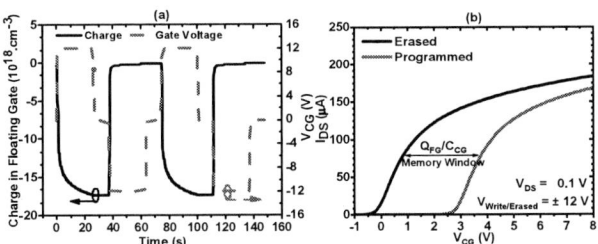

Fig. 1. Floating Gate MOSFET, for read ($Q_{FG} = 0$) (a), write (b), read ($Q_{FG} \neq 0$) (c) and erase (d) operations.

Fig. 2. Charge in floating gate MOSFET as a function of time, in write ($V_{CG}$ = 12 V), read ($V_{CG}$ = 0 V) and erase ($V_{CG}$ = -12) condition (a) and drain current read as a function of control gate voltage, for floating gate programmed and erased (b).

$$\Delta V_{Th} = V_{Th} - V_{Th0} = -\frac{Q_{FG}}{C_{CG}} \quad (1)$$

### III. DEVICE CHARACTERISTICS AND OPERATION

The new proposal of an Ion Sensitive Floating Gate MOSFET based on memory effect, was done using Synopsys Sentaurus [24]. UTBB SOI technology is used to facilitate future manufacturing processes, such as bottom oxide, a thin floating gate and a good interface for top oxide.

The proposed structure can be seen in Fig. 3 (a), where the "metal gate" is composed by an aqueous solution with work function of 4.71 eV, the bottom oxide is the buried oxide of the UTBB wafer ($t_{box} = 25$ nm) ($SiO_2$), the top oxide is tunneling oxide thickness with 6 nm ($SiO_2$), floating gate thickness (silicon <100>, boron $10^{15}$ cm$^{-3}$) with 6 nm and silicon substrate <100> with boron $10^{15}$ atm.cm$^{-3}$. The physics models of simulations are mobility including doping dependence and high-field saturation (velocity saturation), Shockley–Read–Hall recombination with doping-dependent lifetime and band-to-band tunneling model [24]. The charge in solution is apply at the interface between top oxide (tunneling oxide) and solution to control the interfacial potential at the interface of the solution and the oxide.

The Fig. 3 show the operation mode of SOI Ion-Sensitive FGMOSFET base on memory effect, in read operation the transistor has a similar operation with nMOSFET, where current conduction between drain and source is enabled when $V_{cc} > V_{Th0}$ for the floating gate erased (a) (c) and $V_{cc} > V_{Th}$ when programmed (b) (d). However, the oxide between the floating gate and solution is tunneling oxide, so the Fowler-Nordheim tunneling can transports charges from solution to the floating gate or floating gate to solution. The potential applied to tunneling charges from solution to floating gate

negative (write operation) (e) or positive to tunneling charges from floating gate to solution (erase operation) (f).

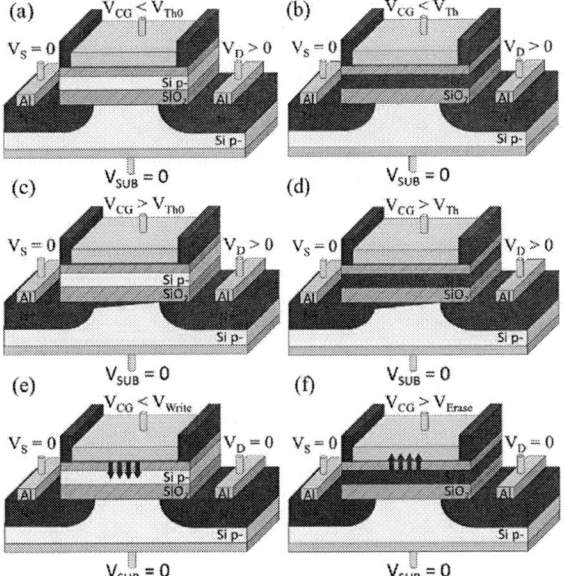

Fig. 3. The operation of proposed SOI Ion-Sesitive FGMOSFET base on memory effect.

### IV. RESULTS AND DISCUSSIONS

The Fig. 4 show the average charge in floating gate of SOI Ion Sensitive FGMOSFET in function of time (s), for write ($V_{CG} \ll 0$ V), retention ($V_{CG} = 0$ V) and erase ($V_{CG} \gg 0$ V) operation, with different concentration of charges at interface tunneling oxide/solution.

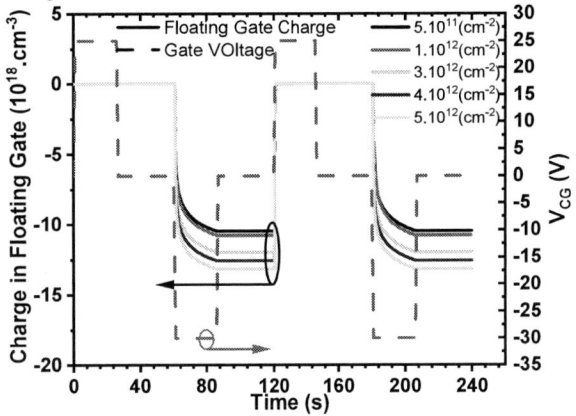

Fig. 4. Charge in floating gate as a function of time, in write, retenption and erase, for different charges at oxide/solution interface.

At the beginning of the simulation, independently of the charge in the solution, there is minimal stored charge ($Q_{FG} \approx 0$) on the floating gate. Consequently, when a positive voltage is applied to the control gate ($V_{Erase}$), there is no change in the floating gate charge. However, over time, with the application of a high negative potential, electrons tunneling from the gate to the floating gate, increasing the average electron concentration in function of time. Removing the voltage applied to the control gate, the electron concentration on the floating gate is trapped in the condition $V_{WRITE} > V_{CG} > V_{Erase}$, defining the write operation. Applying a potential higher than $V_{ERASE}$ the electrons are transported from floating gate to control gate, resulting in $Q_{FG} \approx 0$ under the condition

($V_{Write} > V_{CG} > V_{Erase}$). Increasing the interface charge at the solution-oxide interface facilitates charge (red line in Fig. 3) tunneling to the floating gate, allowing an increase in the average electron charge on the floating gate.

The Fig. 5 shows the charge in floating gate in function of time, for 6 and 4 nm of tunneling oxide. The reduction in tunnel oxide thickness promotes a greater charge accumulation in the floating gate during the writing operation. However, after being charged, the stored charge is lost over time more rapidly than in 6 nm. Although the charge in the floating gate continues to be influenced by the interface charge, the reduction in oxide thickness can cause potential read errors. Thus, the work proposes the use of a 6 nm tunnel oxide to reduce charge losses over time. The Fig. 6 shows the drain current read in function of control gate voltage, after erase and write operation, with different concentration of charges at interface oxide/solution.

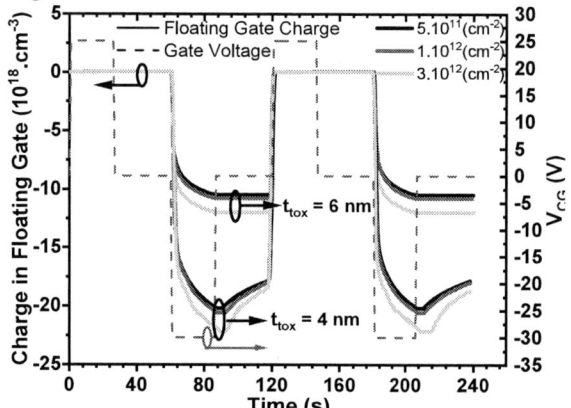

Fig. 5. Charge in floating gate in function of time, in write, retenption and erase, for different interface charges.

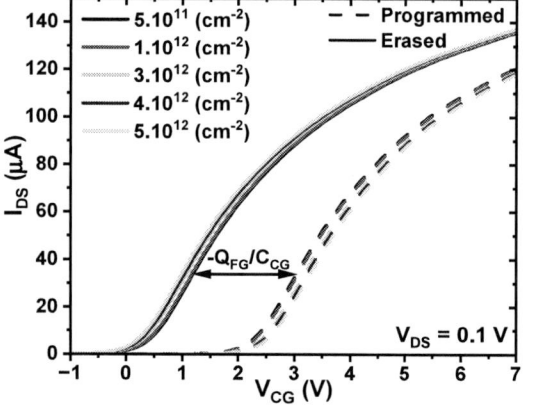

Fig. 6. Drain current as a function of control gate voltage, written and erased, for different charges at oxide/solution interface.

The threshold voltage of the transistor depends on the charge trapped in the floating gate. When the floating gate has no stored charge (erased), it is observed that an increase in positive charge at the oxide/solution interface tends to decrease the transistor's threshold voltage, similar to a conventional ISFET. However, the write operation allows electron storage in the floating gate over time, increasing the transistor's threshold voltage, by $-Q_{FG}/C_{CG}$ as shown in (1). In contrast to the effect of decreasing the transistor's threshold voltage with positive charge at the interface when erased, it was shown in Fig. 4 that increasing interface charge enhances the average electron concentration in the floating gate when programmed, thereby competitively increasing the transistor's threshold voltage with the charge at the interface.

The memory window of the transistor depends on the charge in the floating gate. However, the amount of charge will depend on the loading and unloading of the floating gate, which is a function of the charging and discharging time and the write and erase voltage levels. Fig. 7 illustrates the average charge in the floating gate as a function of time for various write times and constant $V_{Write}$.

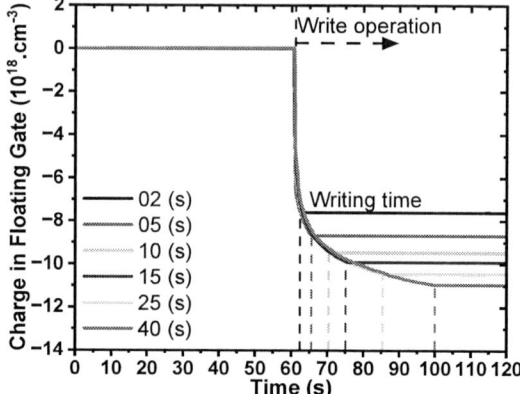

Fig. 7. Floating Gate MOSFET, for read ($Q_{FG} = 0$) (a), write (b), read ($Q_{FG} < 0$) (c) and erase (d) operations.

At the beginning of the write operation, the floating gate is erased ($Q_{FG} \approx 0$). Then the electrons are transported to the floating gate because $V_{CG} < V_{Write}$, following a capacitive loading trend. Thus, the average charge on the floating gate is a function of time and write voltage. Fig. 8 shows the drain current read as a function of control gate voltage, for constant erase time (30 s) and different write times. The threshold voltage of the transistor when erased is constant as ($Q_{FG} \approx 0$) over a long time, but the charge on the floating gate is proportional to the write time, so the threshold voltage of the transistor when programmed increases with time. Thus, Fig. 9 shows the threshold voltage of the transistor as a function of interface charge and write time, programmed and erased, for constant write voltage (30 V) and erase time (30 s).

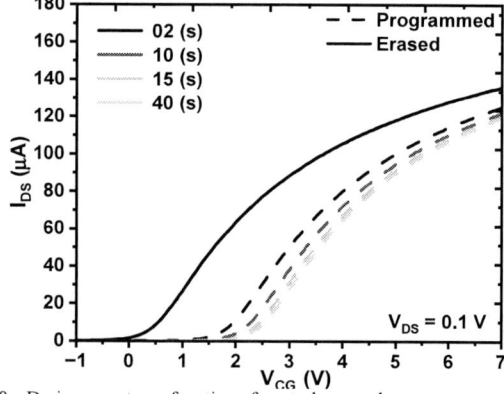

Fig. 8. Drain current as a function of control gate voltage, programmed and erased, for different write times.

The threshold voltages were extracted using the method [25], where the threshold voltage represents the maximum of $d^2 I_{DS}/dV_{CG}^2$. When the transistor is erased, the threshold voltage decreases with increasing interface charge and is not influenced by long erase time. When programmed, the threshold voltage increases with increasing interface charge and with write time. It is observed that the memory window increases with increasing interface charge, thus it is also possible to sense charge at the oxide/solution interface from the difference between memory windows, with a reference memory window. However, the adopted sensitivity definition is (2), where the adopted reference is the threshold voltage of the erased transistor ($Q_{FG} \approx 0$) and with no charge at the oxide/solution interface. Fig. 10. presents the sensitivity in function of interface charge for different write times.

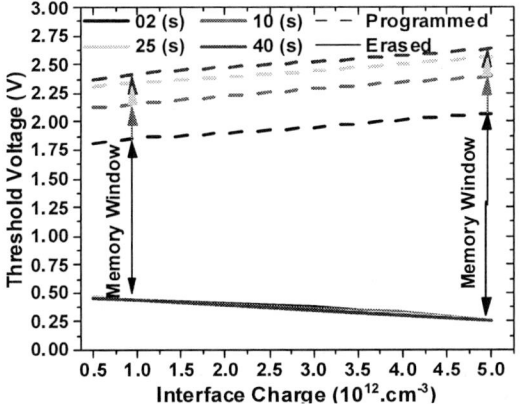

Fig. 9. Threshold voltage as a function of interface charge for different time, programmed and erased.

$$Sensitivity = \frac{V_{Th(Prog)} - V_{Tho(ref)}}{V_{Tho(ref)}}$$

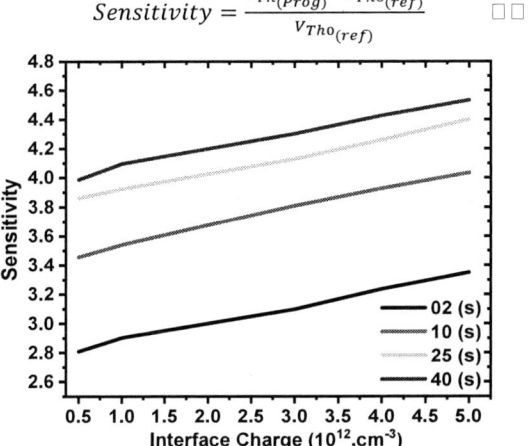

Fig. 10. Sensitivity as a function of interface charge for different write times.

The device exhibits different levels of sensitivity (2) depending on the writing time, thus the difference between the threshold voltages with respect to the reference can be controlled by the write time, for different charges at the interface.

## V. CONCLUSION

In conclusion, the proposed Ion-Sensitive Floating Gate MOSFET, using ultra-thin body and box (UTBB) silicon-on-insulator (SOI) structure, demonstrates promising capabilities for sensing interface charges at the interface oxide/metal-gate that working as a solution with ion concentration in solution, based on a floating gate charge effect for detection, and retaining the charge giving a memory effect, in the interface charge range of $5.10^{11}$ and $5.10^{12}$ cm$^{-2}$.

Through simulations, it was demonstrated the operation and characteristics of the proposed device with 6 and 4 nm of tunneling oxide. the average concentration of charges (electrons) in the floating gate depends on the oxide/solution interface charge, thus a higher interface charge ($5.10^{12}$ cm$^{-2}$) promotes a higher electron concentration ($\sim 1.7.10^{19}$ cm$^{-3}$) than a minor interface charge ($5.10^{11}$ cm$^{-2}$) in the floating gate when charged ($\sim 1.10^{19}$ cm$^{-3}$), increasing the transistor's threshold voltage with the increase of interface charge; however, the natural effect of decreasing threshold voltage with the increase of interface charge is still observed when the transistor is erased. It was found that the loading of the

floating gate follows a capacitive charging trend, therefore the memory window is affected by the write time, along with the control gate voltage applied for the write operation; the longer the write time and the higher the write voltage, the greater the charge in the floating gate, consequently, the larger the memory window, For constant voltage writing, a memory window of 1.42 V and 2 V was obtained for charge times of 2 and 40 seconds, with an interface charge of $1.10^{12}$ cm$^{-2}$.

It was demonstrated that device's sensitivity can be controlled by adjusting the write time, offering an option to detect ion in solution. Furthermore, the memory window read ($V_{Th} - V_{Th0}$) with charges at the interface oxide/solution can be used to sense interface charges from a reference memory window without charges. The proposed device has potential for applications in ion sensing and detection, offering tunable sensitivity and memory properties.

## ACKNOWLEDGMENT

The authors acknowledge CNPq, CAPES and São Paulo Research Foundation - FAPESP (under grant #2020/04867-2) for the financial support.

## REFERENCES

[1] J.Lee , "Review paper: Nano-floating gate memory devices," Electron. Mater. Lett. 7, 175–183, 2011, doi: 10.1007/s13391-011-0901-5.
[2] S. Agarwal, et al., "Using Floating-Gate Memory to Train Ideal Accuracy Neural Networks," IEEE Exploratory, 2019.
[3] J. Colinge,"The SOI MOSFET: from Single Gate to Multigate," Springer, Boston, MA. doi: 10.1007/978-0-387-71752-4_1
[4] S. Valasa, et al."A critical review of advancements in negative capacitance field effect transistors: A revolution in next-generation electronics," Mat. Sci in Semiconductor Processing, Vol 173, 2024.
[5] T. Mikolajick, et. al. "20 Years of reconfigurable field effect transistors: From concepts to future applications," 2021 Solid State Electronics. 186, Doi: doi:10.1016/j.sse.2021.108036.
[6] S. Rahi, S. Tayal, A. Upadhyay, "A review on emerging negative capacitance field effect transistor for low power electronics," Microelectronics Journal, Volume 116, 2021.
[7] Chenbi Li, et al. "Review of the AlGaN/GaN High-Electron-Mobility Transistor-Based Biosensors: Structure, Mechanisms, and Applications," Micromachines 2024.
[8] Ž. Janićijević, et al., "Extended-gate field-effect transistor chemo- and biosensors: State of the art and perspectives," Next Nanotechnology, 2023, doi:10.1016/j.nxnano.2023.100025.
[9] M. Kaisti, et al., "Compact model and design considerations of an ion-sensitive floating gate FET," Sensors and Actuators B: Chemical, Volume 241, 2017, doi:10.1016/j.snb.2016.10.051.
[10] T. Cong, et al., "A simulated fabrication and characterization of a 65nm floating-gate MOS transistor," Ain Shams Engineering Journal, Volume 14 2023, doi:10.1016/j.asej.2022.101917.
[11] Yunjae Kim, et al., "Circuit simulation of floating-gate FET (FGFET) for logic application," Memories - Materials Devices Circuits and Systems, Volume 6, 2023, doi:10.1016/j.memori.2023.100090.
[12] J. Padovese, et al., "Back Enhanced SOI MOSFET as UV light sensor", 33rd Symposium on Microelectronics Technology and Devices (SBMicro), IEEExplorer, 2018, doi: 10.1109/SBMicro.2018.8511328.
[13] L. Yojo, et al., "Optimization of the permittivity-based BE SOI biosensor," IEEE SOI-3D-Subthreshold Microelectronics Technology Unified Conference (S3S), 2018.
[14] L. Yojo, et al., "Influence of biological element permittivity on BE (Back Enhanced) SOI MOSFETs," 33rd Symposium on Microelectronics Technology and Devices (SBMicro), pp. 1-4, 2018.
[15] L. Yojo, et al., "Back enhanced (BE) SOI pMOSFET behavior at high temperatures," 31rd Symposium on Microelectronics Technology and Devices (SBMicro), pp. 1-4, 2016.
[16] P. Duarte, et al., "ISFET Fabrication and Characterization for Hydrogen Peroxide sensing," Journal of Integrated Circuits and Systems, Volume 18, 2023, doi: 10.29292/jics.v18i1.646
[17] O. Gubanova, et al., "A novel extended gate ISFET design for biosensing application compatible with standard CMOS," Materials Science in Semiconductor Processing, Volume 177, 2024.
[18] Cao Shengli, et al., "ISFET-based sensors for (bio)chemical applications: A review," Electrochemical Science Advances, Volume 3, 2022, doi: 10.1002/elsa.202100207
[19] Li-Te Yin, et al., "Separate structure extended gate H+-ion sensitive field effect transistor on a glass substrate," Sensors and Actuators B: Chemical, Volume 71, 2000, doi: 10.1016/S0925-4005(00)00613-4.
[20] M. Hussin, et al., "Simulation and Fabrication of Extended Gate Ion Sensitive Field Effect Transistor for Biosensor Application," vol 339. Springer, Berlin, Heidelberg. doi: 10.1007/978-3-642-35264-5_53
[21] Wang S, et al., "Nonvolatile memory: New floating gate memory with excellent retention characteristics," Adv Electron Mater, 2019.
[22] R. Wunderlich, et al., "Floating Gate Based Field Programmable Mixed-Signal Array," in IEEE Trans on (VLSI) Systems, vol. 21, 2013.
[23] P.Pavan, et al. "Floating Gate Devices: Operation and Compact Modeling," Springer New York, NY, doi: 10.1007/b105299
[24] Synopsys TCAD, Sentaurus Device User Guide, (Version L2016.03).
[25] H. Wong, at al., "Modeling of transconductance degradation and extraction of threshold voltage in thin oxide MOSFET's.", Solid-State Electron, 1987;30:953, doi: 10.1016/0038-1101(87)90132-8.

# AUTHOR INDEX

Agopian, Paula G. D. ..................... 9, 17, 37
Agopian, Paula Ghedini Der .................... 41
Aguiar, Vitor Ângelo P. ...................... 118
Alaferdov, Andrei ............................ 33
Alberton, Saulo Gabriel ..................... 118
Andrade, M. G. C. .......................... 134
Barraud, Sylvain ........................... 100
Bazilio, Willian M. M. ....................... 80
Benevenuti, Fábio ........................... 96
Benvenutti, Julia Willow ..................... 96
Beraldo, R. M. ............................. 64
Bergamaschi, Flávio Enrico .................. 100
Boas, Alexis C. Vilas ....................... 84
Bôas, Alexis Cristiano Vilas ................ 118
Boas, Alexis V. ............................ 45
Bonnaud, Olivier ........................... 13
Bontempo, Leonardo ......................... 29
Bordon, Camila D. S. ....................... 92
Bordon, Camila Dias Da Silva ................ 21
Brito, Francisco ........................... 68
Caetano, Fabio Domingues .................... 33
Camillo, L. M. ............................ 130
Candido, J. .............................. 134
Carvalho, Henrique L. ...................... 138
Carvalho, Marcel Castilho Batista De ......... 72
Cassé, Mikael ............................ 100
Cavalcante, Tássio V. ...................... 118
César, R. R. .............................. 64
Chi, Kung Shao ........................... 104
Choudhari, Asawari .......................... 1
Cicareli, Rodrigo ........................... 57
Cioldin, F. H. ............................. 64
Cirino, Giuseppe A. ........................ 57
Correia, M. M. ........................... 130
Costa, Priscila ............................ 25
Dejous, Corinne ............................. 1
Dias, Lucas Paiva .......................... 25
Diniz, J. A. .............................. 64
Diniz, José Alexandre ....................... 68
Doria, Rodrigo T. ........................... 5
Duarte, Pedro H. .......................... 126
Etcheverry, Louise Patron ................... 49
Fang, Lee Kuan ............................ 88
Faynot, Olivier ........................... 100
Fernandes, Thiago Vecchi .................... 92
Ferreira, Carlos L. ....................... 115
Filho, Sebastião Gomes Dos Santos ......... 29, 72
Finco, Saulo .............................. 118

Galembeck, E. H. S. ....................... 130
Garcia, José Augusto Martins .............. 21, 29
Garcia, Paulo R. ........................... 45
Garcia, Paulo Roberto ...................... 118
Giacomini, Renato C. ....................... 118
Giacomini, Renato Camargo ................... 84
Giacomini, Renato .......................... 45
Gimenez, S. P. ........................... 130
Gomes, Antonio Aurélio De Sousa ............. 84
Gonçalves, Raphael De Carvalho .............. 21
González, Maria E. L. ...................... 61
Grandesi, Guilherme Inácio .................. 45
Guazzelli, Marcilei A. ................... 45, 84
Guazzelli, Marcilei ....................... 118
Guerreiro, Joel Felipe ..................... 53
Hilkner, Henrique .......................... 41
Horiguchi, Naoto ............................ 9
Jakomin, Roberto .......................... 76
Joo, Jerald Sim Mong ....................... 88
Junior, Carlos R. P. Dos Santos ............. 49
Júnior, N. Graziano ....................... 134
Kassab, Luciana R. P. ...................... 92
Kassab, Luciana Reyes Pires ............... 21, 29
Kastensmidt, Fernanda ...................... 96
Kato, Beatrice Sayuri ...................... 21
Kawabata, Rudy M. S. ................ 76, 80, 115
Kawabata, Rudy Massami Sakamoto ............ 108
Kumada, Daniel Keij ........................ 29
Kumada, Daniel Kendji ...................... 21
Leite, Fernando Idalírio De Lima ............ 33
Lima, M. P. Braga De ...................... 130
Martino, Joao A. .............. 9, 17, 37, 126, 138
Martino, Joao Antonio ...................... 41
Matos, Jefferson Almeida ................... 100
Mederos, M. ............................... 64
Medina, Nilberto H. ....................... 118
Melo, Marco Antônio A. .................... 118
Minamisawa, R. A. ......................... 64
Minamisawa, Renato. A. .................... 104
Moreira, Claudio Villela .................. 122
Moreira, Eduardo Ceretta ................... 25
Morelhão, Sérgio Luiz ..................... 108
Mourão, R. T. ............................. 80
Namba, Igor Fernandes ...................... 33
Nunes, Carolina Carvalho Previdi ........... 33
Oliveira, Adhimar F. ....................... 61
Panzo, E. C. ............................. 134
Pavanello, Marcelo Antonio ........ 100, 111, 122

Pavanello, Marcelo ....................................................68
Peixoto, M. A. P. ....................................................130
Penello, Germano Maioli ..........................................108
Pereira, Cleiton Felix ................................................84
Pereira, Pedro Henrique ..........................................108
Perina, Welder F. .....................................................37
Philip, Deborah Debbie Anak ....................................88
Pinto, Francisco Rogelio Palomo ...............................118
Pinto, Luciana D. .....................................................76
Pires, Maurício P. ............................................ 76, 115
Pontes, Fagnaldo Braga ............................................49
Portes, Felipe S. C. .................................................61
Prates, Vinícius Rodrigues ........................................111
Quivy, Alain André ................................................108
Rangel, Ricardo C. .......................................... 126, 138
Rebiere, Dominique ...................................................1
Ribeiro, Arllen D. R. ..................................................9
Rodrigues, Jaime Calçade ........................................111
Rossi, Wagner De .....................................................92
Rouxinol, Francisco ..................................................68
Rua, Marcelo G. .....................................................115
Rube, Maxence ..........................................................1
Sadli, Idris...............................................................1
Santos, Favero Guilherme .........................................49
Santos, Marcos V. Puydinger Dos....................... 53, 104
Santos, Roberto Baginski B. .....................................118
Sebeloue, Martine .....................................................1
Seixas, L. E. ............................................................45
Seixas, Luis Eduardo ..............................................118
Seng, Ng Hong .........................................................88
Silva, Alexander. M. ..................................................76
Silva, Denison Rodrigo Ferreira .................................53
Silva, Elvio C. Dutra E .............................................49
Silva, Everton M. ......................................................5
Silva, Pedro H. Penna Da ..........................................17
Silva, Vanessa C. P. ...................................................9
Sim, Florinna ...........................................................88
Simoen, E. .............................................................134
Soares, Guillermo J. N. ..............................................76
Sousa, Graciana S. ....................................................76
Sousa, Matheus Dias..................................................33
Souza, Lucas Andrade Teixeira De .............................108
Souza, Michelly De ........................................... 100, 111
Souza, Patricia L. De .................................................80
Souza, Patrícia L. ............................................. 76, 115
Spejo, Lucas B. .........................................................53
Spejo, Lucas Barroso ...............................................104
Stolf, Ricardo Germano ..............................................84
Streit, Lívia .............................................................96
Stucchi-Zucchi, Lucas ................................................68
Tamarin, Ollivier........................................................1
Tat, Eddie Chaim Tau ................................................88

Tavares, Fabiele C. ...................................................76
Teixeira, R. C. .........................................................64
Torelly, Guilherme M. ........................................80, 115
Torelly, Guilherme Monteiro.....................................108
Trevisoli, Renan ........................................................5
Valaski, Rogério .......................................................76
Veloso, Anabela..........................................................9
Wetter, Niklaus U. ....................................................92
Yojo, Leonardo Shimizu.............................................49

**IEEE**
445 Hoes Lane
Piscataway, NJ  08854-4141

ISBN 979-8-3315-4064-7